Antioxidants and Reactive Oxygen Species in Plants

Biological Sciences Series

A series which provides an accessible source of information at research and professional level in chosen sectors of the biological sciences.

Series Editor:

Professor Jeremy A. Roberts, Plant Science Division, School of Biosciences, University of Nottingham. UK.

Titles in the series:

Biology of Farmed Fish   Edited by K.D. Black and A.D. Pickering

Stress Physiology in Animals   Edited by P.H.M. Balm

Seed Technology and its Biological Basis   Edited by M. Black and J.D. Bewley

Leaf Development and Canopy Growth   Edited by B. Marshall and J.A. Roberts

Environmental Impacts of Aquaculture   Edited by K.D. Black

Herbicides and their Mechanisms of Action   Edited by A.H. Cobb and R.C. Kirkwood

The Plant Cell Cycle and its Interfaces   Edited by D. Francis

Meristematic Tissues in Plant Growth and Development   Edited by M.T. McManus and B.E. Veit

Fruit Quality and its Biological Basis   Edited by M. Knee

Pectins and their Manipulation   Edited by Graham B. Seymour and J. Paul Knox

Wood Quality and its Biological Basis   Edited by J.R. Barnett and G. Jeronimidis

Plant Molecular Breeding   Edited by H.J. Newbury

Biogeochemistry of Marine Systems   Edited by K.D. Black and G. Shimmield

Programmed Cell Death in Plants   Edited by J. Gray

Water Use Efficiency in Plant Biology   Edited by M.A. Bacon

Plant Lipids – Biology, Utilisation and Manipulation   Edited by D.J. Murphy

Plant Nutritional Genomics   Edited by M.R. Broadley and P.J. White

Plant Abiotic Stress   Edited by M.A. Jenks and P.M. Hasegawa

Gene Flow from GM Plants   Edited by G.M. Poppy and M.J. Wilkinson

Antioxidants and Reactive Oxygen Species in Plants   Edited by N. Smirnoff

Antioxidants and Reactive Oxygen Species in Plants

Edited by

NICHOLAS SMIRNOFF
School of Biological and Chemical Sciences
University of Exeter
UK

Blackwell
Publishing

Editorial Offices:
Blackwell Publishing Ltd, 9600 Garsington Road, Oxford OX4 2DQ, UK
  Tel: +44 (0)1865 776868
Blackwell Publishing Professional, 2121 State Avenue, Ames, Iowa 50014-8300, USA
  Tel: +1 515 292 0140
Blackwell Publishing Asia, 550 Swanston Street, Carlton, Victoria 3053, Australia
  Tel: +61 (0)3 8359 1011

First published 2005 by Blackwell Publishing Ltd

Library of Congress Cataloging-in-Publication Data:

Antioxidants and reactive oxygen species in plants / edited by Nicholas Smirnoff.
      p. cm.
  Includes bibliographical references and index.
  ISBN-13: 978-1-4051-2529-1 (hardback : alk. paper)
  ISBN-10: 1-4051-2529-2 (hardback : alk. paper)
  1. Antioxidants—Physiological effect. 2. Active oxygen—Physiological effect.
3. Plants—Metabolism. I. Smirnoff, N.

QK898.A57A57 2005
572′.42—dc22

                                                                2004027446

ISBN-13: 978-1-4051-2529-1
ISBN-10: 1-4051-2529-2

British Library Cataloguing-in-Publication Data
A catalogue record for this title is available from the British Library

Set in 10/12 pt Times
by Newgen Imaging Systems (P) Ltd, Chennai, India
Printed and bound in India
by Replika Press Pvt. Ltd

For further information on Blackwell Publishing, visit our website:
www.blackwellpublishing.com

Contents

Contributors

Dr Radhika Desikan Centre for Research in Plant Science, University of the West of England, Frenchay Campus, Coldharbour Lane, Bristol BS16 1QY, UK

Professor Karl-Josef Dietz Lehrstuhl für Biochemie und Physiologie der Pflanzen, Fakultät für Biologie W5, Universität Bielefeld, 33501 Bielefeld, Germany

Professor Jürgen Feierabend Institute of Botany, J.W. Goethe-Universität, D-60054 Frankfurt/Main, Germany

Professor Christine H. Foyer Crop Performance and Improvement Division, Rothamsted Research, Harpenden, Herts AL5 2JQ, UK

Professor Stephen C. Fry School of Biological Sciences, University of Edinburgh, The King's Buildings, Mayfield Road, Edinburgh EH9 3JH, UK

Dr Leonardo Gomez Centre for Novel Agricultural Products, Department of Biology, University of York, PO Box 373, York YO10 5YW, UK

Dr Stephen C. Grace Biology Department, University of Arkansas at Little Rock, 2801 South University Avenue, Little Rock, AR 72204-1099, USA

Dr John Hancock Centre for Research in Plant Science, University of the West of England, Frenchay Campus, Coldharbour Lane, Bristol BS16 1QY, UK

Ms Pinja Jaspers Department of Biological and Environmental Sciences, University of Helsinki, FIN-00014 Helsinki, Finland

Dr Mark A. Jones School of Biological and Chemical Sciences, University of Exeter, Washington Singer Laboratories, Perry Road, Exeter EX4 4QG, UK

Professor Jaakko Kangasjärvi   Department of Biological and Environmental
                                     Sciences, University of Helsinki, FIN-00014
                                     Helsinki, Finland

Dr Hannes Kollist   Department of Biological and Environmental
                                     Sciences, University of Helsinki, FIN-00014
                                     Helsinki, Finland

Dr Christian Langebartels   GSF – National Research Center for
                                     Environment and Health, Ingolstädter
                                     Landstrasse 1, D-85764 Oberschleissheim,
                                     Germany

Dr Barry A. Logan   Biology Department, Bowdoin College, 6500
                                     College Station, Brunswick ME 04011, USA

Dr Ron Mittler   Department of Biochemistry, University of
                                     Nevada, MS 200, 1664 North Virginia Street,
                                     Reno, Nevada 89557, USA

Professor Steven Neill   Centre for Research in Plant Science,
                                     University of the West of England, Frenchay
                                     Campus, Coldharbour Lane, Bristol BS16
                                     1QY, UK

Dr Thomas L. Poulos   Department of Molecular Biology and
                                     Biochemistry, University of California, Irvine,
                                     CA 92697-3900, USA

Dr Nicholas Smirnoff   School of Biological and Chemical Sciences,
                                     University of Exeter, Washington Singer
                                     Laboratories, Perry Road, Exeter EX4
                                     4QG, UK

Dr Philippus D.R. van Heerden   School of Environmental Sciences and
                                     Development, Section Botany, North-West
                                     University, Potchefstroom 2520, South Africa

Dr Robert A.M. Vreeburg   School of Biological Sciences, University of
                                     Edinburgh, The King's Buildings, Mayfield
                                     Road, Edinburgh EH9 3JH, UK

Preface

Reactive oxygen species (ROS), which include superoxide, hydrogen peroxide and the hydroxyl radical, are produced during the interaction of metabolism with oxygen. ROS have the potential to cause oxidative damage by reacting with biomolecules. Accordingly, the emphasis of research on ROS has been on the oxidative damage that results from exposure to environmental stresses and on the role of ROS in defence against pathogens. More recently, it has become apparent that ROS have important roles as signalling molecules that contribute to the control of plant development and to the sensing of the external environment. A complex network of enzymatic and small molecule antioxidants control the concentration of ROS and repair oxidative damage. Interest in the function of small molecule antioxidants such as tocopherols (vitamin E), glutathione and ascorbic acid (vitamin C) is increasing, now that the details of their biosynthetic pathways have been uncovered. This research is revealing the complex and subtle interplay between ROS and antioxidants in controlling plant growth, development and response to the environment.

The book covers these new developments through a series of in-depth chapters, but it also provides, in one place, an overview of the subject for those outside the research area. The first six chapters cover the synthesis and function of antioxidants (glutathione, proteins involved in thiol homeostasis, ascorbate, tocopherol, carotenoids, phenolic compounds and catalase). Many of these antioxidants are important in the human diet, which provides additional impetus to understand their metabolism in order to manipulate their synthesis more effectively. The next three chapters consider the role of ROS in signalling, developmental processes and cell wall biochemistry. The final two chapters consider photosynthesis, which is a major source of ROS in leaves, and response to ozone, which has provided a useful model system for understanding the complex interplay of responses made by plants to ROS and oxidative stress. The subject matter is generally focussed on the molecular and biochemical details, as these have provided the basis for the large increase in research activity on ROS and antioxidants over the last decade. The antioxidant network is an ideal candidate for investigation on the genome-wide scale, using functional genomics and systems biology tools, and I hope that this volume will provide a background for this endeavour.

Nicholas Smirnoff

1 Glutathione

Christine H. Foyer, Leonardo D. Gomez and Philippus
D.R. van Heerden

1.1 Introduction

Glutathione (γ-Glu-Cys-Gly) is an information-rich molecule that serves a number of key functions in plant biology. Reduced glutathione (GSH) is a product of primary sulphur metabolism and acts as a transport and storage form of reduced sulphur. Together with cysteine, it forms part of the repertoire of signals that modulate sulphate uptake and assimilation (Kopriva & Rennenberg, 2004). GSH is the substrate for phytochelatin synthesis and is thus crucial to the detoxification of heavy metals such as cadmium and nickel (Freeman *et al.*, 2004). It is also a substrate for glutathione-S-transferases (GSTs) that catalyse numerous conjugation reactions for the removal of xenobiotics, and it catalyses conjugation of glutathione to anthocyanins and other secondary metabolites, which are then transported into the vacuole. GSH interacts with nitric oxide to form S-nitrosoglutathione, which is thought to be a stable transport form for this signal molecule. Recently, other roles in the regulation of growth and development, cell defence, redox signalling and gene expression have been attributed to glutathione. These functions depend on the concentration and/or redox state of the glutathione pool in each cellular compartment (May *et al.*, 1998a; Noctor *et al.*, 1998a,b).

Perhaps best known as an antioxidant, GSH reacts chemically with a range of active oxygen species (AOS). Moreover, enzyme-catalysed reactions link GSH to the detoxification of hydrogen peroxide (H_2O_2) in the ascorbate–glutathione cycle (Foyer & Halliwell, 1976; Noctor & Foyer, 1998). Ascorbic acid (AA) and GSH are the two most important soluble reducing compounds in living cells and are integral in maintaining a net reducing environment. Exposure to environmental stresses results in changes to levels of antioxidants, particularly GSH (Noctor *et al.*, 2000). The initial response of the cellular glutathione pool to stresses such as chilling, salinity and drought is an increase in the glutathione/glutathione disulphide (GSSG/GSH) ratio, followed by a marked increase in total glutathione concentration (Tausz *et al.*, 2004). While this response has been interpreted as overcompensation for a stress-induced increase in oxidative load (Tausz *et al.*, 2004), altered glutathione concentration and GSSG/GSH ratio could act as signals triggering specific cellular responses. For example, one extreme effect is the specific blocking of the cell cycle when ascorbate and glutathione become oxidised, thus, causing the cessation of cell division (Potters *et al.*, 2002, 2004). Since overall plant growth is intimately related to meristem cell division rates, and a major consequence of stress is the cessation of plant growth, it is logical to consider that these processes are linked and that cell cycle arrest in response to abiotic stress leads to growth arrest. The evidence discussed later links GSH and AA to specific cell cycle effects on the G1 phase of the cell cycle, and damaging cellular responses to stress including programmed cell death (PCD) and systemic acquired resistance (SAR). Although AA and GSH are intimately associated with each other through the operation of the ascorbate–glutathione cycle, allowing considerable cross-talk between AA and glutathione-related pathways, these two pools are independently assessed in terms of gene regulation (Noctor *et al.*, 2000; Pastori *et al.*, 2003; Ball *et al.*, 2004) and by cell cycle control mechanisms (Potters *et al.*, 2004).

1.2   The glutathione redox couple and cellular redox potential

Like all other aerobic organisms, plants maintain cytoplasmic thiols in the reduced (–SH) state because of the low thiol-disulphide redox potential imposed by milli-molar amounts of glutathione, which acts, therefore, as a thiol buffer. The glutathione redox couple is also considered to be important in the homeostatic adjustment of the cellular redox potential. In animal cells, substantial evidence implicates redox poten-tial as an important factor determining cell fate, and the glutathione redox couple is the key player (Schäfer & Buettner, 2001). Unlike many other redox couples such as ascorbate/dehydroascorbate and NADPH/NADP, glutathione redox potential depends on both GSH/GSSG and absolute glutathione concentration. The concentration-dependent term of the Nernst equation is second-order with respect to GSH but first-order with respect to GSSG; hence, accumulation of GSH can offset the change in redox potential caused by decreases in the GSH/GSSG ratio (Schäfer & Buettner, 2001).

1.3   Glutathione metabolism

In many reactions, the cysteine thiol group is oxidized to form GSSG (Figure 1.1). Cellular GSH:GSSG ratios are maintained by glutathione reductase (GR), a homodimeric flavoprotein that uses NADPH to reduce GSSG to two GSH. Although transient disulphide bonds do occur during the catalytic cycle of some enzymes, stable protein disulphide bonds are relatively rare except in quiescent tis-sues such as seeds. In contrast to GR, which has a specific function with very restricted substrate specificity, the GST family of proteins has a large range of func-tions. In general, they are characterized by their ability to catalyse a nucleophilic attack of the sulphur atom of GSH (or homologue) on the electrophilic centre of their substrates; however, this activity is not necessary for GST functions. GSTs are best known for their protective function in the removal of compounds that are potentially cytotoxic but some GSTs have a 'ligandin' function that is important in

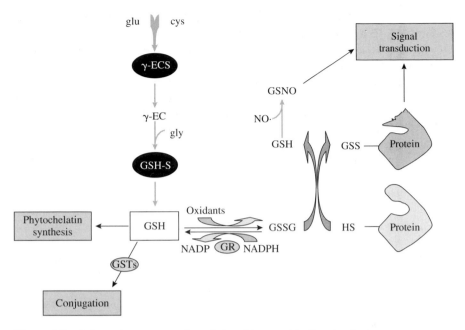

Figure 1.1   Schematic representation of glutathione metabolism in plants.

transport, e.g. in the anthocyanin synthesis pathway (Marrs, 1996). Accumulating evidence suggests that certain GSTs function as flavinoid-binding proteins, e.g. AN9, which is a GST required for efficient anthocyanin export from the cytosol in Petunia (Mueller *et al.*, 2000). Other GSTs have a function in the regeneration of AA from dehydroascorbate. Like glutathione peroxidases (GPXs), certain GSTs are also involved in the detoxification of H_2O_2 and hydroperoxides (Bartling *et al.*, 1993; Cummins *et al.*, 1999; Collinson *et al.*, 2002).

    Although genes encoding plant GPXs have been cloned from a wide variety of plant species (e.g. Eshdat *et al.*, 1997), they are rather different in a number of aspects from the respective enzymes in animals. In contrast to the mammalian H_2O_2-detoxifying GPXs, which involve selenocysteine, the reaction mechanism of plant-type GPXs involves cysteine residues. Some GPX genes encode proteins targeted to the chloroplast (Mullineaux *et al.*, 1998) but most are cytosolic. Like certain GSTs, some GPX genes are strongly induced by oxidative stress (Levine *et al.*, 1994; Willekens *et al.*, 1997). All plant GPX cDNAs, isolated to date, show high homology to the mammalian phospholipid hydroperoxide glutathione peroxidases (PHGPXs), which have a higher affinity to lipid hydroperoxides than to H_2O_2. This is explained by the absence of selenocysteine at the active site, which diminishes the nucleophilic properties of the enzyme and probably accounts for its much lower activity with H_2O_2. It has also been demonstrated that at least two plant PHGPXs probably represent a novel isoform of thioredoxin (TRX) peroxidase, exhibiting much higher oxidation activity for TRX than for glutathione (Herbette *et al.*, 2002). H_2O_2 is the preferred substrate for TRX peroxidase. PHGPXs are induced by a wide range of biotic and abiotic stresses, suggesting that they play a specific role in limiting the extent of lipid peroxidation during stress by removing phospholipid hydroperoxides and H_2O_2. Transgenic tobacco lines overexpressing plant GST/GPX were reported to show enhanced antioxidant capacity and substantial improvement in seed germination and seedling growth under stress (Roxas *et al.*, 1997). The glutatredoxin (GRX)/GR system is also involved in the regeneration of peroxiredoxins (Rouhier *et al.*, 2002).

The breakdown of glutathione could involve GSH, GSSG and glutathione-*S*-conjugates. Different pathways are possible and each might fulfil an essentially different function. Catabolic destruction of GSSG may serve as a detoxification process. GSSG is involved in thiolation reactions forming mixed disulphides with proteins in conditions of oxidative stress. Since this process inactivates many biosynthetic enzymes, the presence of a large GSSG pool is not compatible with many metabolic reactions; catabolism of GSSG would essentially return the system to pre-stress homeostasis. Catabolism of GSH, on the other hand, largely concerns the remobilisation of cysteine, e.g. during seed storage protein synthesis or during periods of sulphur deprivation. This requires the successive breakage of the two peptide bonds. The catabolism of GSH through the γ-glutamyl cycle is well described in animals but poorly defined in plants (Meister, 1988). GSH is degraded in the γ-glutamyl cycle by γ-glutamyl transpeptidase (GGT) and subsequent reactions to produce the component amino acids that can be re-integrated into the cell metabolism. GGT catalyses the reversible hydrolysis of the N-terminal peptide bond, initiating catabolism by removing the γ-linked Glu from GSH, GSSG, glutathione conjugates and other peptides. The Glu moiety is either hydrolysed or donated to an amino acid acceptor or even to another GSH molecule. The second step in catabolism is less well characterised. The Cys–Gly bond is not unique to the glutathione tripeptide and several enzymes, including aminopeptidase M and Cys–Gly dipeptidase, are able to hydrolyse the bond (Meister, 1988). The transpeptidases are part of the γ-glutamyl cycle and as such are involved in amino acid transport in some tissues (Meister, 1988). The γ-glutamyl moiety is metabolised by a γ-glutamylcyclotransferase to oxo-proline which are subsequently converted to glutamate by oxo-prolinase. Homologous activities are also present in plants (Rennenberg *et al.*, 1981; Steinkamp & Renneberg, 1984; Steinkamp *et al.*, 1987). Evidence for a carboxypeptidase and GGT (Schneider *et al.*, 1992; Storozhenko *et al.*, 2002) has been obtained. Transgenic *Arabidopsis* with increased GGT activity showed no changes in tissue glutathione content or GSH/GSSG ratios; hence, the role of this enzyme in glutathione metabolism remains uncertain (Storozhenko *et al.*, 2002). Since GSSG can be considered as a glutathione-*S*-glutathione conjugate, transport of GSSG by the vacuolar conjugate transporter may play a role in removing this species from the cytosol. The failure to detect significant accumulation of glutathione-*S*-conjugates in vacuoles suggests that they are rapidly catabolised in this compartment (Marrs, 1996).

1.4 Biosynthesis and inhibition by L-buthionine-SR-sulphoximine

In all organisms studied to date, GSH biosynthesis occurs from constituent amino acids via a two-step ATP-dependent reaction sequence (Figure 1.2) catalysed by γ-ECS and glutathione synthetase (GSH-S) (Hell & Bergmann, 1988, 1990; Meister, 1988). Genes encoding the above enzymes have been identified from a number of C_3 dicotyledonous plants such as *Arabidopsis, Lycopersicon, Brassica* and *Medicago*. All the plant γ-ECS and GSH-S enzymes described to date display a high degree of sequence homology. In *Arabidopsis thaliana*, γ-ECS is encoded by a single gene, *gsh1*, with a plastid target signal (May & Leaver, 1994). Moreover, recent studies have indicated that most, if not all, of the *A. thaliana* γ-ECS protein is targeted to the chloroplasts (Maughan, 2003; Wachter *et al.*, 2005). There may also be differences between species as there are two *gsh1* genes in the maize genome and one gene encoding GSH-S. Targeting information may suggest that GSH-S has an exclusively cytosolic location (Wachter *et al.*, 2005). However, γ-ECS and GSH-S activities have been shown to be located both inside and outside the chloroplast (Klapheck *et al.*, 1987; Hell & Bergmann, 1988, 1990) and the ability of photosynthetic cells to synthesize GSH in chloroplastic and cytosolic compartments has been confirmed by overexpression studies (Noctor *et al.*, 1996, 1998a; Creissen *et al.*, 1999). The chloroplastic and extra-chloroplastic isoforms possess similar

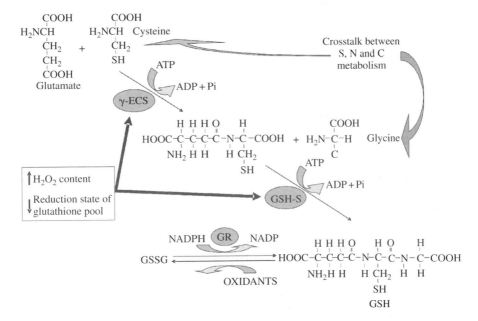

Figure 1.2 The pathway of GSH synthesis showing the required integration of C, N and S metabolism and indicating that increased oxidation either via enhanced H_2O_2 or decreased GSH/GSSG ratios will stimulate enzyme activity and throughput to increase GSH levels.

regulatory properties (Noctor et al., 2002a). Nevertheless, it should be borne in mind that the bulk of the leaf cell γ-EC may be made and hence exported from the chloroplast while the majority of GSH could be made in the cytosol and imported into the chloroplast. Such a system would have important implications for transport, signalling and defence. However, more data is required before any conclusions can be drawn with regard to differential intracellular localisation of the enzymes of GSH synthesis.

In C_4 species such as maize, the situation may even have a greater degree of complexity because of the leaf structure. Cysteine synthesis appears to be exclusively localised in the bundle sheath (BS) cells of maize leaves (Burgener et al., 1998). While an earlier study had shown that GSH-S activity was located primarily in the mesophyll (M) cells (Burgener et al., 1998), a more recent study (Gomez et al., 2004a) using in situ hybridization and immunolocalization of leaf sections showed that γ-ECS and GSH-S transcripts and proteins were similar in M and BS cells and found in most leaf cell types including the epidermis and stomatal guard cells.

The nature and roles of the gsh1 gene products continues to cause much speculation. It appears that there are two gsh1 genes in most species and certainly one is targeted to the chloroplast and one to the cytosol. However, if recent speculation is correct, and most of the γ-EC formation occurs in the chloroplast, then the role of the cytosolic isoform becomes very interesting.

A number of GSH-deficient mutants have been described in the single locus which encodes γ-ECS in A. thaliana. In particular, the analysis of the ROOT-MERISTEMLESS1 (RML1), CADMIUM SENSITIVE2 (CAD2) and the REGULATOR OF APX21-1 (RAX1-1) has provided useful information concerning the effects of glutathione depletion in plants and the mechanisms whereby this control is exerted (Ball et al., 2004). The cad2-1 mutant has a deletion in the region encoding the active site of the γ-ECS protein and this appears to alter the structure of the active site (Cobbett et al., 1998) in such a way that the mutant has only 15–30% of wild-type levels of GSH and, as a result, is rendered cadmium-sensitive (Howden et al., 1995). The rml1 mutant has only about 2% of the leaf GSH of the wild type and cannot form a normal root system (Cheng et al., 1995; Vernoux et al., 2000). The lack of complementation between rml1 and cad2-1 and later cloning of the RML1 gene confirmed that it is allelic with CAD2 (Vernoux et al., 2000). The rax1-1 mutant has about 50% of the wild-type leaf levels of glutathione and is allelic to gsh1 (Ball et al., 2004). Although both cad1 and rax1-1 lack γ-ECS activity and have lower glutathione contents, they have very different leaf transcriptome profiles (Ball et al., 2004). This suggests that at least one of the gsh1 gene products may be involved in signal transduction as well as in biosynthesis. The marked chlorotic phenotype produced by chloroplastic Escherichia coli γ-ECS overexpression in transformed tobacco might be considered to result from perturbed signal transduction rather than direct effects of GSH or γ-EC accumulation per se (Creissen et al., 1999). However, expression of the E. coli γ-ECS and GSH-S in poplar and mustard (Foyer et al., 1995; Strohm et al., 1995; Noctor et al., 1996, 1998a; Pilon-Smits *et al.*, 1999; Zhu *et al.*, 1999) did not produce the chlorotic phenotypes observed in tobacco (Creissen *et al.*, 1999). Overexpression, with targeting of the bacterial enzyme protein to either the chloroplast or cytosol, led to marked increases in enzyme activity. Increases in γ-ECS, but not GSH-S, not only led to large constitutive increases in leaf glutathione (Noctor *et al.*, 1996, 1998a; Creissen *et al.*, 1999) but glutathione was also increased in xylem sap, phloem exudates and roots (Herschbach *et al.*, 2000). Despite enhanced GSH contents in the phloem of poplar, sulphur uptake by the roots was markedly enhanced to meet the requirements of increased demand for sulphur (Herschbach *et al.*, 2000). Incubation of leaf discs with cysteine increased glutathione contents substantially in untransformed and transformed poplars, particularly in the light, suggesting that cysteine supply remains a key limiting factors during glutathione synthesis (Strohm *et al.*, 1995; Noctor *et al.*, 1996, 1997). Such studies have also shown that γ-ECS abundance and cysteine availability are major factors controlling GSH accumulation (Noctor *et al.*, 1996, 1998a,b, 2002a; Xiang & Oliver, 1998; Xiang & Bertrand, 2000; Xiang *et al.*, 2001). However, the activities of enzymes involved in sulphate reduction and assimilation are not greatly changed in transgenic plants with high γ-ECS activities (Hartmann *et al.*, 2004). Similarly, micro-array analysis of plants undergoing sulphate starvation did not reveal effects on γ-ECS or GSH-S transcripts suggesting little co-ordination of the pathways of GSH synthesis in response to sulphur availability (Hirai *et al.*, 2003; Maruyama-Nakashita *et al.*, 2003).

Relatively little is known about the co-ordinate regulation of expression of *gsh1* and *gsh2* in plants, but it is clear that GSH and GSSG per se exercise little or no control over transcription (Xiang & Oliver, 1998). Similarly, while the application of H_2O_2 increased tissue GSH (as did the catalase inhibitor, amino triazole), it did not affect *gsh1* or *gsh2* transcript abundance (Xiang & Oliver, 1998). Moreover, when *Arabidopsis* cell cultures were exposed to oxidative stress (by the addition of either amino triazole, menadione or fenchlorazole), cellular γ-ECS activity and glutathione content increased, but γ-ECS mRNA levels were unchanged. The abundance of *gsh1* and *gsh2* transcripts was increased by cadmium in *Brassica juncea* (Schäfer *et al.*, 1998) and by both cadmium and copper in *Arabidopsis* (Xiang & Oliver, 1998). Jasmonic acid (JA) also increased *gsh1* and *gsh2* transcripts and a common signal transduction pathway may be involved (Xiang & Oliver, 1998). Indeed, the application of methyl jasmonate to *Arabidopsis* plants induced a fast transient increase in the abundance of transcripts encoding a number of enzymes of sulphur assimilation in addition to *gsh1* and *gsh2* transcripts (Harada *et al.*, 2000). Although transcript abundance was increased by heavy metals and JA, oxidative stress was required for the translation of the transcripts, implicating regulation at the post-transcriptional level and a possible role for factors such as H_2O_2 or modified GSH/GSSG ratios in de-repressing translation of the existing mRNA (Xiang & Oliver, 1998). Stress-induced increases in glutathione, such as those observed in plants deficient in catalase, have shown that glutathione accumulation is preceded or accompanied by a marked decrease in the reduction state of the pool (Smith *et al.*, 1984; Willekens *et al.*, 1997). A similar response was elicited by exposing poplar leaves to ozone (Sen Gupta *et al.*, 1991). The 5' untranslated region (5' UTR) of the *gsh1* gene was found to interact with a repressor-binding protein that was released upon addition of H_2O_2 or changes in the GSH/GSSG ratio (Xiang & Bertrand, 2000). A redox-sensitive 5' UTR-binding complex is thus suggested to control γ-ECS mRNA translation in *A. thaliana* (Xiang & Bertrand, 2000).

Post-translational regulation of γ-ECS through end-product inhibition by GSH is a crucial factor in controlling GSH concentration in animals and plants (Jez *et al.*, 2004). Tobacco and parsley γ-ECS was found to be subject to GSH inhibition, in a manner that is competitive with respect to glutamate (Hell & Bergmann, 1990; Schneider & Bergmann, 1995). However, since a range of environmental triggers enhance cellular GSH contents, it would appear that feedback inhibition of γ-ECS by GSH can be overcome by other regulatory mechanisms. Moreover, overexpression of either γ-ECS (Noctor *et al.*, 1996) or serine acetyltransferase (Harms *et al.*, 2000) greatly enhances tissue GSH contents. The consensus sequences indicate that γ-ECS proteins house multiple putative phosphorylation sites but there is no evidence, to date, that any of these are functionally significant. There is some evidence to suggest that rat γ-ECS is regulated by protein phosphorylation (Sun *et al.*, 1996) but this has not yet been found in studies on γ-ECS from plants. May *et al.* (1998b) concluded that protein factors are involved in post-translational control of γ-ECS and are required for full activity. In the animal enzyme system, a smaller regulatory subunit acts to increase the catalytic potential of the larger catalytic subunit by increasing its K_i value for GSH and decreasing the K_m for glutamate, thereby alleviating feedback control and allowing the enzyme to operate under *in vivo* conditions (Huang *et al.*, 1993). It is, nevertheless, clear that even in the absence of the smaller subunit, the large catalytic subunit is capable of effective catalysis, since overexpression of this polypeptide alone yielded increased glutathione levels in transfected human cells (Mulcahy *et al.*, 1995). Highest glutathione contents were, however, obtained by dual overexpression of both subunits (Mulcahy *et al.*, 1995). While protein factors have not been identified in plants, and there is as yet no evidence for control of γ-ECS by phosphorylation, several enzymes in plants are controlled by interactions between phosphorylation status and factors such as 14-3-3 proteins – regulatory components found in several compartments of the plant cell (DeLille *et al.*, 2001).

The glutamate analogue methionine sulphoximine (MSO) is an inhibitor of γ-ECS and glutamine synthetase but MSO analogues such as L-buthionine-SR-sulphoximine (BSO) are much more specific and effective inhibitors of γ-ECS (Griffith & Meister, 1979). Phosphorylated BSO is very similar to the glutamylphosphate cysteine adduct which binds to the active site (Griffith, 1982), but once phosphorylated by γ-ECS, BSO binds irreversibly to the γ-ECS protein, preventing γ-EC synthesis. This inhibitor has thus been widely used to deplete cellular GSH levels (Hell & Bergmann, 1990; Vernoux *et al.*, 2000). Analysis of *A. thaliana* T-DNA insertion and EMS mutants selected by their ability to grow on BSO concentrations that are inhibitory to the wild type have indicated that different mechanisms can confer BSO resistance (Maughan, 2003). In particular, the T-DNA insertion in one BSO resistant mutant (*brt1*) resides within a putative transport protein and is a novel component influencing glutathione metabolism (Maughan, 2003). Although this mutant lacks a visible phenotype, it has double the levels of tissue glutathione. BRT1 is a novel family consisting of three genes in *A. thaliana* with homologues in other organisms including the malaria parasite (*Plasmodium falciparum*), where the homologue, PfCRT, confers resistance to chloroquine, a widely used antimalarial drug (Maughan, 2003). The apparent conservation at a functional level between the BRT1 and PfCRT proteins may be important in advancing our present concepts of the regulation of glutathione metabolism in both organisms (Maughan, 2003).

1.5   Glutathione and the cell cycle

Both oxidants and antioxidants have roles in the regulation of the cell cycle. Low levels of ROS are synchronously generated with the cell cycle in mammals indicating that ROS are closely connected with the cell cycle. Moreover, small perturbations of these ROS levels prevented normal progression of the cell cycle, indicating that distinct signalling between ROS levels and the cell cycle exists and is biologically significant (Meijer & Murray, 2001; Dewitte & Murray, 2003). Also, direct connections exist between antioxidants and cell cycle control (especially the G1 checkpoint) in animals. For example, GSH depletion in mice fibroblasts results in decreased Cyclin D1 and increased p27 (a CDK inhibitor) protein levels. Glutathione also regulates progression through the plant cell cycle (Vernoux *et al.*, 2000). The first evidence that GSH was involved in cell cycle control was obtained by Earnshaw and Johnson (1985), who showed that carrot cells stopped dividing when GSH was depleted from the growth media. Synchronized tobacco BY-2 suspension cells are sensitive to GSH depletion by BSO treatment (Vernoux *et al.*, 2000). Moreover, BSO causes a specific arrest in cell division at the G1 checkpoint (Vernoux *et al.*, 2000). Thus, cell cycles that have already been initiated were completed, but cycling cells were blocked the next time they passed through the G1 phase, showing that cell division processes were not blocked in a general sense, but rather a specific checkpoint in G1 was activated in response to low GSH levels. Arrest was accompanied by changes in the expression of cell cycle related genes. Depletion of either ascorbate or glutathione causes the cessation of cell division (Potters *et al.*, 2002) but these antioxidants do not have compensatory functions in cell cycle regulation (Potters *et al.*, 2004).

Recent results have shown that down-regulation of the enzyme poly(ADP ribose) polymerase (PARP) provides substantial protection against a range of abiotic stresses in both monocot and broad leaf crops (De Block *et al.*, 2004) allowing plants to continue growing under conditions where controls not only stop growing but exhibit extreme stress responses including necrosis. This raises the possibility that PARP may be a key player linking stress responses to growth and division at the cellular level.

The evidence from manipulation of PARP levels in divergent plant species confirms that cessation of growth is not a necessary adaptive response to stress, but rather a result of an unnecessary cellular response to stress. Moreover, it demonstrates that broad-spectrum resistance to abiotic stress is both possible and beneficial to crops, and also suggests that PARP plays a key role in stress effects. Since PARP is closely linked with cell cycle control in mammals, it provides clues as to one of the best potential links into the plant cell cycle.

1.6   Glutathione in leaves and its relationship to chilling tolerance

Although glutathione can be synthesised in many plant organs, green leaves are the major sites of synthesis. It has long been appreciated that chloroplasts contain high concentrations of glutathione (3–5 mM) (Foyer & Halliwell, 1976; Smith *et al.*, 1985). Moreover, Klapheck *et al.* (1987) calculated that, in pea leaves, 35% of the leaf glutathione was localised in the chloroplasts. The global glutathione concentration outside the chloroplast in wheat leaves has been calculated to be between 1 and 4 mM (Noctor *et al.*, 2002a). Plant cells fed with cysteine or GSH show a strong decrease in sulphur uptake (Kreuzwieser & Rennenberg, 1998). Evidence suggests that GSH accumulation has feedback effects on the pathway of primary sulphur assimilation. For example, adenosine 5′phosphosulphate reductase (APR) activity and transcript abundance were decreased following addition of GSH in *Arabidopsis* roots (Vauclare *et al.*, 2002) and the activity of ATP sulphurylase was also lower when *Arabidopsis* plants were supplied with GSH (Lappartient *et al.*, 1999). Analysis of phloem sap indicated that GSH (rather than cysteine) was the signal lowering sulphur assimilation in *Brassica napus* (Lappartient & Touraine, 1996) and poplar (Herschbach *et al.*, 2000). A simple regulatory mechanism has been proposed with regard to the regulation of demand-driven sulphur assimilation by positive signals such as *O*-acetylserine and negative signals such as GSH (Kopriva & Rennenberg, 2004).

Plants vary greatly in their ability to tolerate low growth temperature (chilling stress). This can be seen in particular plants of tropical and subtropical origin that are affected most severely by chilling temperatures. Species can be classified according to the threshold below which chilling injury is observed. Chilling-sensitive plants such as soybean usually show decreased photosynthesis (Van Heerden *et al.*, 2003), growth, reproductive development and yield when exposed to temperatures of between 10 and 15°C. Some of the detrimental effects of chilling stress may be attributed to increased oxidation resulting from AOS accumulation (Prasad *et al.*, 1994; Prasad, 1996). Chilling increases photoinhibition and hence increases the production of singlet oxygen by photosystem II and it also favours enhanced lipid hydroperoxidation that leads to membrane damage (Yoshimura *et al.*, 2004). Strong evidence for a relationship between tissue glutathione and chilling tolerance has been obtained in the chilling-sensitive C_4 plant, maize. Chilling-sensitive maize cultivars increase leaf GSH concentrations when grown for long periods at sub-optimal growth temperatures, but have lower GSH/GSSG ratios under these conditions than more tolerant cultivars (Hodges *et al.*, 1996). Experiments in the field linked chilling tolerance in these cultivars to their capacity to accumulate leaf GSH and enhance GR activities (Leipner *et al.*, 1999). Interestingly, increasing the glutathione contents of maize leaves by the addition of herbicide safeners and other metabolites that stimulate GSH accumulation also enhanced chilling tolerance (Kocsy *et al.*, 2000). Pre-treatment of mung bean with H_2O_2 also led to increased foliar glutathione content and this correlated with improved chilling tolerance (Yu *et al.*, 2002). Our own experiments with maize grown under field conditions suggest that the quantitative trait loci (QTLs) for glutathione content overlap, at least to some extent, with key QTLs for photosynthesis and yield at chilling temperatures (authors, unpublished data).

Chilling temperatures in the absence of illumination (dark chilling) also enhanced GR activity in soybean genotypes with superior chilling tolerance, especially when dark chilling and drought stress was applied simultaneously (Van Heerden & Krüger, 2002). All chilling sensitive species, however, do not respond in the same fashion; e.g. chilling stress in tobacco resulted in an accumulation of oxidised glutathione and reduced activity of GR and other antioxidant enzymes (Gechev *et al.*, 2003). In the case of certain soybean genotypes, prolonged exposure to simultaneously applied dark chilling and drought stress also resulted in loss of GR activity when compared to short-term stress exposure (Van Heerden & Krüger, 2002). Chilling tolerance in maize is accompanied by increased γ-ECS activity and GSH accumulation in leaves (Kocsy *et al.*, 2000, 2001). Sulphur assimilation enzymes as well as the enzymes involved in GSH synthesis are increased in chilled maize plants (Kopriva *et al.*, 2001). Chilling young maize plants for 48 h, strongly induced γ-ECS mRNA (Gomez *et al.*, 2004b; Figure 1.3) – an effect reversed during recovery. However, the chilling-induced increase in γ-ECS transcripts was not accompanied by enhanced total leaf γ-ECS protein or extractable activity in these circumstances (Gomez *et al.*, 2004b). Moreover, chilling increased γ-ECS transcripts and protein in the BS but not in the M cells. This finding is surprising because maize leaves contain no GR in the BS cells and GSSG has to be transported to the M cells, where GR is abundant, for regeneration. A pertinent question concerns why GSH synthesis is induced in the one tissue of the maize leaf that cannot keep the glutathione pool reduced. The answer may lie in the requirement to keep a lower GSH/GSSG ratio for signalling purposes under chilling conditions. Increased BS γ-ECS was correlated with a two-fold increase in both leaf cysteine and γ-glutamylcysteine but leaf total glutathione significantly increased only in the recovery period, when the GSH/ GSSG ratio decreased three-fold (Gomez *et al.*, 2004b). These changes may be related to roles of glutathione in plant growth and development but these remain poorly defined. High leaf glutathione contents stimulate early flowering in *Arabidopsis* (Ogawa *et al.*, 2002). Moreover, enhanced leaf glutathione promoted bolting in *Eustoma grandiflorum* with or without vernalisation, and led to early leaf senescence (Yanagida *et al.*, 2002).

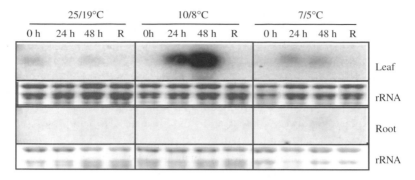

Figure 1.3   The expression of γ-ECS and GSH-S in leaves and roots of maize plants exposed to chilling temperatures. Northern blot analysis was performed to determine maize leaf γ-ECS and GSH-S transcript abundance in plants grown at optimal temperatures (25/19°C, 0 h); after 24 or 48 h exposure to 10/8°C or 7/5°C day/night temperatures; or after transfer from each chilling temperature treatment back to optimal growth temperatures for 2 days (recovery, R). Leaf and root RNA samples (20 μg) were loaded on to each lane and blots were hybridised with homologous radio-labelled probes under high stringency conditions. The rRNA bands stained with ethidium bromide were used as loading controls.

There is also evidence linking PHGPX to chilling tolerance in plants (Ursini *et al.*, 1995). During chilling stress, transgenic tobacco plants expressing a PHGPX-like protein in either the cytosol (TcGPX) or chloroplasts (TpGPX) showed suppressed production of malondialdehyde (MDA), compared to wild-type plants (Yoshimura *et al.*, 2004). The photosynthetic capacities of the TcGPX and TpGPX plants also remained high during chilling exposure as compared to those of the wild-type plants.

1.7   Glutathione and homoglutathione in the regulation of root and root nodule development

The *A. thaliana rml1* mutant, which has tissue GSH levels that are less than 2% of the wild type, has normal embryonic development (within a heterozygote parent) (Cheng *et al.*, 1995; Vernoux *et al.*, 2000). Following germination, however, it is unable to initiate cell division and, consequently, it cannot maintain a root meristem (Cheng *et al.*, 1995). In contrast, shoot development appears to be normal. Moreover, the *cad2-1* mutant with 15–30% of wild-type GSH levels is less sensitive than the wild type to root growth inhibition induced by BSO. The mutant phenotype can be rescued by addition of exogenous GSH (but not other reducing agents), indicating the importance of GSH in cell division in the root meristem. Furthermore, wild-type plants treated with BSO to deplete GSH resemble *rml1* plants. Cell division is inhibited in the roots of both *rml1* mutants and wild-type plants treated with BSO (Vernoux *et al.*, 2000). Glutathione is therefore required for normal root development.

The root cytosolic glutathione concentration, estimated using *in situ* imaging techniques, is between 1.8 and 4 mM (Fricker & Meyer, 2001). Little information is available on the regulation of glutathione synthesis in roots. Plastids isolated from very young maize roots were reported to contain about half the total γ-ECS activity and a small proportion of the root GSH-S activity (Rüegsegger & Brunold, 1993). In our own experiments with maize, neither γ-ECS nor GSH-S transcripts could be detected by Northern blot analysis in roots (Figure 1.4). While exposure to chilling temperatures induced a large increase in leaf γ-ECS transcripts, γ-ECS transcripts were below the level of detection in all conditions (Figure 1.3). We were able to detect γ-ECS transcripts in maize roots using RT-PCR (Figure 1.5). This analysis revealed that in contrast to leaf γ-ECS transcripts, root γ-ECS transcripts actually decreased in response to chilling (Figure 1.5) suggesting that the roots became even more dependent on the shoots for GSH than under optimal growth conditions. While GSH-S transcripts were detected in leaves using RT-PCR methods, we were unable to detect any GSH-S transcripts in roots in any condition (Figure 1.5).

The symbiotic association between the roots of legumes and soil rhizobia results in the formation of specialized organs known as root nodules, whose main function is symbiotic nitrogen fixation (SNF). Under physiological conditions, respiration associated with SNF is a major source of nodule AOS production (Matamoros *et al.*, 2003). AOS are involved in all stages of nodule development, from initiation to senescence (Becana *et al.*, 2000; Baudouin *et al.*, 2004; Frendo *et al.*, 2004). GSH is abundant in nodules (Figure 1.6) where it performs a number of functions that are critical for optimal SNF (Iturbe-Ormaetxe *et al.*, 2002). Recent evidence shows that nodules will not form on roots if GSH synthesis is blocked by addition of BSO, suggesting that, like the root meristem, the nodule meristem is unable to develop in the absence of GSH (Frendo *et al.*, 2005).

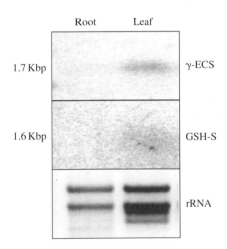

Figure 1.4 Expression of γ-ECS and GSH-S in leaves and roots of maize plants grown under optimal temperatures. Plants were grown for 21 days under a 25/19°C day/night temperature regime. Root and leaf samples (1 g) were used for RNA extraction and subsequent Northern blot analysis. Homologous probes, generated by cloning the maize cDNA for both enzymes (Gomez *et al.*, 2004b), were labelled with P[32] and hybridised under high stringency conditions. The rRNA bands stained with ethidium bromide were used as loading controls.

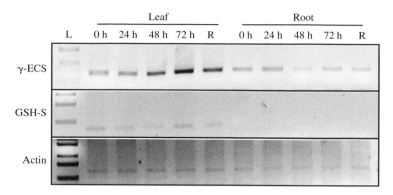

Figure 1.5   Semi-quantitative RT-PCR estimation of relative γ-ECS and GSH-S transcript abundance in maize leaves and roots. Plants were grown at 25/19°C day/night temperatures (0 h) and then transferred to chilling temperatures (10/8°C day/night temperatures) for three consecutive days (24 h, 48 h and 72 h). After the chilling treatment, plants were returned to optimal growth temperatures for 24 h. Reverse transcription was performed using 1 μg of total RNA extracted from leaves or roots. Linear amplification was obtained after 25 PCR cycles using an annealing temperature of 50°C. Actin was used as internal transcription control and the specific primers were 5′-GAGAAGATGACTCAGATC-3′ (sense) and 5′-ATCCTTCCTGATATCGAC-3′ (antisense). For γ-ECS, the primers used were 5′-GAGATGAGAGGTGCTGATGGT-3′ (sense) and 5′-GTTGCCACTTGGTCTCGTAC-3′ (antisense). For GSH-S, the primers used were 5′-AAGCGGAAGGCCAGGTTCT-3′ (sense) and 5′-CTGACAAGCGTTTCCCTCAA-3′ (antisense).

Legumes often contain homologues of GSH such as hydroxymethylglutathione (hmGSH; γ-Glu-Cys-Ser) and homoglutathione (hGSH; γ-Glu-Cys-βAla) in addition to GSH. hmGSH was first described by Klapheck *et al.* (1992) and reported to exist in numerous cereal species. On the other hand, hGSH has been detected only in leguminous plants (Matamoros *et al.*, 1999) and was first found in leaves of soybean and common bean (Price, 1957). In some legumes, hGSH may partially or completely replace GSH. In nodules of soybean, common bean and mung bean, hGSH is the dominant tripeptide, whereas GSH is predominant in nodules of pea, alfalfa and cowpea (Matamoros *et al.*, 2003). Many of the roles ascribed to GSH are also performed by hGSH. In species such as mung bean, hGSH is the major storage form of reduced sulphur (Macnicol & Bergmann, 1984). hGSH has also been implicated in plant defense against heavy metals and as a scavenger of AOS (Dalton *et al.*, 1986). The ability of plant species to make homologues of glutathione depends on the specificity of the synthetases involved. Specific legume GSH-Ss use either glycine to form GSH or β-alanine to form hGSH. Recent work in *Medicago truncatula* suggests that separate genes encode GSH-S and homoglutathione synthetase (hGSH-S) and that the divergence in specificity has arisen by gene duplication after the evolutionary divergence of the *Leguminaceae* (Frendo *et al.*, 1999). The two genes are very homologous and are found on the same fragment of genomic DNA. In a consideration of the distribution of the biosynthetic enzymes in legume nodules, Becana *et al.* (2000) have suggested that γ-ECS is plastidic, hGSH-S is

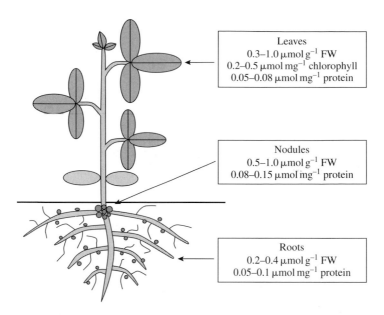

Figure 1.6 Typical values for total glutathione contents in leaves and roots, and root nodules under optimal growth conditions. For simplicity, we have shown a legume here but the values for leaves and roots equally apply to other plant species. In all tissues, the glutathione pool is reduced by more than 90%.

cytosolic and GSH-S isoforms exist in both the cytosol and mitochondria in several legume species. hGSH is synthesized in the cytosol and subsequently taken up by the mitochondria as well as the bacteroids, which suggest that hGSH is important for both symbiotic partners (Iturbe-Ormaetxe *et al.*, 2001).

1.8 Transport and transporters

Photosynthetic cells produce and export glutathione which is supplied to other organs through the phloem (Herschbach *et al.*, 2000). It has long been appreciated that pho-toautotrophic tobacco cells export glutathione into the culture medium faster than heterotrophically grown cells (Bergmann & Rennenberg, 1978). GSH is also readily taken up by cells in culture and by leaf protoplasts (Schneider *et al.*, 1992; Jamaï *et al.*, 1996) (for a review of the functions of intercellular glutathione transport systems in plants, see Foyer *et al.*, 2001). Although glutathione concentrations in the vacuole are thought to be low, it is clear that the compound enters via transport of conjugates. Transport from the cytosol to the vacuole occurs via a Mg-ATP glutathione-*S*-conjugate transporter which is up-regulated along with GSTs upon exposure to xenobiotics (Martinoia *et al.*, 1993; Li *et al.*, 1995). The glutathione-*S*-conjugates formed with anthocyanin and medicarpin are also transported into the vacuole by a specific glutathione-*S*-conjugate transporter (Li *et al.*, 1995), where they are further metabolised (Marrs, 1996). To date, there is little information on glutathione transport across the chloroplast envelope. As suggested earlier, the bulk of the leaf cell γ-EC may be synthesised in the chloroplast and hence requires a transporter for export to the cytosol where the majority of the cellular GSH might be produced for immediate use or export to the roots through the phloem.

We have previously described the kinetics of ^{35}S-labelled GSH uptake into chloroplasts, which suggest the active uptake of glutathione from the external medium (Noctor et al., 2000, 2002a). Such data suggest that GSH and GSSG are transported by common systems, although GSH appears to be the preferred metabolite. Further work is required to identify the chloroplast envelope transporter. It remains to be determined whether GSSG and GSH are transported by the same transporter but it would appear that the chloroplast transporter protein system is rather different from that on the plasmalemma, which is essentially a peptide transporter. In A. thaliana, there are nine orthologues of the yeast oligopeptide transporter (OPT) family (Koh et al., 2002). These are integral membrane proteins whose function in plants is the transport of small peptides including GSH (Koh et al., 2002). However, while heterologous expression of the A. thaliana OPT genes in Saccharomyces cerevisiae has confirmed that most are able to transport small peptides, only some exhibit glutathione transport capabilities (Koh et al., 2002). To date, no specific high-affinity GSH-transporters have been identified in plants.

Significant GSH uptake was observed in broad bean protoplasts (Jamaï et al., 1996) and wheat protoplasts (Noctor et al., 2002a). Two phases of GSH uptake were observed in wheat and in photoheterotrophic tobacco cells (Schneider et al., 1992). The bean protoplast plasmalemma transports GSSG at higher rates than GSH (Jamaï et al., 1996), consistent with a predominant physiological role in the recovery of glutathione oxidised in the apoplast, where there is little GR activity and no NAD(P)H. The two species had distinct kinetics, GSH showing a single saturable phase with K_m 0.4 mM whereas GSSG showed two saturable phases (Jamaï et al., 1996). Protoplast transport of the two species was similarly dependent on pH and both resulted in decreased acidification of the extracellular medium in leaf pieces (Jamaï et al., 1996). A high-affinity glutathione transporter has recently been cloned from yeast, with a K_m value close to that of the high-affinity system characterised in tobacco cells and wheat chloroplasts (54 μM) (Bourbouloux et al., 2000). Negligible inhibition was observed with other peptides or amino acids, but glutathione conjugates and GSSG both competed significantly with GSH. A database search of amino acid sequences identified homologues from other organisms, including five from Arabidopsis, which had 38–51% identity with the cloned yeast gene (Bourbouloux et al., 2000).

1.9   Glutathione and signalling

The GSH/GSSG couple is well suited to the role of redox sensor, indicative of the general cellular thiol-disulphide redox balance, and producing profound effects on metabolism and gene expression. Application of exogenous glutathione can elicit changes in the transcription of genes encoding cytosolic Cu, Zn superoxide dismutase, GR and 2-cys peroxiredoxins and pathogenesis-related protein-1 (Hérouart *et al.*, 1993; Wingsle & Karpinski, 1996; Baier & Dietz, 1997; Gomez *et al.*, 2004b). Glutathione-inducible hypersensitive elements have been identified in the proximal region of the chalcone synthase (CHS) promoter (Dron *et al.*, 1988). In animal cells, redox regulation of the transcription factor NFKB involves glutathione. This regulation is important for T cell function since glutathione augments the activity of T cell lymphocytes (Suthanthiran *et al.*, 1990).

It has long been known that defense genes are among those induced by GSH (Dron *et al.*, 1988; Wingate *et al.*, 1988). In particular, thiol-disulphide status appears to be crucial for expression of pathogenesis-related (PR) proteins, which are induced by salicylic acid (SA) and are involved in SAR. Increases in SA trigger reduction of disulphide bonds located on both the regulatory protein NPR1 and on certain TGA transcription factors with which NPR1 interacts (Després *et al.*, 2003; Mou *et al.*, 2003). This reduction stimulates both the translocation of NPR1 from the cytosol to the nucleus and the physical interaction of NPR1–TGA1 that is necessary for activation of PR gene transcription (Després *et al.*, 2003; Mou *et al.*, 2003). The redox dependence of the pathway suggests that any biotic or abiotic stimulus that can perturb the cellular redox state will up-regulate the same set of defense genes via the NPR1 pathway (Mou *et al.*, 2003). Redox-linked effects explain, e.g. PR gene expression in response to UV-B exposure (Green & Fluhr, 1995) or in catalase-deficient mutants (Willekens *et al.*, 1997) where extensive, specific oxidation of the glutathione pool and greatly enhanced glutathione accumulation occurs in certain conditions (May & Leaver, 1993; Noctor *et al.*, 2000). Indeed, it is noteworthy that oxidation of the leaf glutathione pool followed by enhanced glutathione synthesis is also seen in plant–pathogen interactions (Vanacker *et al.*, 2000; Mou *et al.*, 2003). In this situation, SAR induction involves an early burst of ROS and a transient or more sustained increase in cellular redox state (more oxidised), followed by a sharp decrease in cellular redox potential (more reduced) as a result of accumulation of antioxidants such as GSH. Interestingly, recent data with *Arabidopsis* shows that a specific cytosolic Trx is co-expressed with PR-1 in response to stress, including pathogen attack (Laloi *et al.*, 2004).

The simultaneous presence of pro- and anti-oxidants may be important in the NPR1–TGA interaction (Després *et al.*, 2003; Mou *et al.*, 2003). The glutathione couple shows a complex response in plant–pathogen interactions (Vanacker *et al.*, 2000; Noctor *et al.*, 2002a), where an oxidative burst and rapid induction of glutathione accumulation precede maximal induction of transcripts encoding phenylpropanoid metabolism enzymes (Zhang *et al.*, 1997; Vanacker *et al.*, 2000). We have recently shown that the addition of GSH or GSSG to tobacco leaves triggers calcium release (Gomez *et al.*, 2004b). Moreover, H_2O_2-induced expression of *GST1* involved a biphasic pattern of calcium release in *Arabidopsis* seedlings (Rentel & Knight, 2004). The addition of BSO enhanced the calcium signature and led to higher levels of *GST1* transcripts indicating that GSH was involved in the signal transduction process (Rentel & Knight, 2004). This evidence suggests that glutathione interacts with at least two key signalling factors: H_2O_2 and Ca^{2+}. It is probably also involved in transmission of oxidative stress signals, e.g. through GSSG-driven or enzyme-catalysed protein glutathionylation, reactions that may be reversed and regulated by specific enzymes such as glutaredoxins (Johansson *et al.*, 2004; Lemaire, 2004).

1.10   Conclusions and perspectives

The evidence provided shows that glutathione is a high abundance metabolite in plants (Figure 1.6) that has many diverse and important functions. In particular, glutathione is required for mitosis and root development. Although much work has focused on the regulation of GSH biosynthesis, it is only recently that molecular genetics has revealed that γ-ECS may fulfill a signalling function as well as a catalytic role (Ball *et al.*, 2004). In addition, the intra-cellular distribution of the biosynthetic enzymes has now been questioned (Wachter *et al.*, 2005) and more experimentation is required to determine whether the chloroplast and cytosol enzymes fulfill different roles in GSH production. GSH and GSSG have distinct roles in signal transduction. These metabolites trigger calcium signalling and are also involved in the H_2O_2-mediated signal transduction cascades. Many questions with regard to the transport of γ-EC, GSH and GSSG remain to be answered but it appears likely that the chloroplast transport system is rather different from that of other membranes. While peptide transporters have long been considered as candidates for plasma-membrane glutathione transport other types of GSH transporters may exist.

Recent evidence suggests that the enzymes of GSH synthesis and metabolism are induced together in response to stress (Mittova *et al.*, 2003). This suggests that there is considerable overlap in the signal transduction cascades. For example, common pathways may induce the genes encoding the enzymes of GSH synthesis and those involved in the induction of GPX homologous genes by singlet oxygen (Op den Camp *et al.*, 2003) and GST1 by H_2O_2 (Rentel & Knight, 2004).

References

Baier, M. and Dietz, K.-J. (1997) 'The plant 2-Cys peroxiredoxin BAS1 is a nuclear-encoded chloroplast protein: its expressional regulation, phylogenetic origin, and implications for its specific physiological function in plants', *The Plant Journal* 12, 179–190.

Ball, L., Accotto, G., Bechtold, U. *et al.* (2004) 'Evidence for a direct link between glutathione biosynthesis and stress defense gene expression in *Arabidopsis*', *The Plant Cell* 16, 2448–2462.

Bartling, D., Radzio, R., Steiner, U. and Weiler, E.W. (1993) 'A glutathione-S-transferase with glutathione-peroxidase activity from *Arabidopsis thaliana* – molecular cloning and functional characterization', *European Journal of Biochemistry* 216, 579–586.

Baudouin, E., Frendo, P., Le Gleuher, M. and Puppo, A. (2004) 'A *Medicago sativa* haem oxygenase gene is preferentially expressed in root nodules', *Journal of Experimental Botany* 55, 43–47.

Becana, M., Dalton, D.A., Moran, J.F., Iturbe-Ormaetxe, I., Matamoros, M.A. and Rubio, M.C. (2000) 'Reactive oxygen species and antioxidants in legume nodules', *Physiologia Plantarum* 109, 372–381.

Bergmann, L. and Rennenberg, H. (1978) 'Efflux and production of glutathione in suspension cultures of *Nicotiana tabacum*', *Zeitschrift für Planzenphysiologie Bd* **88**, 175–185.

Bourbouloux, A., Shahi, P., Chakladar, A., Delrot, S. and Bachhawat, A.K. (2000) 'Hgt1p, a high affinity glutathione transporter from the yeast *Saccharomyces cerevisiae*', *Journal of Biological Chemistry* **275**, 13259–13265.

Burgener, M., Suter, M., Jones, S. and Brunold, C. (1998) 'Cysteine is the transport metabolite of assimilated sulfur from bundle-sheath to mesophyll cells in maize leaves', *Plant Physiology* **116**, 1315–1322.

Cheng, J.C., Seeley, K.A. and Sung, Z.R. (1995) 'RML1 and RML2, *Arabidopsis* genes required for cell proliferation at the root tip', *Plant Physiology* **107**, 365–376.

Cobbett, C.S., May, M.J., Howden, R. and Rolls, B. (1998) 'The glutathione-deficient, cadmium-sensitive mutant, *cad2-1*, of *Arabidopsis thaliana* is deficient in γ-glutamylcysteine synthetase', *The Plant Journal* **16**, 73–78.

Collinson, E.J., Wheeler, G.L., Garrido, E.O., Avery, A.M., Avery, S.V. and Grant, C.M. (2002) 'The yeast glutaredoxins are active as glutathione peroxidases', *Journal of Biological Chemistry* **277**, 16712–16717.

Creissen, G., Firmin, J., Fryer, M. *et al.* (1999) 'Elevated glutathione biosynthetic capacity in the chloroplasts of transgenic tobacco plants paradoxically causes increased oxidative stress', *The Plant Cell* **11**, 1277–1291.

Cummins, I., Cole, D.J. and Edwards, R. (1999) 'A role for glutathione transferases functioning as glutathione peroxidases in resistance to multiple herbicides in black grass', *The Plant Journal* **18**, 285–292.

Dalton, D.A., Russel, S.A., Hanus, F.J., Pascoe, G.A. and Evans, H.J. (1986) 'Enzymatic reactions of ascorbate and glutathione that prevent peroxide damage in soybean root nodules', *Proceedings of the National Academy of Sciences USA* **83**, 3811–3815.

De Block, M., Verduyn, C., De Brouwer, D. and Cornelissen, M. (2004) 'Generating stress tolerant crops by economizing energy consumption', *Pflanzenschutz-Nachrichten Bayer* **57**, 105–110.

DeLille, J.M., Sehnke, P.C. and Ferl, R.J. (2001) 'The *Arabidopsis* 14-3-3 family of signaling regulators', *Plant Physiology* **126**, 35–38.

Després, C., Chubak, C., Rochon, A. *et al.* (2003) 'The *Arabidopsis* NPR1 disease resistance protein is a novel cofactor that confers redox regulation of DNA binding activity to the basis domain/leucine zipper transcription factor TGA1', *The Plant Cell* **15**, 2181–2191.

Dewitte, W. and Murray, J.A.H. (2003) 'The plant cell cycle', *Annual Review of Plant Biology* **54**, 235–264.

Dron, M., Clouse, S.D., Dixon, R.A., Lawton, M.A. and Lamb, C.J. (1988) 'Glutathione and fungal elicitor regulation of a plant defense gene promoter in electroporated protoplasts', *Proceedings of the National Academy of Sciences USA* **85**, 6738–6742.

Earnshaw, B.A. and Johnson, M.A. (1985) 'The effect of glutathione on development in wild carrot suspension cultures', *Biochemical and Biophysical Research Communications* **133**, 988–993.

Eshdat, Y., Holland, D., Faltin, Z. and Ben-Hayyim, G. (1997) 'Plant glutathione peroxidases', *Physiologia Plantarum* **100**, 234–240.

Foyer, C.H. and Halliwell, B. (1976) 'The presence of glutathione and glutathione reductase in chloroplasts: a proposed role in ascorbic acid metabolism', *Planta* **133**, 21–25.

Foyer, C.H., Souriau, N., Perret, S. *et al.* (1995) 'Overexpression of glutathione reductase but not glutathione synthetase leads to increases in antioxidant capacity and resistance to photoinhibition in poplar trees', *Plant Physiology* **109**, 1047–1057.

Foyer, C.H., Theodoulou, F.L. and Delrot, S. (2001) 'The functions of intercellular and intracellular glutathione transport systems in plants', *Trends in Plant Science* **6**, 486–492.

Freeman, J.L., Persans, M.W., Nieman, K. *et al.* (2004) 'Increased glutathione biosynthesis plays a role in nickel tolerance in Thlaspi nickel hyperaccumulators', *The Plant Cell* **16**, 2176–2191.

Frendo, P., Gallesi, D., Turnbull, R., Van de Sype, G., Hérouart, D. and Puppo, A. (1999) 'Localisation of glutathione and homoglutathione in *Medicago truncatula* is correlated to a differential expression of genes involved in their synthesis', *The Plant Journal* **17**, 215–219.

Frendo, P., Harrison, J., Norman, C. *et al.* (2005) 'Glutathione and homoglutathione play a critical role in the nodulation process of *Medicago truncatula*', *Molecular Plant–Microbe Interactions* **18**, 254–259.

Fricker, M.D. and Meyer, A.J. (2001) 'Confocal imaging of metabolism in vivo: pitfalls and possibilities', *Journal of Experimental Botany* **52**, 631–640.

Gechev, T., Willekens, H., Van Montagu, M. *et al.* (2003) 'Different responses of tobacco antioxidant enzymes to light and chilling stress', *Journal of Plant Physiology* **160**, 509–515.

Gomez, L., Vanacker, H., Buchner, P., Noctor, G. and Foyer, C.H. (2004a) 'Regulation of glutathione metabolism during the short-term chilling response of maize leaves', *Plant Physiology* **134**, 1662–1671.

Gomez, L.D., Noctor, G., Knight, M. and Foyer, C.H. (2004b) 'Regulation of calcium signaling and gene expression by glutathione', *Journal of Experimental Botany* **55**, 1851–1859.

Green, R. and Fluhr, R. (1995) 'UV-B-induced PR-1 accumulation is mediated by active oxygen species', *The Plant Cell* **7**, 203–212.

Griffith, O.W. (1982) 'Mechanism of action, metabolism, and toxicity of buthione sulfoximime and its higher homologs, potent inhibitors of glutathione synthesis', *Journal of Biological Chemistry* **257**, 13704–13712.

Griffith, O.W. and Meister, A. (1979) 'Potential and specific inhibition of glutathione synthesis by buthionine sulfoximine (S-n-butyl homocysteine sulfoximime)', *Journal of Biological Chemistry* **254**, 7558–7560.

Harada, E., Kusano, T. and Sano, H. (2000) 'Differential expression of genes encoding enzymes involved in sulfur assimilation pathways in response to wounding and jasmonate in *Arabidopsis thaliana*', *Journal of Plant Physiology* **156**, 272–276.

Harms, K., Von Ballmoos, P., Brunold, C., Höfgen, R. and Hesse, H. (2000) 'Expression of a bacterial serine acetyltransferase in transgenic potato plants leads to increased levels of cysteine and glutathione', *The Plant Journal* **22**, 335–343.

Hartmann, T., Hönicke, P., Wirtz, M., Hell, R., Rennenberg, H. and Kopriva, S. (2004) 'Sulfate assimilation in poplars (Populus tremula x P. alba) overexpressing γ-glutamyl synthetase in the cytosol', *Journal of Experimental Botany* **55**, 837–845.

Hell, R. and Bergmann, L. (1988) 'Glutathione synthetase in tobacco suspension cultures: catalytic properties and localization', *Physiologia Plantarum* **72**, 70–76.

Hell, R. and Bergmann, L. (1990) 'γ-Glutamylcysteine synthetase in higher plants: catalytic properties and subcellular localization', *Planta* **180**, 603–612.

Herbette, S., Lenne, C., Leblanc, N., Julien, J.-L., Drevet, J.R. and Roeckel-Drevet, P. (2002) 'Two GPX-like proteins from *Lycopersicon esculentum* and *Helianthus annuus* are antioxidant enzymes with phospholipids hydroperoxide glutathione peroxidase and thioredoxin peroxidase activities', *European Journal of Biochemistry* **269**, 2414–2420.

Hérouart, D., Van Montagu, M. and Inzé, D. (1993) 'Redox-activated expression of the cytosolic copper/zinc superoxide dismutase gene in *Nicotiana*', *Proceedings of the National Academy of Sciences USA* **90**, 3108–3112.

Herschbach, C., Van der Zalm, E., Schneider, A., Jouanin, L., De Kok, L.J. and Rennenberg, H. (2000) 'Regulation of sulfur nutrition in wild-type and transgenic poplar over-expressing γ-glutamylcysteine synthetase in the cytosol as affected by atmospheric H₂S', *Plant Physiology* **124**, 461–473.

Hirai, M.Y., Fujiwara, T., Awazuhara, M., Kimura, T., Noji, M. and Saito, K. (2003) 'Global expression profiling of sulfur-starved *Arabidopsis* by DNA macroarray reveals the role of O-acetyl-l-serine as a general regulator of gene expression in response to sulfur nutrition', *The Plant Journal* **33**, 651–663.

Hodges, M.D., Andrews, C.J., Johnson, D.A. and Hamilton, R.I. (1996) 'Antioxidant compound responses to chilling stress in differentially sensitive inbred maize lines', *Physiologia Plantarum* **98**, 685–692.

Howden, R., Andersen, C.R., Goldsbrough, P.B. and Cobbett, C.S. (1995) 'A cadmium-sensitive, glutathione-deficient mutant of *Arabidopsis thaliana*', *Plant Physiology* **107**, 1067–1073.

Huang, C.S., Chang, L.S., Anderson, M.E. and Meister, A. (1993) 'Catalytic and regulatory properties of the heavy subunit of rat kidney γ-glutamylcysteine synthetase', *Journal of Biological Chemistry* **268**, 19675–19680.

Iturbe-Ormaetxe, I., Heras, B., Matamoros, M.A., Ramos, J., Moran, J.F. and Becana, M. (2002) 'Cloning and functional characterization of homoglutathione synthetase from pea nodules', *Physiologia Plantarum* **115**, 69–73.

Iturbe-Ormaetxe, I., Matamoros, M.A., Rubio, M.C., Dalton, D.A. and Becana, M. (2001) 'The antioxidants of legume nodule mitochondria', *Molecular Plant–Microbe Interactions* **14**, 1189–1196.

Jamaï, A., Tommasini, R., Martinoia, E. and Delrot, S. (1996) 'Characterization of glutathione uptake in broad bean leaf protoplasts', *Plant Physiology* **111**, 1145–1152.

Jez, J.M., Cahoon, R.E. and Chen, S. (2004) 'Arabidopsis thaliana glutamate-cysteine ligase: functional properties, kinetic mechanism, and regulation of activity', *Journal of Biological Chemistry* **279**, 33463–33470.

Johansson, C., Lillig, C.H. and Holmgren, A. (2004) 'Human mitochondrial glutaredoxin reduces S-glutathionylated proteins with high affinity accepting electrons from either glutathione or thioredoxin reductase', *Journal of Biological Chemistry* **279**, 7537–7543.

Klapheck, S., Chrost, B., Starke, J. and Zimmermann, H. (1992) 'γ-Glutamylcysteinylserine – a new homologue of glutathione in plants of the family *Poaceae*', *Botanica Acta* **105**, 174–179.

Klapheck, S., Latus, C. and Bergmann, L. (1987) 'Localisation of glutathione synthetase and distribution of glutathione in leaf cells of *Pisum sativum* L.', *Journal of Plant Physiology* **131**, 123–131.

Kocsy, G., Von Ballmoos, P., Rüegsegger, A., Szalai, G., Galiba, G. and Brunold, C. (2001) 'Increasing the glutathione content in a chilling-sensitive maize genotype using safeners increased protection against chilling-induced injury', *Plant Physiology* **127**, 1147–1156.

Kocsy, G., Von Ballmoos, S., Suter, M. *et al.* (2000) 'Inhibition of glutathione synthesis reduces chilling tolerance in maize', *Planta* **211**, 528–536.

Koh, S., Wiles, A.M., Sharp, J.S., Naider, F.R., Becker, J.M. and Stacey, G. (2002) 'An oligopeptide transporter gene family in *Arabidopsis*', *Plant Physiology* **128**, 21–29.

Kopriva, S. and Rennenberg, H. (2004) 'Control of sulphate assimilation by glutathione synthesis: interactions with N and C metabolism', *Journal of Experimental Botany* **55**, 1831–1842.

Kopriva, S., Jones, S., Koprivova, A. *et al.* (2001) 'Influence of chilling stress on the intercellular distribution of assimilatory sulfate reduction and thiols in *Zea mays*', *Plant Biology* **3**, 24–31.

Kreuzwieser, J. and Rennenberg, H. (1998) 'Sulphate uptake and xylem loading of mycorrhizal beech roots', *New Phytologist* **140**, 319–329.

Laloi, C., Mestres-Ortega, D., Marco, Y., Meyer, Y. and Reichheld, J.P. (2004) 'The *Arabidopsis* cytosolic thioredoxin *h5* gene induction by oxidative stress and its W-Box-mediated response to pathogen elicitor', *Plant Physiology* **134**, 1006–1016.

Lappartient, A.G. and Touraine, B. (1996) 'Demand-driven control of root ATP sulphurylase activity and SO_4^{2-} uptake in intact canola. The role of phloem-translocated glutathione', *Plant Physiology* **111**, 147–157.

Lappartient, A.G., Vidmar, J.J., Leustek, T., Glass, A.D.M. and Touraine, B. (1999) 'Inter-organ signaling in plants: regulation of ATP sulphurylase and sulfate transporter genes expression in roots mediated by phloem-translocated comound', *The Plant Journal* **18**, 89–95.

Leipner, J., Fracheboud, Y. and Stamp, P. (1999) 'Effect of growing-season on the photosynthetic apparatus and leaf antioxidative defenses in two maize genotypes of different chilling tolerance', *Environmental and Experimental Botany* **42**, 129–139.

Lemaire, S.D. (2004) 'The glutaredoxin family of oxygenic photosynthetic organisms', *Photosynthesis Research* **79**, 305–318.

Levine, A., Tenhaken, R., Dixon, R. and Lamb, C. (1994) 'H_2O_2 from the oxidative burst orchestrates the plant hypersensitive disease resistance response', *Cell* **79**, 583–593.

Li, Z.-S., Zhao, Y. and Rea, P.A. (1995) 'Magnesium adenosine 5'-triphosphate-energized transport of glutathione-*S*-conjugates by plant vacuolar membrane vesicles', *Plant Physiology* **107**, 1257–1268.

Macnicol, P.K. and Bergmann, L. (1984) 'A role for homoglutathione in organic sulfur transport to the developing mung bean seed', *Plant Science* **36**, 219–223.

Marrs, K.A. (1996) 'The functions and regulation of glutathione S-transferases in plants', *Annual Review of Plant Physiology and Plant Molecular Biology* **47**, 127–158.

Martinoia, E., Grill, E., Tommasini, R., Kreuz, K. and Amrhein, N. (1993) 'An ATP-dependent glutathione-*S*-conjugate "export" pump in the vacuolar membrane of plants', *Nature* **364**, 247–249.

Maruyama-Nakashita, A., Inoue, E., Watanabe-Takahashi, A., Yarnaya, T. and Takahashi, H. (2003) 'Transcriptome profiling of sulfur-responsive genes in *Arabidopsis* reveals global effects of sulfur nutrition on multiple metabolic pathways', *Plant Physiology* **132**, 597–605.

Matamoros, M.A., Dalton, D.A., Ramos, J., Clemente, M.R., Rubio, M.C. and Becana, M. (2003) 'Biochemistry and molecular biology of antioxidants in the rhizobia-legume symbiosis', *Plant Physiology* **133**, 499–509.

Matamoros, M.A., Moran, J.F., Iturbe-Ormaetxe, I., Rubio, M.C. and Becana, M. (1999) 'Glutathione and homoglutathione synthesis in legume root nodules', *Plant Physiology* **121**, 879–888.

Maughan, S. (2003) 'Selection and characterization of *Arabidopsis* mutants resistant to a glutathione biosynthesis inhibitor', PhD thesis, University of Melbourne, Australia.

May, M.J. and Leaver, C.J. (1993) 'Oxidative stimulation of glutathione synthesis in *Arabidopsis thaliana* suspension cultures', *Plant Physiology* **130**, 621–627.

May, M.J. and Leaver, C.J. (1994) '*Arabidopsis thaliana* γ-glutamylcysteine synthetase is structurally unrelated to mammalian, yeast and *E. coli* homologues', *Proceedings of the National Academy of Sciences USA* **91**, 10059–10063.

May, M.J., Vernoux, T., Leaver, C., van Montagu, M. and Inzé, D. (1998a) 'Glutathione homeostasis in plants: implications for environmental sensing and plant development', *Journal of Experimental Botany* **49**, 649–667.

May, M.J., Vernoux, T., Sánchez-Fernández, R., van Montagu, M. and Inzé, D. (1998b) 'Evidence for posttranscriptional activation of γ-glutamylcysteine synthetase during plant stress responses', *Proceedings of the National Academy of Sciences USA* **95**, 12049–12054.

Meijer, M. and Murray, J.A.H. (2001) 'Cell cycle controls and the development of plant form', *Current Opinion in Plant Biology* **4**, 44–49.

Meister, A. (1988) 'Glutathione metabolism and its selective modification', *Journal of Biological Chemistry* **263**, 17205–17208.

Mittova, V., Kiddle, G., Theodoulou, F.L. *et al.* (2003) 'Co-ordinate induction of glutathione biosynthesis and glutathione-metabolising enzymes is correlated with salt tolerance in tomato', *FEBS Letters* **554**, 417–421.

Mou, Z., Fan, W. and Dong, X. (2003) 'Inducers of plant systemic acquired resistance regulate NPR1 function through redox changes', *Cell* **27**, 935–944.

Mueller, L.A., Goodman, C.D., Silady, R.A. and Walbot, V. (2000) 'AN9, a petunia glutathione S-transferase required for anthocyanin sequestration, is a flavonoid-binding protein', *Plant Physiology* **123**, 1561–1570.

Mulcahy, R.T., Bailey, H.H. and Gipp, J.J. (1995) 'Transfection of complementary DNAs for the heavy and light subunits of human γ-glutamylcysteine synthetase results in an elevation of intracellular glutathione and resistance to melphalan', *Cancer Research* **55**, 4771–4775.

Mullineaux, P.M., Karpinski, S., Jimenez, A., Cleary, S.P., Robinson, C. and Creissen, G.P. (1998) 'Indentification of cDNAs encoding plastid-targeted glutathione peroxidase', *The Plant Journal* **13**, 375–379.

Noctor, G. and Foyer, C.H. (1998) 'Simultaneous measurement of foliar glutathione, γ-glutamylcysteine and amino acids by high-performance liquid chromatography: comparison with two other assay methods for glutathione', *Analytical Biochemistry* **264**, 98–110.

Noctor, G., Arisi, A.-C.M., Jouanin, L. and Foyer, C.H. (1998a) 'Manipulation of glutathione and amino acid biosynthesis in the chloroplast', *Plant Physiology* **118**, 471–482.

Noctor, G., Arisi, A.-C.M., Jouanin, L., Kunert, K.J., Rennenberg, H. and Foyer, C.H. (1998b) 'Glutathione: biosynthesis and metabolism explored in transformed poplar', *Journal of Experimental Botany* **49**, 623–647.

Noctor, G., Arisi, A.-C.M., Jouanin, L., Valadier, M.-H., Roux, Y. and Foyer, C.H. (1997) 'The role of glycine in determining the rate of glutathione synthesis in poplars. Possible implications for glutathione production during stress', *Physiologia Plantarum* **100**, 255–263.

Noctor, G., Gomez, L., Vanacker, H. and Foyer, C.H. (2002a) 'Glutathione homeostasis and signaling: the influence of biosynthesis, compartmentation and transport', *Journal of Experimental Botany* **53**, 1283–1304.

Noctor, G., Strohm, M., Jouanin, L., Kunert, K.J., Foyer, C.H. and Rennenberg, H. (1996) 'Synthesis of glutathione in leaves of transgenic poplar (*Populus tremula x P. alba*) overexpressing γ-glutamylcysteine synthetase', *Plant Physiology* **112**, 1071–1078.

Noctor, G., Veljovic-Jovanovic, S., Driscoll, S., Novitskaya, L. and Foyer, C.H. (2002a) 'Drought and oxidative load in wheat leaves: a predominant role for photorespiration?', *Annals of Botany* **89**, 841–850.

Noctor, G., Veljovic-Jovanovic, S. and Foyer, C.H. (2000) 'Peroxide processing in photosynthesis: antioxidant coupling and redox signaling', *Proceedings of the Royal Society of London, B* **355**, 1465–1475.

Ogawa, K., Hatano-Iwasaki, A., Tokuyama, M. and Iwabuchi, M. (2002) A possible role of glutathione as an electron source for flowering in *Arabidopsis thaliana*, *PS2001 Proceedings*, S20-30, CSIRO Publishing, Melbourne.

Op den Camp, R.G., Przybyla, D., Ochsenbein, C. *et al.* (2003) 'Rapid induction of distinct stress responses after the release of singlet oxygen in *Arabidopsis*', *The Plant Cell* **15**, 2320–2332.

Pastori, G.M., Kiddle, G., Antoniw, J. *et al.* (2003) 'Leaf vitamin C contents modulate plant defense transcripts and regulate genes that control development through hormone signaling', *The Plant Cell* **15**, 939–951.

Pilon-Smits, E.A.H., Ahu, Y.L., Pilon, M. and Terry, N. (1999) Overexpression of glutathione synthesizing enzymes enhances cadmium accumulation in *Brassica juncea, Proceedings of the 5th International Conference on the Biogeochemistry of Trace Elements*, Vienna, pp. 890–891.

Potters, G., De Gara, L., Asard, H. and Horemans, N. (2002) 'Ascorbate and glutathione: guardians of the cell cycle, partners in crime?', *Plant Physiology and Biochemistry* **40**, 537–548.

Potters, G., Horemans, N., Bellone, S. *et al.* (2004) 'Dehydroascorbate influences the plant cell cycle through a glutathione-independent reduction mechanism', *Plant Physiology* **134**, 1479–1487.

Prasad, T.K. (1996) 'Mechanisms of chilling-induced oxidative stress injury and tolerance in developing maize seedlings: changes in antioxidant system, oxidation of proteins and lipids, and protease activities', *The Plant Journal* **10**, 1017–1026.

Prasad, T.K., Anderson, M.D., Martin, B.A. and Steward, C.R. (1994) 'Evidence for chilling-induced oxidative stress in maize seedlings and a regulatory role for hydrogen peroxide', *The Plant Cell* **6**, 65–74.

Price, C. (1957) 'A new thiol in legumes', *Nature* **180**, 148–149.

Rennenberg, H., Steinkamp, R. and Kesselmeier, J. (1981) '5-oxo-prolinase in *Nicotiana tabacum*: catalytic properties and subcellular localization', *Physiologia Plantarum* **62**, 211–216.

Rentel, M.C. and Knight, M.R. (2004) 'Oxidative stress-induced calcium signaling in *Arabidopsis*', *Plant Physiology* **135**, 1471–1479.

Rouhier, N., Gelhaye, E. and Jacquot, J.P. (2002) 'Glutaredoxin-dependent peroxiredoxin from poplar: protein–protein interaction and catalytic mechanism', *Journal of Biological Chemistry* **277**, 13609–13614.

Roxas, V.P., Smith, R.K., Allen, E.R. and Allen, R.D. (1997) 'Overexpression of glutathione *S*-transferase/glutathione peroxidase enhances the growth of transgenic tobacco seedlings during stress', *Nature Biotechnology* **15**, 988–991.

Rüegsegger, A. and Brunold, C. (1993) 'Localisation of γ-glutamylcysteine synthetase and glutathione synthetase activity in maize seedlings', *Plant Physiology* **101**, 561–566.

Schäfer, F.Q. and Buettner, G.H. (2001) 'Redox environment of the cell as viewed through the redox state of the glutathione disulfide/glutathione couple', *Free Radical Biology and Medicine* **30**, 1191–1212.

Schäfer, H.J., Haag-Kerwer, A. and Rausch, T. (1998) 'cDNA cloning and expression analysis of genes encoding GSH synthesis in roots of the heavy-metal accumulator *Brassica juncea* L.: evidence for Cd-induction of a putative mitochondrial γ-glutamylcysteine synthetase isoform', *Plant Molecular Biology* **37**, 87–97.

Schneider, S. and Bergmann, L. (1995) 'Regulation of glutathione synthesis in suspension cultures of parsley and tobacco', *Botanica Acta* **108**, 34–40.

Schneider, S., Martini, N. and Rennenberg, H. (1992) 'Reduced glutathione (GSH) transport in cultured tobacco cells', *Plant Physiology and Biochemistry* **30**, 29–38.

Sen Gupta, A., Alscher, R.G. and McCune, D. (1991) 'Response of photosynthesis and cellular antioxidants to ozone in *Populus* leaves', *Plant Physiology* **96**, 650–655.

Smith, I.K., Kendall, A.C., Keys, A.J., Turner, J.C. and Lea, P.J. (1984) 'Increased levels of glutathione in a catalase-deficient mutant of barley (*Hordeum vulgare* L.)', *Plant Science Letters* **37**, 29–33.

Smith, I.K., Kendall, A.C., Keys, A.J., Turner, J.C. and Lea, P.J. (1985) 'The regulation of the biosynthesis of glutathione in leaves of barley (*Hordeum vulgare* L.)', *Plant Science* **41**, 11–17.

Steinkamp, R. and Rennenberg, H. (1984) 'γ-glutamyltranspeptidase in tobacco suspension cultures: catalytic properties and subcellular localization', *Physiologia Plantarum* **61**, 251–256.

Steinkamp, R., Schweihofen, B. and Rennenberg, H. (1987) 'γ-glutamylcyclotransferase in tobacco suspension cultures: catalytic properties and subcellular localization', *Physiologia Plantarum* **69**, 499–503.

Storozhenko, S., Belles-Boix, E., Babiychuk, E. *et al.* (2002) 'γ-glutamyl transpeptidase in transgenic tobacco plants. Cellular localization, processing, and biochemical properties', *Plant Physiology* **128**, 1109–1119.

Strohm, M., Jouanin, L., Kunert, K.J. *et al.* (1995) 'Regulation of glutathione synthesis in leaves of transgenic poplar (*Populus tremula* x *P. alba*) overexpressing glutathione synthetase', *The Plant Journal* **7**, 141–145.

Sun, W.M., Huang, Z.Z. and Lu, S.C. (1996) 'Regulation of γ-glutamylcysteine synthetase by protein phophorylation', *Biochemical Journal* **320**, 321–328.

Suthanthiran, M., Anderson, M.E., Sharma, V.K. and Meister, A. (1990) 'Glutathione regulates activation-dependent DNA synthesis in highly purified normal human T lymphocytes stimulated via the CD2 and CD3 antigens', *Proceedings of the National Academy of Sciences USA* **87**, 3343–3347.

Tausz, M., Sircelj, H. and Grill, D. (2004) 'The glutathione system as a stress marker in plant ecophysiology: is a stress-response concept valid?', *Journal of Experimental Botany* **55**, 1855–1862.

Ursini, F., Mariorino, M., Brigelius-Flohé, R. *et al.* (1995) 'Diversity of glutathione peroxidase', *Methods in Enzymology* **252**, 38–53.

Vanacker, H., Carver, T.L.W. and Foyer, C.H. (2000) 'Early H_2O_2 accumulation in mesophyll cells leads to induction of glutathione during the hypersensitive response in the barley-powdery mildew interaction', *Plant Physiology* **123**, 1289–1300.

Van Heerden, P.D.R. and Krüger, G.H.J. (2002) 'Separately and simultaneously induced dark chilling and drought stress effects on photosynthesis, proline accumulation and antioxidant metabolism in soybean', *Journal of Plant Physiology* **159**, 1077–1086.

Van Heerden, P.D.R., Krüger, G.H.J., Loveland, J.E., Parry, M.A.J. and Foyer, C.H. (2003) 'Dark chilling imposé metabolic restrictions on photosynthesis in soybean', *Plant Cell and Environment* **26**, 323–337.

Vauclare, P., Kopriva, S., Fell, D. *et al.* (2002) 'Flux control of sulphate assimilation in *Arabidopsis thaliana*: Adenosine 5′-phosphosulphate reductase is more susceptible to negative control by thiols than ATP sulphurylase', *The Plant Journal* **31**, 729–740.

Vernoux, T., Wilson, R.C., Seeley, K.A. *et al.* (2000) 'The ROOT MERISTEMLESS1/CADMIUM SENSITIVE2 gene defines a glutathione-dependent pathway involved in initiation and maintenance of cell division during postembryonic root development', *The Plant Cell* **12**, 97–110.

Wachter, A., Wolf, S., Steininger, H., Bogs, J. and Raush, T. (2005) 'Differential targeting of GSH1 and GSH2 is achieved by multiple transcription initiation: implications for the compartmentation of glutathione biosynthesis in the *Brassicaceae*', *The Plant Journal* **41**, 15–30.

Willekens, H., Chamnongpol, S., Davey, M. *et al.* (1997) 'Catalase is a sink for H_2O_2 and is indispensable for stress defense in C_3 plants', *EMBO Journal* **16**, 4806–4816.

Wingate, V.P.M., Lawton, M.A. and Lamb, C.J. (1988) 'Glutathione causes a massive and selective induction of plant defense genes', *Plant Physiology* **87**, 206–210.

Wingsle, G. and Karpinski, S. (1996) 'Differential redox regulation by glutathione of glutathione reductase and CuZn-superoxide dismutase gene expression in *Pinus sylvestris* L. needles', *Planta* **198**, 151–157.

Xiang, C. and Bertrand, D. (2000) Glutathione synthesis in *Arabidopsis*: multilevel controls coordinate responses to stress, in *Sulfur Nutrition and Sulphur Assimilation in Higher Plants* (ed. C. Brunold, H. Rennenberg, L.J. De Kok, I. Stulen and J.C. Davidian), Paul Haupt, Bern, Switzerland, pp. 409–412.

Xiang, C. and Oliver, D.J. (1998) 'Glutathione metabolic genes coordinately respond to heavy metals and jasmonic acid in *Arabidopsis*', *The Plant Cell* **10**, 1539–1550.

Xiang, C., Werner, B.L., Christensen, E.M. and Oliver, D.J. (2001) 'The biological functions of glutathione revisited in *Arabidopsis* transgenic plants with altered glutathione levels', *Plant Physiology* **126**, 564–574.

Yanagida, M., Mino, M., Iwabuchi, M. and Ogawa, K. (2002) 'Vernalization-induced bolting involves regulation of glutathione biosynthesis in *Eustoma grandiflorum*', *Plant Cell Physiology* **43**, s43.

Yoshimura, K., Miyao, K., Gaber, A. *et al.* (2004) 'Enhancement of stress tolerance in transgenic tobacco plants overexpressing *Chlamydomonas* glutathione peroxidase in chloroplasts or cytosol', *The Plant Journal* **37**, 21–33.

Yu, C.W., Murphy, T.M., Sung, W.W. and Lin, C.H. (2002) 'H_2O_2 treatment induces glutathione accumulation and chilling tolerance in mung bean', *Functional Plant Biology* **29**, 1081–1087.

Zhang, L., Robbins, M.P., Carver, T.L.W. and Zeyen, R.J. (1997) 'Induction of phenylpropanoid gene transcripts in oat attacked by *Erysiphe graminis* at 20°C and 10°C', *Physiological and Molecular Plant Pathology* **51**, 15–33.

Zhu, Y., Pilon-Smits, E.A.H., Jouanin, L. and Terry, N. (1999) 'Overexpression of glutathione synthetase in *Brassica juncea* enhances cadmium tolerance and accumulation', *Plant Physiology* **119**, 73–79.

2 Plant thiol enzymes and thiol homeostasis in relation to thiol-dependent redox regulation and oxidative stress

Karl-Josef Dietz

2.1 Introduction: plant sulfur and thiol contents

Plant shoots contain about 30 μmol sulfur per gram dry weight when grown with an adequate sulfur supply (Epstein, 1965). Assuming 10% dry weight of plant tissue, this figure corresponds to 3.3 mM of sulfur. At low sulfur supply, the majority of S is present in the organic sulfur pool. For example, in *Lemna* grown at 0.32 μM sulfate in the medium, 5% S was contained in sulfate and 95% S in organic sulfur compounds. Within the protein fraction of *Lemna* and *Chlorella*, respectively, methionine made up about 60%, and cysteine about 40% of the S-containing amino acids. The latter was equivalent to 0.8–1.1 mM protein-thiols (Giovanelli *et al.*, 1980). In the same experiment, the glutathione concentrations varied between 77 and 205 μM, and cysteine concentrations between 4 and 20 μM. Usually, leaf glutathione concentrations exceed root glutathione (GSH) levels by a factor of about 5 (Bergmann & Rennenberg, 1993). Protein thiols mostly surpass non-protein thiols by a factor of about 2–4. However, upon elicitation with heavy metals, the dipeptidyl transferase phytochelatin synthase is activated through posttranslational regulation that involves metal ion binding, and phytochelatins [$(\gamma$-EC$)_n$X, where $n = 3$–8 and X corresponds to G and A in most cases] are synthesized from glutathione by repeated transfer of γ-glutamyl-cysteine residues to a glutathione primer (Grill *et al.*, 1989). In the presence of certain heavy metals such as Cd, Cu and Zn, total non-protein thiol concentrations may increase by factors of 10, and may equal or even exceed protein thiol levels, particularly in roots (Finkemeier *et al.*, 2003). Sulfur availability may become limiting and the glutathione pool depleted. Under normal growth conditions, glutathione is the predominant and largest homogeneous thiol buffer of the cells, whereas the protein-bound thiol pool (the 'thiol proteome'), naturally, is extremely diversified. Nevertheless, single protein thiols may be abundant, e.g. the 2-cysteine peroxiredoxin (2-Cys Prx) with two functional thiol groups has been reported to account for 0.6% of the chloroplast protein (König *et al.*, 2002). Assuming 10 mg protein per milligram chlorophyll and 40 μl chloroplast volume per milligram chlorophyll (Winter *et al.*, 1993), the 2-Cys Prx is present at stromal concentrations of about 60 μM. Thiols of ribulose-1,5-bisphosphate carboxylase subunits are present in millimolar concentrations. In relation to the topic of oxidative defence, it should be concluded here that protein

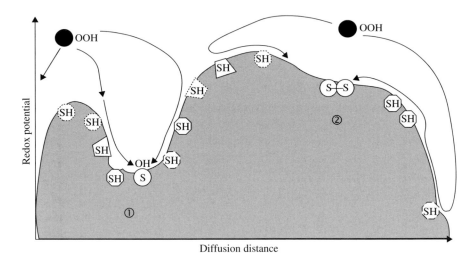

Figure 2.1 Thiols as antioxidants. In principle, all thiols have an antioxidant capacity and are targets to oxidation, for instance, by peroxides. However, the energy distribution (redox potential) and specificity of the active sites alters the efficiency of oxidation, thereby creating sinks such as peroxiredoxins that protect other thiols from rapid oxidation. The scheme depicts a thiol landscape through which the peroxide diffuses. Specific thiols efficiently react with the peroxide and oxidize to sulfenic acid (1) or disulfide derivatives (2).

thiol concentrations are high in plasmatic compartments of the cell, and protein thiols are likely to be efficient targets for unwanted oxidation. An elaborate antioxidant defence system serves in two ways: first by offering sinks such as peroxiredoxins and glutathione peroxidase for preferential attack of reactive oxygen species (Figure 2.1) and second by maintenance of highly active mechanisms to restore functional thiols following their accidental oxidation.

Obviously, dithiol–disulfide transitions have an immense physiological importance in regulating plant cell function. This chapter summarizes some of the knowledge related to the chemical reactivity of organic thiols in cell metabolism and extends to previous reviews and comments that have focused on redox regulation (e.g. Pastori & Foyer, 2002; Dietz, 2003a; Pfannschmidt, 2003; Dietz & Scheibe, 2004). Additionally, emphasis is placed on the role of thiol proteins in antioxidant defence. The author apologizes to all colleagues for compiling a somewhat biased description of the thiol world of plant cells and for not including all the important findings due to lack of space.

2.2   The redox potential and its relation to the redox proteome

The midpoint redox potential E_m of a redox-pair defines its tendency to gain electrons. The more negative the E_m of a redox element, the more is the energy required for its reduction. Since the various thiol pools do not exchange freely, it is impossible to describe the cellular redox state in a comprehensive manner. However, there exist indicators such as the redox state of the glutathione pool that may be used to describe the redox environment of cells. The midpoint redox potential of glutathione is -0.23 V (Segel, 1976). The redox potential of glutathione within the cell may be more negative, since the glutathione pool often is rather reduced. Furthermore, since two GSH molecules form one molecule of oxidized glutathione (glutathione disulphide, GSSG), the redox potential of the glutathione couple is also dependent on the total glutathione concentration (Schafer & Buettner, 2001), decreasing with decreasing concentration of total glutathione (GSH + GSSG). Schafer and Buettner (2001) analysed the state of the glutathione couple that appeared to be related to the biological status of yeast cells. Dividing cells had a redox state of GSH/GSSG of about -0.24, differentiating cells of about -0.2 and apoptopic cells of -0.17 V. The redox midpoint potential of protein thiol groups strongly depends on the microenvironment defined by amino acid residues neighbouring the cysteinyl thiol. The most common method of determining the redox potential of proteins is realized in redox buffers adjusted through defined ratios of GSH/GSSG or dithiothreitol/threitoldisulfide (DTT_{red}/DTT_{ox}). Accurate determination of E_m is problematic if a significant fraction of mixed disulfide intermediates between the protein and the redox buffer is formed. For that reason, Åslund et al. (1997) conducted redox potential determinations by directly quantifying the reduced and oxidized fraction of *Escherichia coli* redox proteins following HPLC separation. Midpoint redox potentials of two *E. coli* glutaredoxins Grx 1 and Grx 3 with a standard redox site containing a C-P-Y-C-motif were found to be -233 and -198 mV, respectively. The differences in E_m correlated with a 15-fold higher activity of Grx 1 than Grx 3 as hydrogen donor to ribonucleotide reductase (Åslund et al., 1996). This example shows the importance of redox potential of active site thiol groups in driving efficient disulfide–dithiol exchange reactions. Quite negative redox potentials of regulatory sulfhydryl groups are found in some redox regulated chloroplast enzymes such as fructose-1,6-bisphosphatase. E_m of NADP-dependent malate dehydrogenase of sorghum was reported with -0.33 mV (Schürmann & Jacquot, 2000) and that of chloroplast 2-Cys peroxiredoxin with -0.32 mV (Horling et al., 2003). The redox state of thioredoxins is coupled to the NADP system with an E_m of -0.32 (Segel, 1976). The redox potential of the NADP system in the cell may be significantly more negative when the NADP pool becomes more reduced (Dietz, 2003). In addition, the midpoint redox potential of thioredoxins is considerably more negative than that of glutaredoxins. Accordingly, thioredoxins are the more powerful thiol reductants in the cells. E_m of chloroplast Trx (thioredoxin) f and Trx m is -0.29 and -0.3 V at pH 7, respectively (Schürmann & Jacquot, 2000). Even more negative E_m values of At-Trx m1-4, At-Trx-f1 and At-Trx-x were reported by Collin et al. (2003) using a fluorescence labelling approach with monobromobimane at pH 7.9, where E_m ranged between -0.368 and -0.351 V, except for Trx x with an E_m of -0.336. As expected, E_m of cytosolic Trx h2 and h3 was less negative with values of -0.283 and -0.294 V, respectively (Bréhélin et al., 2004).

The redox state of thiols depends on the relative rates of reduction and oxidation of the various interacting components of the cellular redox environment and the kinetic constraints governing their exchange reaction. Specificity and irreversible thermodynamics are important features of the redox networks similar to general metabolism. Usually, only specific pairs of redox proteins catalyse the thiol–disulfide exchange reaction. Reduction of disulfides is catalysed by dithiol–disulfide exchange proteins and reductases, respectively. Many of the enzymes involved are represented in gene families whose members are present in plants in higher numbers than in non-photosynthetic organisms (Dietz, 2003), suggesting that thiol-dependent redox regulation is highly important and more diversified in plants than in animals, fungi and bacteria. Furthermore, the redox milieu and redox network differ among the sub-cellular compartments of eukaryotic cells, i.e. the plastid, mitochondrion, cytosol and nucleus. The diversity of thiol–disulfide exchange proteins shows that distinct pathways of electron transfer exist, which may separate reductive processes involved in regulation and in thiol restoration.

2.3 Oxidation of thiol groups

Thiol groups can undergo successive reactions of oxidation producing disulfide – (Equation 2.1), sulfenic acid – (Equation 2.2), sulfinic acid – (Equation 2.3) and, finally, sulfonic acid – (cysteic acid; Equation 2.4) derivatives of cysteinyl group.

$$4RSH + O_2 \rightarrow 2RSSR + 2H_2O \qquad (2.1)$$

$$RSH + R'OOH \rightarrow [RSOH] + R'OH \qquad (2.2)$$

$$[RSOH] + R'OOH \rightarrow RSO_2H + R'OH \qquad (2.3)$$

$$RSO_2H + R'OOH \rightarrow RSO_3H + R'OH \qquad (2.4)$$

$$[RSOH] + R'SH \rightarrow RSSR' + H_2O \qquad (2.5)$$

The sulfenic acid [RSOH] is highly reactive and does not permit isolation except in very special cases (Capozzi & Modena, 1974). Usually, the sulfenic acid derivative reacts with another cysteinyl thiol to yield a disulfide and water (Equation 2.5). The oxidation of thiols by hydrogen peroxide (H_2O_2), alkylhydroperoxides as well as peroxyacids proceeds in most cases via the corresponding disulfide. In the presence of excess and strong oxidant, the sulfenic acid form and also the disulfide can be oxidized further to form sulfinic and sulfonic acid derivatives (Figure 2.2). This is interesting since dithiol formation is considered to protect thiols from further oxidation. The sulfinic acid form is usually slowly autoxidized in an oxygen-containing environment or rapidly oxidized by H_2O_2. The dithiol–disulfide transition is easily reversed, and, consequently, the thiol–disulfide exchange constitutes the most commonly encountered thiol-related redox reaction and will be discussed in detail later. Oxidation of cysteinyl sulfur to sulfinic and sulfonic acid was believed to be irreversible. Recently, a novel enzyme activity was identified that reduces

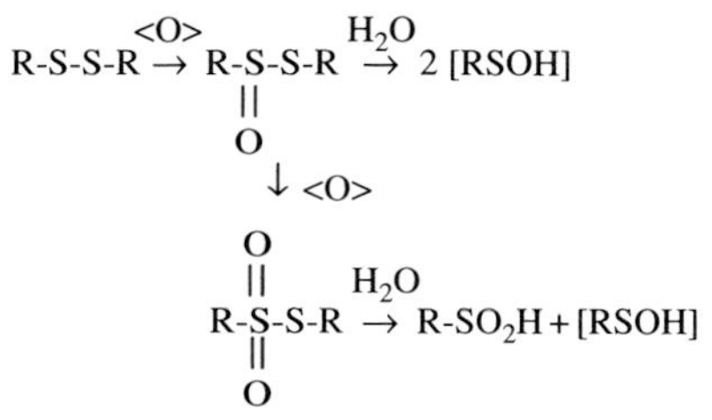

$$R\text{-}S\text{-}S\text{-}R \xrightarrow{<O>} \underset{\underset{O}{\overset{\|}{}}}{R\text{-}S\text{-}S\text{-}R} \xrightarrow{H_2O} 2\,[RSOH]$$

$$\downarrow <O>$$

$$\underset{\underset{O}{\overset{\|}{}}}{\overset{\overset{O}{\overset{\|}{}}}{R\text{-}S\text{-}S\text{-}R}} \xrightarrow{H_2O} R\text{-}SO_2H + [RSOH]$$

Figure 2.2 Oxidation of organic disulfides. These reactions will only occur in limited and specific cases. $<O>$ indicates a strong oxidant.

over-oxidized protein thiols. The enzyme was identified by its activity in reducing sulfinic acid derivatives of the active site thiol of yeast peroxiredoxin Tsa1. The enzyme was termed sulfiredoxin (Biteau *et al.*, 2003). A homologous gene is present in the *Arabidopsis* genome.

2.4   C-X-X-C and C-X-X-S motifs in redox proteins

The most commonly found motif in redox proteins is the C-X-X-C motif, where two cysteinyl groups are spaced by two other residues. C-X-X-C is a motif employed in the formation, reduction and isomerization of disulfide bonds. Within the motif, serine and threonine may substitute for the N-terminal and, interestingly, also the C-terminal Cys (Fomenko & Gladyshev, 2003). The basic and modified motifs are found in redox proteins such as thioredoxins, glutaredoxins, peroxiredoxins and disulfide isomerases. The N-terminally located Cys residue has a lower pK and serves as a nucleophilic group to attack disulfides or sulfenic acid derivatives of substrate proteins. A mixed disulfide is formed between the redox protein and the thiol substrate. In a second step, the resolving C-terminal Cys group of the redox protein attacks the intramolecular disulfide (Figure 2.3). The reduced form of the substrate protein is released and a disulfide bridge in the redox protein formed. Fomenko and Gladyshev (2003) searched three bacterial and the yeast genomes for 'true' variants of the C-X-X-C motif. Peroxiredoxins (Prx) belong to the thioredoxin-fold proteins, possessing T-X-X-C or S-X-X-C motifs. Catalytic function of Prx as redox proteins involves the second cysteinyl residue of the basic motif usually located in the N-terminal part of the protein (Dietz, 2003). The same variants of the C-X-X-C motif are found in Trx and glutathione-*S*-transferase (GST). Fomenko and Gladyshev (2003) conclude from their *in silico* study that not necessarily the N-terminal Cys residue but also the C-terminal Cys residue of the motif can function as nucleophilic group to attack disulfides. Furthermore, regularly occurring modifications of the motif to C-X-X-S as well as S-X-X-C in Grx- and Trx-like proteins, and to T-X-X-C in peroxiredoxins, as well as to C-X-X-T in glutathione peroxidase suggest that the hydroxyl groups of the serine and threonine residue

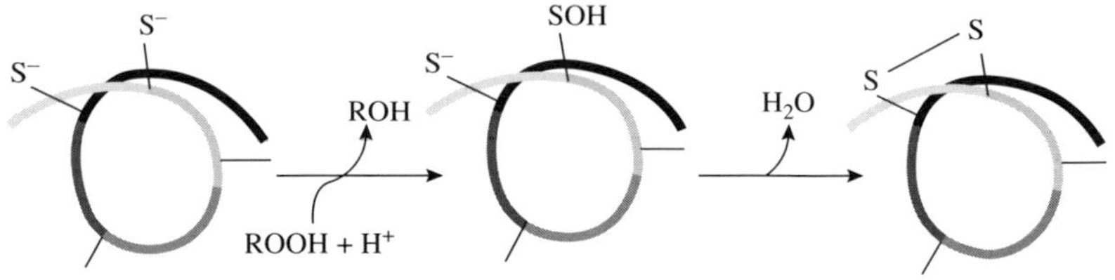

Figure 2.3 Principle structure of the C-X-X-C motif. Schematic representation of an active site as α-helical loop that forms a disulfide bridge upon oxidation.

may function in stabilizing the thiolate anion within the catalytic centre (Fomenko & Gladyshev, 2003).

2.5   The principle reactions that maintain thiol-redox homeostasis

Unavoidably in an oxygen-enriched atmosphere, thiol groups are finally oxidized by oxygen, reactive oxygen species or other oxidizing reagents, or substrates, and need to be re-reduced in order to be maintained in the reduced state. For that process, a continuous electron flux from appropriate donors through the protein thiols to oxygen as final acceptors has to be established (Table 2.1). Basically, the electron flux through the electron transport systems is fuelled by two reduction mechanisms coupled either to oxidation of NADPH or ferredoxin. As with all irreversible reaction sequences, the redox state of individual components of the electron transport chain depends on the relative rates of reduction and oxidation reactions. The difference in redox potential between the initial electron donors NAD(P)H and ferredoxin and the final acceptor O_2 is large enough to allow for almost full reduction as well as complete oxidation of protein thiols depending on the kinetic constraints. If the dithiol–disulfide interchange converts active proteins into inactive ones, or vice versa, this mechanism can be employed for both, the realization of switch on-/off-mechanisms as initially suggested for enzymes in the Calvin cycle during light/dark transitions (Scheibe, 1991), and fine-tuning as proposed for many of the recently identified novel targets. Thiol–disulfide exchange proteins with high specificity for special target proteins are required. In other cases, thiol reduction may not be part of regulation but may just represent a necessity to avoid irreversible inactivation and degradation of enzymes and proteins if they have been oxidized through oxidative processes. The latter process requires non-specific disulfide reductases; in fact, for all proteins that contain sensitive target cysteinyl groups.

NADPH-dependent thioredoxin reductases (NTRs) are present in the cytosol and mitochondrion; a ferredoxin-dependent thioredoxin reductase exists in the chloroplast (Walters & Johnson, 2004). The *Arabidopsis* genome encodes three NADPH-dependent thioredoxin reductases (NTR-A, B, C). The initially identified

Table 2.1 Energization of thiol-redox pathways by coupling to photosynthetic electron transport via ferredoxin and ferredoxin-dependent thioredoxin reductase and to glycolysis/respiration via NADPH

Donor →	Non-thiol reductant →	Thiol reductant →	Thiol–disulfide exchange →	Final acceptor
Photosynthesis	Fd → FTR	Trx	Protein-thiols	$O_2/H_2O_2/ROOH$
Glucose	NADPH → NTR	Trx	Protein-thiols	$O_2/H_2O_2/ROOH$
Glucose	NADPH → GR	GSH → Grx	Protein-thiols	$O_2/H_2O_2/ROOH$

NTR-B efficiently reduces cytosolic h-type Trx (Jacquot *et al.*, 1994). NTR-A was shown to be translated from two different mRNAs; the shorter transcript produces a cytosolic form of NTR-A, whereas the longer and less-abundant transcript is translated to a gene product imported by mitochondria (Laloi *et al.*, 2001). The third NTR-C contains an N-terminal extension that is likely to mediate posttranslational import into plastids. The identification of the sequence suggests that plastids possess an NADPH-dependent NTR in addition to ferredoxin-dependent thioredoxin reductase (Bréhélin *et al.*, 2004). This system would allow reduction of thioredoxins in the dark independent of photosynthetic electron transport. Glutathione reductases (GRs) constitute the third line of electron flow into the thiol system and maintain the glutathione pool reduced. Thiol/disulfide exchange reactions link individual elements of the thiol/disulfide proteome. At present, evidence accumulates indicating that the thiol/disulfide proteome uses specific electron transfer pathways to code regulatory information in specific context.

Glutathionylation is the formation of mixed disulfides between protein thiols and GSH in the presence of oxidants. Binding of GSH to protein thiols introduces (i) a bulky residue of about 310 Da, and (ii) a negative charge with low pK-value into the protein structure. For this reason, glutathionylation affects protein conformation and thereby may regulate protein activities on the one hand and protect thiol groups from easy oxidation (but not generally, see earlier) to sulfinic and sulfonic acid on the other. Human T-lymphocytes exposed to oxidants such as H_2O_2 and diamide showed a strong increase in glutathionylation of cellular proteins as visualized by 35S-GSH binding. In a proteomics approach, *de novo* glutathionylated proteins were marked and identified by mass spectrometry (Fratelli *et al.*, 2002). Protein disulfide isomerase, aldolase, adenylate kinase, cyclophilins and cytochrome *c* oxidase are examples of enzymes that were shown to be glutathionylated in diamide-treated cells. Stress proteins, many cytoskeletal proteins, signalling molecules and proteins of posttranscriptional regulation were also identified indicating the widespread occurrence of glutathionylation in oxidatively stressed lymphocytes (Fratelli *et al.*, 2002). Recently, glutathionylation was also detected in oxidatively stressed suspension cultured *Arabidopsis* cells (Ito *et al.*, 2003). One particular target enzyme of glutathionylation, i.e. triose phosphate isomerase is inactivated upon glutathionylation. Additional and more comprehensive studies in plants are needed to evaluate the significance of this process in protection and regulation of plant metabolism in response to oxidative stress.

2.6   Enzymes involved in thiol–disulfide interconversion

Thioredoxins were identified as the first thiol–disulfide exchange proteins in plants (Schürmann & Jacquot, 2000). Thioredoxins share a common tertiary protein structure – the so-called thioredoxin fold (Martin, 1995). The same domain structure is found in many other thiol-proteins such as the glutaredoxins, GSTs, glutathione peroxidases, peroxiredoxins and protein disulfide isomerase. These enzymes with thioredoxin fold and variants of C-X-X-C motifs frequently catalyse the thiol–disulfide exchange between proteins. In addition to thioredoxins and glutaredoxins, experimental data have accumulated showing that other proteins have acquired the same function in cellular thiol homeostasis as well, e.g. some GSTs and cyclophilins. Apparently, each cell orchestrates a whole set of thiol/disulfide exchange proteins to maintain a functional redox environment.

2.6.1   Thioredoxins

Thioredoxins (Trx) are small heat-stable polypeptides of 12–14 kDa with a typical W-C-G/P-P-C motif. They function as dithiol–disulfide exchange catalysts (protein disulfide oxidoreductase) and link the reductive power of the NADPH- and ferredoxin-system, repectively, to thiol reduction of target proteins. The *Arabidopsis* genome encodes at least 19 classical Trx and more than 20 atypical Trx (Trx-like proteins), the former with targeting information to mitochondria (Trx o1), chloroplasts (Trx f1, 2; Trx m1–4; Trx y1, 2; Trx x), or predicted to reside in the cytosol (Trx m1–8) (Bréhélin *et al.*, 2004). The location of Trx o2 remains unknown; an intron in the presequence results in an import-incompetent protein (Laloi *et al.*, 2001). Some of the Trx-like polypeptides are predicted to enter the secretory pathway. Knowledge on Trx structure, function and evolution has been summarized recently (Jacquot *et al.*, 1997; Schürmann and Jacquot, 2000; Baumann and Juttner, 2002; Bréhélin *et al.*, 2004). The most exciting progress during the last years has been the identification of new members of the Trx family in the chloroplast and the mitochondrion. The chloroplast Trx x is closely related to a prokaryotic thioredoxin from Synechocystis PCC 6803 (Slr1139) (Mestres-Ortega & Meyer, 1999). A detailed biochemical characterization of chloroplast Trx-proteins revealed specificity in activation efficiency of malate dehydrogenase and regeneration capacity for 2-Cys Prx (Collin *et al.*, 2003). Thioredoxins activated malate dehydrogenase in the order Trx f1 > f2 > m4 > m1 > m2, while Trx m3 and Trx x were hardly effective (Collin *et al.*, 2003). Chloroplast fructose-1,6-bisphosphatase was specifically activated by f-type Trx. A completely different picture emerged for regeneration of 2-Cys Prx, where Trx x allowed for the highest rates of reduction of t-butylhydroperoxide (Figure 2.4). Trx m3 was inactive in the peroxidase assay, whereas all other Trx

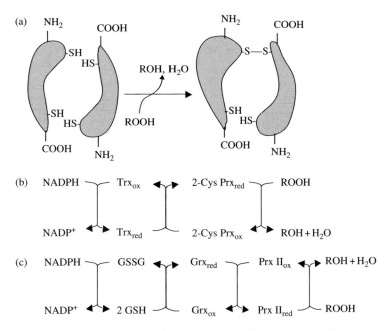

Figure 2.4 Mechanism of peroxiredoxin mediated detoxification of peroxides. (a) Interaction of the N- and C-terminal Cys residue of 2-Cys Prx. (b) Coupling of 2-Cys Prx function to thioredoxin. (c) Coupling of cytosolic type II Prx to glutaredoxin.

revealed intermediate activities with 2-Cys Prx (Collin *et al.*, 2003). Specificity of cytosolic Trx h isoforms has been investigated by complementation of a yeast mutant deficient in both endogenous cytosolic Trx genes (Mouaheb *et al.*, 1998). All five Trx h-isoforms complemented the cell cycle defect of the yeast mutant and allowed for growth on methionine sulfoxide in the medium. The latter depends on reduction of methionine sulfoxide by methionine sulfoxide reductase. Some differences were observed in complementation efficiency of these defects. Interestingly, AtTrx h2 restored growth on sulfate containing medium, whereas AtTrx h3 enabled the yeast cells to detoxify H_2O_2. These distinct results convincingly prove specificity in Trx targeting to substrate proteins. In contrast to chloroplast Trx, knowledge on interacting partners of cytosolic Trx h is still scarce.

2.6.2 Glutaredoxins

Glutaredoxins (Grxs) also belong to the family of thioredoxin fold proteins and share common properties with thioredoxins. They are small thiol–disulfide exchange polypeptides present in plants as a large family comprised of three subgroups. The first plant glutaredoxins were cloned and isolated, respectively, from rice aleurone layers (Minakuchi *et al.*, 1994), spinach leaves (Morell *et al.*, 1995) and sieve tubes of *Ricinus communis* (Szederkenyi *et al.*, 1997). Lemaire (2004) has analyzed the annotated *Arabidopsis* genome for Grx and Grx-like proteins and identified 30 putative members. The three subgroups are distinguished on the basis of sequence-conserved cysteinyl residues, i.e. the classical *E. coli*-Grx motif C-P-Y-C represented in 6 members of 111–179 amino acids length, the C-G-F-S motif present in 4 members of variable length between 169 and 488 amino acids, and the higher plant-specific Grx form with a C-C-M-C/S motif with 20 putative genes encoding polypeptides of 99–150 amino acids (Lemaire, 2004). Lemaire extended his *in silico* analysis to the genomes of *Chlamydomonas* and *Synechocystis* and included the predicted subcellular localization. The *Chlamydomonas* genome encodes 2 Grx with C-P-Y-C and 4 with C-G-F-S motif, the blue-green algae Synechocystis 2 and 1 of each, respectively. Most Grx appear to reside in the cytosol; however, two might be secreted, and four posttranslationally imported into mitochondria or plastids.

Glutaredoxins have two known functions; they catalyse dithiol–disulfide exchange reactions similar to thioredoxins and reduce thiols of glutathionylated proteins, respectively. Initially, the dithiol–disulfide oxidoreductase activity was attributed to the Grx with C-P-Y-C motif; however, recently, a Grx with C-G-F-S motif was suggested to function as disulfide reductases employing a second cysteinyl group found in *Saccharomyces cerevisiae*, Grx5 (Tamarit *et al.*, 2003), being also conserved in three out of the four members of that specific group in plants (Lemaire, 2004). Poplar glutaredoxin is probably the best characterized plant Grx so far (Rouhier *et al.*, 2003). Poplar Grx functioned as reductant of *E. coli* ribonucleotide reductase and *E. coli* 3′-phosphoadenylylsulfate reductase, however, at low rates. In contrast to their yeast counterparts Grx1 and Grx2 (Wheeler & Grant, 2004), poplar Grx did not directly reduce H_2O_2 and alkylhydroperoxide *in vitro* (Rouhier *et al.*, 2003). However, Bréhélin *et al.* (2003) observed efficient regeneration of type II peroxiredoxins by Grx. In a system containing glutathione, NADPH, *E. coli* glutathione reductase and *Arabidopsis thaliana* Grx (Figure 2.4), the cytosolic Prx IIB, but not the plastidic Prx IIE, was able to reduce H_2O_2 (Bréhélin *et al.*, 2003). The At-Trx h2 was ineffective as reductant of At-Prx IIB. Earlier, in 2001, Rouhier *et al.* had shown that both poplar Grx and Trx were able to regenerate poplar Prx in a peroxide detoxification assay. These significant differences hint to specificity in the Grx- and Trx-dependent redox pathways. Such donor–acceptor specificity must be expected in the light of the large size and sequence diversity of the respective gene families.

Similar to separate redox 'cofactors', glutaredoxins function as electron donors in the reduction of activated sulfate in conjunction with 5′-adenylylsulfate reductase (APS). Interestingly, plant APS contains both functional domains, the sulfate reductase and glutathione oxidase, on one hybrid polypeptide. The reductase domain is located in the N-terminus and a glutaredoxin-like domain in the C-terminus (Bick *et al.*, 1998). With a K_M of 0.6 mM, glutathione efficiently supported APS-catalysed sulfate reduction. This example demonstrates that glutaredoxins can be used as domain modules in the evolution of new protein functions.

2.6.3   Omega and lambda-GSTs

GSTs constitute an interesting protein family demonstrating the divergence of protein function during evolution and differentiation of redox proteins. In plants, all GSTs described to date form dimers from 25 kDa subunits. They have evolved into four major classes namely phi-, tau-, zeta- and theta-GSTs that mainly function in xenobiotic detoxification and as glutathione peroxidase, but also in less defined and more diversified reactions such as stress signalling and protein binding (Marrs, 1996). These GSTs contain a conserved catalytic Cys residue needed for glutathione binding and a Ser residue involved in stabilizing the thiolate of the glutathione in the peroxidation, conjugation or isomerization reaction. Constitutively expressed GSTs possess high GST but low glutathione peroxidase activities, the latter being an activity of glutathione-dependent reduction of peroxides. In a converse manner, xenobiotic-induced GSTs have both activities at similar level (Dixon et al., 1998). At least in some cases, these GSTs are induced in conjunction with multiple drug resistance pumps needed for safe deposition of glutathione-conjugated xenobiotics into secretory compartments (Theodoulou et al., 2003). Recently, two additional groups of GSTs were identified. Omega-GSTs encode a small gene family of four putative dehydroascorbate reductases (DHARs) in A. thaliana with C-P-F-C/S motif, denominated At-DHAR1–4, where At-DHAR1 and 2 are well represented in the EST databases, whereas At-DHAR3 was present only once and At-DHAR4 was missing (Dixon et al., 2002). DHAR regenerates ascorbic acid from dehydroascorbic acid using glutathione as electron donor. DHAR regenerates the electron donor ascorbic acid in the Asada–Halliwell–Foyer cycle, where H_2O_2 is detoxified by ascorbate peroxidase, linking thiol antioxidants with ascorbic acid (see Chapters 1, 3 and 4). The catalytic mechanism of dehydroascorbate reduction by DHAR involves an intermediate S-glutathionylation (Cys20 in At-Dhar1), whereas an intramolecular disulfide bridge was promoted in the presence of GSSG in At-Dhar3. The sixth subgroup of so-called lambda GSTs is comprised of two members, At-GST-L1 and the putatively plastidic form At-GST-L2. At-Gstl did not show any DHAR activity but showed GSH-dependent thiol transferase activity suggesting that lambda-GSTs might function in dethiolation of S-glutathionylated proteins that accumulate during oxidative stress (Dixon et al., 2002). The cytosolic forms At-dhar1 and At-gstl-1 were induced by stress, whereas the chloroplastic forms At-dhar3 and At-gstl-2 were expressed constitutively.

2.6.4   Protein disulfide isomerases

Protein disulfide isomerases (PDIs) belong to the superfamily of proteins with thioredoxin folds and catalyse the exchange reaction of dithiols to disulfide. Initially, they were identified as proteins converting thiols to disulfide bridges in polypeptides of the secretory pathway including proteins of the extracellular space and storage vacuoles such as seed storage proteins (Shorrosh et al., 1993; Freedman et al., 1994). Recent data on reactivation of reductively inactivated acidic phospholipase A212 revealed a chaperone function of PDI in addition to the thiol–disulfide exchange reaction (Yao *et al.*, 1997). Catalytic and binding domains interact in these functions (Freedman *et al.*, 2002). Accordingly, PDI have been identified not only in the endoplasmic reticulum but also in chloroplasts (Trebitsh *et al.*, 2001) and mitochondria (Sweetlove *et al.*, 2002). The *Arabidopsis* genome encodes seven PDIs. In plasmatic compartments, they are likely to function in reduction of aberrantly formed or regulatory disulfide bridges. PDIs possess thioredoxin-like domains with conserved cysteinyl motifs that interchange from dithiol to disulfide form. In dependence on the redox potential gradients, PDIs may either accept or donate electrons to target proteins. Sweetlove *et al.* (2002) hypothesized that the mitochondrial PDI could be involved in reducing and regulating alternative oxidase (AOX). AOX is active in its reduced form and inactive in its oxidized form. Thereby, activity of AOX and the energy dissipation through the alternative respiratory electron pathway could be tuned in relation to the electron pressure in the mitochondria. In line with this hypothesis, AOX was not identified as a target of thioredoxin in a proteomic approach (Balmer *et al.*, 2004b). In animal mitochondria, PDI has been shown to be associated with the membrane (Rigobello *et al.*, 2001). PDI is also involved in light-dependent regulation of chloroplast translation as exemplified by the reaction centre protein of photosystem II, D1, encoded by the *psbA* gene. Chloroplast PDI (RB60) reduces RNA-binding protein RB47 in dependence on the electron pressure of the ferredoxin/thioredoxin system and activates translation in the light (Kim & Mayfield, 1997). Conversely, ADP-dependent phosphorylation of RB60 inactivates the psbA-protein complex and inhibits translation in the dark (Trebitsh *et al.*, 2000). Oxidizing conditions in the chloroplast will immediately feed back on translation and photosystem II repair cycle.

2.7   Peroxiredoxins, thiol/disulfide proteins in antioxidant defence

The previous section has outlined properties of thiol/disulfide exchange proteins whose primary function acquired during evolution was likely linked to thiol homeostasis of the cell, counteracting thiol oxidation. Apparently, also early during evolution, Prxs specialized in H_2O_2, alkyl hydroperoxide and peroxinitrite detoxification. In eukaryotes, Prx appears to have evolved additional functions in redox signalling (Hofmann *et al.*, 2002; Dietz *et al.*, 2004). Prxs belong to the proteins with thioredoxin fold. The *Arabidopsis* genome contains 10 *prx* genes belonging to the four different groups, (i) the 1-Cys Prx, (ii) 2-Cys Prx, (iii) Prx Q and (iv) type II Prx. The active site Cys is located in a T-P-X-C motif (Dietz *et al.*, 2004). It reacts with a broad range of peroxide substrates ranging from H_2O_2 (molecular weight as low as 34), butylhydroperoxide, cumene hydroperoxide, peroxinitrite, fatty acid peroxides to complex phosphatidylcholine peroxides ($M_r > 700$) (Cheong *et al.*, 1999; König *et al.*, 2003).

2.7.1   1-Cys Prx

The higher plant 1-Cys Prx was initially identified in barley aleurone and embryo tissue (Aalen *et al.*, 1994). Highest transcript amounts were measured during late stages of seed development. Protein abundance was related to seed dormancy. 1-Cys Prx was localized to the nucleus (Stacy *et al.*, 1999). In the presence of a mixed function oxidation system containing Fe^{2+}, O_2 and dithiothreitol (DTT), the 1-Cys Prx protected DNA from degradation and glutamine synthase from oxidative inactivation. This effect is explained by scavenging of H_2O_2 by Prx to avoid hydroxyl radical formation by reduction of H_2O_2 in the presence of Fe^{2+}. In the assay, Prx is regenerated by DTT. The 1-Cys Prx lacks the resolving thiol and, at present, the reductant regenerating functional 1-Cys Prx *in vivo* is yet unknown. Using full length and truncated *1-cys prx* promoter : β-glucuronidase gene fusions, Haslekas *et al.* (2003a) localized an antioxidant responsive element (ARE) in the promoter region required for early *1-cys prx* expression during seed development, while an abscisic-acid-responsive element (ABRE) was involved in expression from the stage of bent cotyledons on. Using this approach, expression was also observed in cotyledons, meristems and stem branching tissue. The promoter responded to oxidative stress induced by application of hydroquinone. Recently, the hypothesis that 1-Cys Prx is related to dormancy has been tested using *Arabidopsis* plants with increased and decreased levels of *1-cys prx* expression (Haslekas *et al.*, 2003b). No correlation was found between 1-Cys Prx expression and the duration of dormancy required before germination in non-stressed plants. Interestingly, under osmotic, salt and oxidative stress conditions, seeds with high 1-Cys Prx levels germinated with a delay as compared to wild-type plants. Haslekas *et al.* (2003b) suggested that 1-Cys Prx might be involved in sensing stress and inhibiting germination under adverse growth conditions. In the resurrection plant *Xerophyta viscosa*, which is a monocotyledonous desiccation-tolerant plant from Southern Africa, a 1-Cys Prx is expressed in all vegetative tissues following stress and stimulus treatments such as dehydration, heat, high light, salinity and abscisic acid application (Mowda *et al.*, 2002). Since the predicted gene product carries a nuclear localization sequence (NLS) and was shown to be present in the nucleus of dehydrated *X. viscosa* tissue sections, the authors proposed that Xv 1-Cys Prx has a specialized role in combating oxidative stress in the nucleus under conditions of water deprivation (Mowda *et al.*, 2002).

2.7.2   2-Cys Prx

The 2-Cys Prx was first cloned from the growing zone of leaves and shown to be posttranslationally targeted to the chloroplast (Baier and Dietz, 1997). The 2-Cys Prx has been studied most intensely on the levels of biochemical activity, effect of knock-down in transgenic plants, structure and possible regulatory role as redox sensor. Here, these results are summarized only briefly. For a more comprehensive treatment, the reader is referred to recent reviews (Dietz *et al.*, 2002a, 2004; Rouhier & Jacquot, 2002; Dietz, 2003a). The functional unit of the 2-Cys Prx is a homodimer, associated in a head-to-tail arrangement of both subunits. Each fully reduced dimer can detoxify two peroxide substrate molecules, and is simultaneously converted to the partially and then fully oxidized form. Regeneration is achieved by reaction of the oxidized form with reduced thioredoxin as successively evidenced in assays containing *E. coli* Trx (Cheong *et al.*, 1999), chloroplast Trx f and m (König *et al.*, 2002) and, most efficiently, chloroplast Trx x, respectively (Collin *et al.*, 2003). The fully reduced as well as the over-oxidized dimers arrange to higher aggregates of doughnut-shaped decamers (Wood *et al.*, 2002, 2003). These decamers have been detected in protein separations by size exclusion chromatography (König *et al.*, 2002, 2003) and by electron microscopy (Harris *et al.*, 2001). The decamers attach to the thylakoid membrane. It was hypothesized that these conformational changes have a sensor function in plants and membrane attachment could trigger modifications of the photosynthetic activity (Dietz *et al.*, 2004).

The 2-Cys Prx can function as peroxidase in the water–water cycle alternative to ascorbate peroxidase (Dietz *et al.*, 2002a). Superoxide radicals are produced in the Mehler reaction at the acceptor site of photosystem I, when the ferredoxin pool is highly reduced due to a lack of acceptors. Superoxide anions are dismutated to H_2O_2 by superoxide dismutase. H_2O_2 is scavenged by ascorbate peroxidase (Apx) and two molecules of ascorbate are converted to monodehydroascorbate, that may disproportionate to one molecule of ascorbic acid and one molecule of dehydroascorbate. The ascorbate–glutathione cycle regenerates ascorbic acid on the expense of NADPH or glutathione (Section 2.6.3; Chapters 1 and 3; Noctor & Foyer, 1998). Glutathione is reduced by NADPH-dependent glutathione reductase. Following H_2O_2 detoxification by Prx, the generated Prx disulfide is reduced by Trx coupled to the electron transport chain via ferredoxin–thioredoxin reductase or NADPH–thioredoxin reductase (Dietz *et al.*, 2002a). Apparently, photosynthesis does not rely on a single mechanism to detoxify H_2O_2 but maintains a set of alternative pathways, the relative contribution of which to the overall rates still needs to be established.

Development of transgenic antisense *Arabidopsis* with decreased levels of 2-Cys Prx was delayed during the first six weeks after germination, but recovered during later stages with concomitant accumulation of 2-Cys Prx to wild-type levels (Baier & Dietz, 1999). During the phase of impaired development, the ascorbate pool was more oxidized in the antisense lines than in the wild-type plant, whereas the glutathione pool was unaffected (Baier *et al.*, 2000). In this phase of development, gene transcripts related to ascorbate homeostasis were up-regulated. These observations concur with the hypothesis that Prx- and Apx-dependent detoxification of photosynthetically produced H_2O_2 function as alternative pathways.

Gene expression of *2-cys prx* is coupled to cues related to leaf development, and to redox and hormonal control. Antioxidants such as thiols and particularly ascorbic acid decreased *2-cys prx* transcripts (Baier & Dietz, 1997; Horling *et al.*, 2001, 2003). Ascorbate induced rapid shutdown of transcription. The signalling pathway involved a staurosporin-sensitive protein kinase (Horling *et al.*, 2001). Using a reporter gene fused to the *2-cys prx* promoter, transcript regulation was traced back to promoter regulation. High electron pressure in the photosynthetic electron transport chain in the absence of acceptor stimulated promoter activity, whereas saturating CO_2 decreased the activity (Baier *et al.*, 2004). The regulatory network is further modulated by abscisic acid that acts as suppressor of *2-cys prx* expression. The integration of the complex promoter regulation in a physiological context is not understood.

2.7.3 Prx Q

Prx Q was identified in the higher plant *Sedum lineare* on the basis of its similarity to a bacterial protein co-purifying with bacterioferritin (BCP: bacterioferritin-comigratory protein) (Kong *et al.*, 2000). Prx Q possesses two conserved cysteinyl residues that form an intramolecular disulfide between Cys54 and Cys59 (amino acid counting of *Arabidopsis* PrxQ) upon oxidation. It should be noted that site-directed mutagenesis of the second Cys residue produced a variant that was still functional towards peroxide substrates (Jeong *et al.*, 2000). Rouhier *et al.* (2004b) presented a detailed biochemical analysis of the poplar homologue. Prx Q runs as monomer during electrophoretic separation, both under reducing and non-reducing conditions. When six different thioredoxins were compared for regeneration efficiency of Pt Prx Q, the chloroplast Pt Trx y and the two cytosolic Pt Trxh3 and Pt Trxh1 allowed high rates of H_2O_2 detoxification. Grx was ineffective. The bulky cumene hydroperoxide and H_2O_2 served as good substrates, whereas t-butylhydroperoxide revealed a three-fold lower turnover number (Rouhier *et al.*, 2004b). Mutation of the first Cys (Cys46 → Ser in poplar) abolished peroxidase function. Interestingly, modification of the second Cys 51-Ser produced a variant that could be re-reduced by Grx as well as Trx. The redox midpoint potential E_m was −0.325 V at pH 7 and dropped with 59 mV per pH unit. Predicted chloroplast location of Prx Q (Horling *et al.*, 2003) was confirmed by transient expression of Prx Q::GFP (green fluorescent protein) in tobacco leaves and by immunohistochemical methods (Rouhier *et al.*, 2004b). A strong signal was seen in chloroplasts of guard cells. Finally, these authors detected a significant up-regulation of Prx Q protein upon infection of poplar leaves with the rust fungus *Melampsora larici-populina*. This regulation could be related to reactive oxygen species produced during the compatible interaction. In a test with *Arabidopsis* leaf slices, H_2O_2 has been described as strong inducer of *prx Q* expression (Horling *et al.*, 2003). In the same study, employing transcript quantification by RT-PCR, *prx Q* transcript amounts behaved mostly similar to the chloroplast *2-cys prx A* and *B*, i.e. with a decrease after application of ascorbic acid and upon darkening, suggesting the role of Prx Q in the context of photosynthesis. Western blot analysis of isolated thylakoids with anti Prx-Q-antibody revealed a significant attachment of the protein to the membrane (Dietz *et al.*, 2004). The specificity or redundancy of antioxidant and/or sensing function, respectively, of Prx Q, 2-Cys Prx A and 2-Cys Prx B should be the focus of future work. In this context, it is noteworthy that the genome of *Synechocystis PCC 6803* encodes five peroxiredoxins, a 2-Cys Prx, a 1-Cys Prx, a type II Prx and two Prx with significant similarity to Prx Q with some sequence modifications. Sequence similarity of higher plant 2-Cys Prx is highest with the cyanobacterial orthologue suggesting an endosymbiotic origin (Baier & Dietz, 1997). Since Prx Q is missing in animals, a similar evolutionary and functional aspect in the context of photosynthesis may exist for Prx Q.

2.7.4   Type II Prx

Type II Prx form the largest subgroup of Prx in *A. thaliana* (Dietz *et al.*, 2002). They are localized to diverse subcellular compartments, i.e. the mitochondrion (Prx IIF), the plastid (Prx IIE) or reside in the cytosol (Prx II A–D). They are characterized by two conserved Cys residues spaced through 24 amino acids. Mutation of Thr48 in the catalytic T48-P-T-C motif to Val in a poplar Prx II abolished dithiothreitol-, Grx- and Trx-dependent H_2O_2- and t-butylhydroperoxide detoxification, but, interestingly, did not fully inhibit cumene hydroperoxide reduction (Rouhier *et al.*, 2004a). The authors suggested that this mutation altered the active site structure but not the regeneration of Pt Prx II. In a dithiothreitol-based assay, Prx IIB, C, E and F showed significantly higher rates of H_2O_2 reduction than 2-Cys Prx (Horling *et al.*, 2002, 2003). Midpoint redox potentials were in the range of -0.28 to -0.31 V – the more negative value was measured for the mitochondrial isoform (Horling *et al.*, 2003). Cytosolic Prx IIB was efficiently regenerated by Grx (Bréhélin *et al.*, 2003). Promoter–reporter gene fusions expressed in non-stressed *A. thaliana* have provided insight in cell- and tissue-specific expression of At-Prx-II-isoforms (Bréhélin *et al.*, 2003). pAt Prx IIC::GUS was highly expressed in flower buds and mature pollen, pAt Prx IID::GUS in germinating pollen, pollinic tissue and fertilized ovules, pAt Prx IIE in stamen, young flowers, embryo sac and the albumen of older seeds from green to yellow siliques. The results were confirmed by RT-PCR (Bréhélin *et al.*, 2003). Leaf transcript levels of Prx IIC and Prx IID were specifically and strongly up-regulated in response to phosphate deficiency indicating severe redox imbalance under these conditions (Kandlbinder *et al.*, 2004). Also, in leaves, Prx IIC transcript amounts have been described as a sensitive indicator of any redox imbalance in the cell: changes in light intensity, ascorbic acid, salt application, H_2O_2, diamide and t-butylhydroperoxide transiently or permanently up-regulated *prx-IIC*-transcript amounts. The complexity of *prx*-gene regulation at tissue and cell level, and in response to environmental stimuli, as well as the precise function of Prx isoforms are not yet understood and presently investigated (Finkemeier *et al.*, 2003; Lamkemeyer *et al.*, 2003). Preliminary results with knock down/out mutants of the various Prx indicate that modified Prx activity affects both developmental programmes such as stomatal patterning and root growth under stress, but also hardening to environmental stress (Petra Lamkemeyer, Iris Finkemeier, Karl-Josef Dietz, unpublished). Prx function as cytoplasmic thiol sinks in peroxide detoxification is clearly established. Their possible function as redox sensors similar to animals (Hofmann *et al.*, 2002) still needs to be assessed in plants (Dietz *et al.*, 2004).

2.8   The thiol proteome of plants

Recently, a number of studies have aimed at identifying novel proteins regulated by thiol–disulfide interchange reactions. Basically, four approaches have been applied successfully:

   (i)   affinity chromatography with redox proteins mutated in the active centre in order to form mixed disulfides, i.e. covalent bonding between donor and acceptor proteins,
  (ii)   affinity chromatography with wild-type redox proteins to purify protein partners electrostatically interacting with redox proteins,
 (iii)   fluorescence labelling of protein extracts before and after reduction with redox proteins. The latter method was also combined with alkylation reactions of reduced thiols and chromatography, and
  (iv)   identification of glutathionylated proteins by using radioisotopically labelled glutathione.

As described above, thiol–disulfide exchange proteins often possess two cysteinyl groups: the primary thiol in the catalytic centre of the reductase reacts with the disulfide of the target substrate; a mixed disulfide is formed between both proteins; a second, the so-called resolving thiol of the reductase attacks the mixed disulfide; the substrate is released in the reduced dithiol state and the reductase is oxidized. Mutation of the resolving cysteinyl of the reductase still allows the first reaction steps to proceed. However, the mixed disulfide state is frozen due to the lack of the resolving thiol group. This strategy has been used successfully to identify protein targets of chloroplast thioredoxins, e.g. 2-Cys peroxiredoxin, peroxiredoxin Q and cyclophilin (Motohashi *et al.*, 2001). Chloroplast Trx m has been used as bait to identify novel thiol-redox proteins of mitochondria (Balmer *et al.*, 2004b). The authors argued that Trx m is similar to the mitochondrial Trx o and should allow identification of real partners also interacting in mitochondria. The set of proteins identified with Trx m was similar to that isolated with the monocysteinic Trx 1 h as bait. Nevertheless, in the light of the proposed specificity of redox interactions, it would be important to verify the results employing a homologous system of compartment-specific proteins. Interestingly, this work allowed to identify mito-chondrial aldehyde dehydrogenase, catalase, formate dehydrogenase, Grx-like protein, Prx IIF, phospholipid hydroperoxide reductase and superoxide dismutase as targets of Trx and underscored the importance of thiol regulation and thiol reactions in stress defence (Balmer *et al.*, 2004b). In an attempt to identify proteins electrostatically interacting with Trx f, Balmer *et al.* (2004a) immobilized Trx f on a chromatographic matrix, passed proteins from stroma extract at low salt through the column and eluted bound polypeptides with high salt buffer containing 200 mM NaCl. In addition to the known interactor of Trx f, i.e. fructose-1, 6-bisphosphate, they identified Calvin cycle enzymes, translation-related proteins, ATP-dependent clp protease as well as other proteins. Interestingly, antioxidant enzymes were not observed, indicating that Trx f is either not the preferred partner of antioxidant enzymes such as peroxiredoxins or that the interaction does not involve a persistent association of these proteins permitting isolation by non-covalent interaction. Yano *et al.* (2001) devised a different strategy for identifying proteins targeted by thioredoxins. Oxidized peanut seed proteins were incubated in the presence of a system containing NADPH/NTR/Trx h, followed by labelling with monobromobimane, 2-D separation and mass spectrometry. Trx h-reduced proteins were identified as three allergens, desiccation-related protein and seed maturation protein. All these experiments aimed at identifying Trx-related proteins. Lee *et al.* (2004) extended the analysis to the whole disulfide proteome of *Arabidopsis*. Free protein thiols were blocked by alkylation during extraction. Protected protein disulfides were reduced, thiol proteins isolated by thiol-affinity chromatography and analysed by mass spectrometry. The method can be applied to both soluble and membrane-bound proteins (Lee *et al.*, 2004) and allowed identification of, amongst other targets, violaxanthin deepoxidase and subunits of the photosystems as possible targets of redox regulation. The results from these proteomics approaches have advanced our knowledge on the complexity of the plant redox network significantly. After further refinement, this methodology will certainly help to dissect the specificity of redox interchange also in the context of oxidative stress situations.

In a recent work, a mutated poplar glutaredoxin was employed to isolate putative plant Grx targets. Among the 94 identified targets, Rouhier *et al.* (2004c) found interaction of Grx with proteins involved in oxidative stress response, nitrogen, sulfur and carbon metabolism, translation and signalling. It is tempting to hypothesize that the interactions reflect a metabolic relationship in glutathionylation or deglutathionylation reactions.

2.9 Thiol homeostasis in subcellular compartments

Thiol homeostasis of chloroplasts has been described in depth as viewed through the activity of regulated target enzymes (Scheibe , 1991; Scheibe, 2004), interacting partners of the thioredoxin system (Motohashi *et al.*, 2001; Collin *et al.*, 2003), the activity of photosynthesis (Johnson, 2003), the transcriptional (Link, 2003; Pfannschmidt, 2003) as well as the translational activities (Trebitsh *et al.*, 2000). Thiol homeostasis of the cytosol is less understood. Some elements of both thiol-regulatory networks have been addressed in this review. By use of new molecular and genomics tools, the thiol network of subcellular compartments such as the mitochondrion and the nucleus have become accessible. Mitochondria contain thioredoxins of the type o and a

NADPH/thioredoxin-reductase. In a screening for proteins that form mixed disulfides with a site-directed mutagenized chloroplast thioredoxin, 50 potential targets were identified and linked to 12 processes in mitochondrial metabolism (Balmer *et al.*, 2004b). The authors compared extracts from mitochondria isolated from pea and spinach leaves and potato tubers, respectively. Photorespiration involves reactions in the chloroplast, the peroxisome and the mitochondrion. Both mitochondrial enzymes involved in photorespiration, i.e. glycine decarboxylase as well as serine hydroxymethyl transferase, were pulled out as Trx-binding proteins from mitochondrial extracts, independent of the tissue used as source for the isolation of mitochondria. Since photorespiration produces NADH in the mitochondrion, this mechanism may function as a feed-back activation to tune glycine decarboxylation to photorespiratory activity. In addition to the redox regulation, other regulatory mechanisms must be active to prevent depletion of the glycine pool. Seven enzymes of the citric acid cycle, the α-, β- and δ-subunits of the mitochondrial ATP synthase, elements of the respiratory electron transport chain, enzymes of the sulfur, nitrogen and lipid metabolism (Balmer *et al.*, 2004b) were also identified as Trx-binding partners. This set of Trx targets possibly coordinates energy and redox metabolism with each other and with metabolic pathways of the chloroplast, respectively. In addition, stress-related proteins such as Grx-like, Prx, superoxide dismutase and alcohol dehydrogenase were identified as Trx-interacting proteins of mitochondria. Interestingly, the study also gave an indication of how oxidative stress could trigger apoptosis through oxidation of the mitochondrial porin. Balmer *et al.* (2004b) suggest that Trx maintains the porin in a reduced state. If the reduction capacity of the mitochondria drops significantly, the porin might become oxidized and be converted into the permeability transition pore releasing cytochrome *c* and initiating apoptosis. It will be necessary to investigate the physiological significance of all these findings. They may either describe a part of the redox regulatory circuits or defence mechanisms to avoid accumulation of disulfides in mitochondria. The suggestion of the authors that Trx acts as redox sensor, regulator and interorganellar coordinator will have to be validated in further experiments. The redox environment of the nucleus certainly is understood the least. Hirota and colleagues (1997) identified a thioredoxin in the nucleus of aleurone cells and linked it to redox-dependent transcriptional activation. As mentioned above, 1-Cys Prx is targeted to the nucleus (Stacy *et al.*, 1999; Mowda *et al.*, 2002). Information on the redox metabolism and the thiol proteome of the nucleus, as well as on regulatory consequences of thiol–disulfide transitions on nuclear activities is scarce, but urgently needed.

Figure 2.5 illustrates compartment-specific differences in the redox networks of illuminated chloroplast and cytosol. Due to coupling to ferredoxin in the light, chloroplast metabolism can exploit regulatory thiols with a broader range of midpoint redox potentials than the exclusively NADPH-linked cytosol. Furthermore, thioredoxin-linked targets have more negative redox potentials than glutaredoxins. This has implications for the specific function of Trx and Grx in plant metabolism (see earlier).

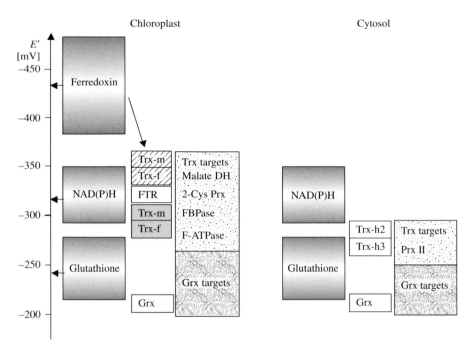

Figure 2.5 Illustration of the reduction capacities of chloroplast and cytosol. The redox state of the illuminated chloroplast is linked to ferredoxin that is a very strong reductant and can efficiently reduce redox partners with high reduction potential. This is used for efficient regulation of target enzymes. In the chloroplast, alkalization of the stroma further shifts the redox potentials to more negative values (shaded boxes). The thiol systems of all other compartments are linked to NADPH and show less negative midpoint redox potentials. See text for details. Abbreviations: Cyt: cytochrome, Trx: thioredoxin, FTR: ferredoxin thioredoxin reductase, FBPase: fructose-1,6-bisphosphatase, Grx: glutaredoxin (E_m of *E. coli*).

2.10   Thiol-dependent redox regulation of gene expression

Thiol-dependent regulation of gene expression is well established in prokaryotes, yeast (Paget & Buttner, 2003) and animals, but is only beginning to be investigated in plants. Initially identified in H_2O_2 challenged *Salmonella typhimurium*, OxyR was described as a transcriptional activator that functions as redox switch in bacteria. The thiol form of OxyR is inactive, whereas the disulfide form was shown to bind to the promoter and initiate transcription (Storz *et al.*, 1990). More than 20 antioxidant genes including the peroxiredoxin AhpC, glutaredoxin I (grxA) and disulfide reductase were activated through that pathway (Zheng *et al.*, 2001). A set of site-specific mutations such as Cys199 → Ser and Cys208 → Ser appeared to concur with the hypothesis that dithiol–disulfide transition is a key regulatory mechanism in OxyR-dependent gene regulation (Zheng *et al.*, 1998). However, the simplicity of the model and the function as an on-/off-switch was questioned by recent results of Kim *et al.* (2002) who proposed a multiple activation state model (Paget & Buttner, 2003). The mechanism relies on chemical modification of Cys199 through S-nitrothiolation and glutathionylation, respectively. These thiol modifications of OxyR are hypothesized to tune different levels of OxyR-dependent transcription. Glutaredoxin GrxA reduces the disulfide form of OxyR. Since *grxA*-gene is activated by OxyR, a homeostatic feed-back-loop is constructed to inhibit OxyR when glutathione in conjunction with GrxA is signalling restoration of redox homeostasis. Redox regulators such as OhrR, Hsp33, σ^R-RsrA, PerR in various bacteria and Yap1 in *S. cerevisiae*, the latter being a bZIP-transcription factor, constitute other examples of well-characterized thiol-based redox regulators at the level of gene transcription (Paget & Buttner, 2003). Yap1 senses diamide stress and H_2O_2 levels, respectively (for review, see Paget & Buttner, 2003), through dithiol oxidation and disulfide bridge formation, however, between distinct cysteinyl residues. In the presence of diamide, Yap1 forms a disulfide bridge between Cys598 and a more C-terminally located Cys. The oxidized form accumulates in the nucleus and activates the genes of the Yap1-regulon through the Yap1 responsive element with consensus sequence -T-T/G-A-C-T-A-A- (Delaunay *et al.*, 2000, 2002). Reduced Yap1 is exported from the nucleus. Interestingly, H_2O_2 sensing is mediated by a glutathione peroxidase Gpx3 (Orp1) that was identified as an interacting partner of Yap1. Upon H_2O_2 stress, the glutathione peroxidase is suggested to form a sulfenic acid intermediate that reacts with Cys598 of Yap1 to form a mixed disulfide. Cys303 of Yap1 resolves the mixed disulfide liberating reduced Gpx3; and the oxidized Yap1 (disulfide bridge between Cys303 and Cys598) is translocated to the nucleus and activates gene expression as before. These examples from fungi and bacteria elucidate two possible levels of redox control of gene expression by regulating binding of transcription factors:

(i) redox-dependent modification of conformation that alters protein binding affinity to the target DNA sequence; and

(ii) redox regulation of subcellular distribution where either the oxidized or reduced form is imported into the nucleus activating gene transcription.

Recently, evidence has emerged that these mechanisms of transcriptional control linked to thiol-redox state also exist in plants. Two transcription factors of the homeodomain family HD-ZipII (*Helianthus annuus* HAHR1, belonging to the glabra2 family) contain a set of conserved cysteinyl residues with the motifs -C-X-X-C-, -C-E-X-L-K/R-R/K-C- and -C-P-S-C-. Site directed modification of single Cys residues within these motifs alters their affinity to the DNA target (Tron *et al.*, 2002). Glutathione was not an efficient reductant, whereas DTT as well as thioredoxin fully activated the factor (Tron *et al.*, 2002). GLABRA is expressed in the epidermis and known to affect development of trichomes, root hairs and seed coat mucilage (Lu *et al.*, 1996; Hung *et al.*, 1998). The physiological relationship between changing redox state of the cells, altered HD-ZipII binding to target genes and specific changes in the developmental programme, e.g. by altered root hair formation still needs to be established.

The *Arabidopsis* genome contains more than 1000 transcription factors. From studies with animals and bacteria, TFs are known to be targets of redox regulation in their control of gene transcription. Amino acid sequences of plant transcription factors should be searched for conserved thiols that could be involved in such a direct thiol/disulfide-dependent activation.

2.11   Linking thiol regulation to metabolic and developmental pathways

Thiol proteins function in reduction reactions of metabolic pathways and in cellular signal transduction and regulatory processes. Electron flow from thiols to oxidized substrates has been established to be involved in reduction of ribonucleotides by ribonucleotide reductase (Åslund *et al.*, 1994; Rouhier *et al.*, 2003), 3'-phosphoadenylylsulfate by PAPS reductase (Lillig *et al.*, 1999; Koprivova *et al.*, 2002), as well as H_2O_2 and alkyl hydroperoxides through peroxiredoxins (Choi *et al.*, 1999; Rouhier & Jacquot, 2002; König *et al.*, 2003) and glutathione peroxidases. This small set of metabolic reduction reactions in plant cells is contrasted by a long list of thiol-dependent regulatory targets characterized by biochemical means, some of which have been addressed above. An even longer list of putative targets of thiol regulation is presently being worked out using thiol proteomics approaches in conjunction with genome wide searches. In many of these cases, thiol–disulfide transition directly mediates a functional switch from an active to an inactive state, or vice versa (Meyer *et al.*, 1999; Baumann & Juttner, 2002; Dietz *et al.*, 2002, 2003b). Increasing evidence suggests that redox regulation is also indirectly coupled to regulatory pathways by means of redox regulated chaperones and adapter proteins. Cyclophilins and FKBP (FK506-binding protein) are emerging examples of putative redox-controlled signalling tools in plant cells: these enzymes, originally described in animals as binding proteins of the immunosuppressive cyclosporin A and FK506, have a peptidyl–prolyl isomerase activity and constitute large gene families in *A. thaliana* (Romano *et al.*, 2004). Interestingly 2–4 cysteinyl groups are conserved in some cyclophilins and have been shown to function as electron donor, e.g. to regenerate Prx (Lee *et al.*, 2004). Cyclophilins can be proposed as proteins transmitting redox signals to other signal transduction pathways.

In illuminated photosynthetic cells, the photosynthetic apparatus is probably the most elaborate mechanism of environmental sensing (Dietz *et al.*, 2002). Any imbalance between energy input through light absorption and acceptor availability will increase electron pressure in the electron transport chain and at the acceptor site of photosystem I. Over-reduction of ferredoxin will fuel electron transfer to oxygen liberating reactive oxygen species. A set of regulatory reactions such as state transitions, violaxanthin cycle and repair cycle are known to modulate and optimize photosynthetic performance (Kruse, 2001). This is described in more detail in Chapter 10. Recently, a thiol-based regulation of thylakoid electron transport was suggested by Johnson (2003). The initial observation with isolated thylakoids concerned the inhibition of oxygen evolution when the redox potential of a DTT buffer was adjusted to values below -0.3 V (at pH 7.35; $E_m = -0.328$) and -0.36 V (at pH 8.01; $E_m = -0.384$). Johnson (2003) proposed a model where increasing electron pressure will reduce the thioredoxin pool via ferredoxin; at very negative redox potentials, the thioredoxin will reduce regulatory disulfides presumably of the cytochrome b_6f-complex as well as of malate dehydrogenase. The latter is known to function in cycling of dicarboxylic acids for export of excess reducing equivalents to the cytosol (Scheibe, 2004). Reduction of the regulatory disulfide of the electron transport efficiently down-regulates electron transport activity. Export of malate from the chloroplast to the cytoplasm and further on to the mitochondrion was suggested to function as interorganellar signalling mechanism (Balmer *et al.*, 2004b). Oxidation of malate in the mitochondrion produces NADPH, that is used for thioredoxin reduction via NADPH-dependent thioredoxin reductase, followed by reduction of mitochondrial target proteins. To this end, the activity of both organelles could be coordinated by redox regulation linked to thiol proteins.

2.12 Outlook

Redox regulation and, specifically, the involvement of thiol/disulfide transitions in plant development and adaptation to oxidative stress are the research topic of a rapidly evolving field. The availability of complete genome sequences, identification of involved gene families and the increasing number of novel players, as well as transgenic and proteomics approaches have promoted our knowledge exponentially during the last years. However, despite increasing awareness of thiol-dependent redox regulation and antioxidant defence in all aspects of cell function, a set of crucial questions and topics need to be addressed: (i) how and to what extent is specificity encoded in the redox signalling and reduction mechanisms mediated by the members of the gene families of Trx, Grx, PDI, Prx and other thiol enzymes? (ii) In addition to analysing thiol/disulfide exchange reactions, should not more emphasis be placed on redox-dependent modification of protein–protein interactions that could trigger appropriate responses ('redox-dependent interactome')? (iii) Can nuclear factors be identified, that alter binding activity to DNA elements or subcellular localization similar to factors described in yeast and animal systems? Such factors could be involved in direct redox-dependent regulation of gene expression and represent key players in plant cell signalling. (iv) What is the significance of nitrothiolation on redox regulation in plants? (v) Finally, cross-talk between regulatory pathways linked to reactive oxygen and reactive nitrogen species is another important emerging topic that will have to be addressed in the near future in order to better understand plant acclimation to environmental conditions.

Acknowledgements

The cited own work was supported by the DFG (For 387, Di 346).

References

Aalen, R.B., Opsahl-Fernstad, H.G., Linnestad, C. and Olsen, O.A. (1994) 'Transcripts encoding an oleosin and a dormancy-related protein are present in both the aleurone layer and the embryo of developing barley (*Hordeum vulgare* L.) seeds', *The Plant Journal* **5**, 385–396.

Åslund, F., Berndt, K.D. and Holmgren, A. (1997) 'Redox potentials of glutaredoxins and other thiol-disulfide oxidoreductases of the thioredoxin superfamily determined by direct protein–protein redox equilibria', *Journal of Biological Chemistry* **272**, 30780–30786.

Åslund, F., Ehn, B., Miranda-Vizuete, A., Pueyo, C. and Holmgren, A. (1994) 'Two additional glutaredoxins exist in *Escherichia coli*: glutaredoxin 3 is hydrogen donor for ribonucleotide reductase in a thioredoxin/glutaredoxin 1 double mutant', *Proceedings of the National Academy of Sciences USA* **91**, 9813–9817.

Åslund, F., Nordstrand, K., Berndt, K.D. *et al.* (1996) 'Glutaredoxin-3 from *Escherichia coli*. Amino acid sequence, 1H AND 15N NMR assignments, and structural analysis', *Journal of Biological Chemistry* **271**, 6736–6745.

Baier, M. and Dietz, K.J. (1997) 'The plant 2-Cys peroxiredoxin BAS1 is a nuclear-encoded chloroplast protein: its expressional regulation, phylogenetic origin, and implications for its specific physiological function in plants', *The Plant Journal* **12**, 179–190.

Baier, M. and Dietz, K.J. (1999) 'Protective function of chloroplast 2-cysteine peroxiredoxin in photosynthesis. Evidence from transgenic Arabidopsis', *Plant Physiology* **119**, 1407–1414.

Baier, M., Noctor, G., Foyer, C.H. and Dietz, K.J. (2000) 'Antisense suppression of 2-Cys peroxiredoxin in *Arabidopsis* specifically enhances the activities and expression of enzymes associated with ascorbate metabolism but not glutathione metabolism', *Plant Physiology* **124**, 823–832.

Baier, M., Ströher, E. and Dietz, K.J. (2004) 'The photosynthetic electron transport and ABA balance 2-Cys peroxiredoxin A promoter activity', *Plant Cell Physiology* **45**(8), 997–1006.

Balmer, Y., Koller, A., del Val, G., Schürmann, P. and Buchanan, B. (2004a) 'Proteomics uncovers proteins interacting electrostatically with thioredoxin in chloroplasts', *Photosynthesis Research* **79**, 275–280.

Balmer, Y., Vensel, W.H., Tanaka, C.K. *et al.* (2004b) 'Thioredoxin links redox to the regulation of fundamental processes of plant mitochondria', *Proceedings of the National Academy of Sciences USA* **101**, 2642–2647.

Baumann, U. and Juttner, J. (2002) 'Plant thioredoxins: the multiplicity conundrum', *Cellular Molecular Life Science* **59**, 1042–1057.

Bergmann, L. and Rennenberg, H. (1993) Glutathione metabolism in plants, in *Sulfur Nutrition and Assimilation in Higher Plants* (ed. L.J. De Kok, I. Stulen, H. Rennenberg, C. Brunold and W.E. Rauser), SPB Academic Publishing, The Hague, pp. 109–124.

Bick, J.A., Åslund, F., Chen, Y.C. and Leustek, T. (1998) Glutaredoxin function for the carboxyl-terminal domain of the plant-type 5′-adenylylsulfate reductase', *Proceedings of the National Academy of Sciences USA* **95**, 8404–8409.

Biteau, B., Labarre, J. and Toledano, M.B. (2003) 'ATP-dependent reduction of cysteine–sulphinic acid by *S. cerevisiae* sulphiredoxin', *Nature* **425**, 980–984.

Bréhélin, C., Laloi, C., Setterdahl, A.T., Knaff, D.B. and Meyer, Y. (2004) 'Cytosolic, mitochondrial thioredoxins and thioredoxin reductases in *Arabidopsis thaliana*', *Photosynthesis Research* **79**, 295–304.

Bréhélin, C., Meyer, E.H., de Souris, J.P., Bonnard, G. and Meyer, Y. (2003) 'Resemblance and dissemblance of *Arabidopsis* type II peroxiredoxins: similar sequences for divergent gene expression, protein localization, and activity', *Plant Physiology* **132**, 2045–2057.

Capozzi, G. and Modena, G. (1974) Oxidation of thiols, in *The Chemistry of Thiol Group*, Part 2, Chapter 17 (ed. S. Patai), John Wiley & Sons, London, pp. 785–839.

Cheong, N.E., Choi, Y.O., Lee, K.O. *et al.* (1999) 'Molecular cloning, expression, and functional characterization of a 2Cys-peroxiredoxin in Chinese cabbage', *Plant Molecular Biology* **40**, 825–834.

Choi, H.Y.O., Cheong, N.E., Lee, K.O. *et al.* (1999) 'Cloning and expression of a new isotype of the peroxiredoxin gene in Chinese cabbage and its comparison to 2-Cys peroxiredoxin isolated from the same plant', *Biochemical and Biophysical Research Communication* **258**, 768–771.

Collin, V., Issakidis-Bourguet, E., Marchand, C. *et al.* (2003) 'The *Arabidopsis* plastidial thioredoxins: new functions and new insights into specificity', *Journal of Biological Chemistry* **278**, 23747–23752.

Delaunay, A., Isnard, A.D. and Toledano, M.B. (2000) 'H_2O_2 sensing through oxidation of the Yap1 transcription factor', *EMBO Journal* **19**, 5157–5166.

Delaunay, A., Pflieger, D., Barrault, M.B., Vinh, J. and Toledano, M. (2002) 'A thiol peroxidase is an H_2O_2-acceptor and redox transducer in gene activation', *Cell* **111**, 471–481.

Dietz, K.J. (2003a) 'Plant peroxiredoxins', *Annual Review of Plant Biology* **54**, 93–107.

Dietz, K.J. (2003b) 'Redox control, redox signaling and redox homeostasis in plant cells', *International Review of Cytology* **228**, 141–193.

Dietz, K.J., Horling, F., König, J. and Baier, M. (2002a) 'The function of the chloroplast 2-cysteine peroxiredoxin in peroxide detoxification and its regulation', *Journal of Experimental Botany* **53**, 1321–1329.

Dietz, K.J., Link, G., Pistorius, E. and Scheibe, R. (2002b) 'Redox regulation in oxygenic photosynthesis', *Progress in Botany* **63**, 207–245.

Dietz, K.J. and Scheibe, R. (2004) 'Redox regulation: an introduction', *Physiologia Plantarum* **120**, 1–3.

Dietz, K.J., Stork, T., Finkemeier, I. *et al.* (2005) The role of peroxiredoxins in oxygenic photosynthesis of cyanobacteria and higher plants: peroxide detoxification or redox sensing?, in *Photoprotection, Photoinhibition, Gene Regulation, and Environment* (ed. B. Demmig-Adams, W. Adams and A. Mattoo), Kluwer Academic Press, in press.

Dixon, D.P., Cummins, I., Cole, D.J. and Edwards, R. (1998) 'Glutathione-mediated detoxification systems in plants', *Current Opinion in Plant Biology* **1**, 258–266.

Dixon, D.P., Davis, B.G. and Edwards, R. (2002) 'Functional divergence in the glutathione transferase superfamily in plants', *Journal of Biological Chemistry* **277**, 30859–30869.

Epstein, E. (1965) Mineral metabolism, in *Plant Biochemistry* (ed. J. Bonner and J.E. Varner), Academic Press, London, Orlando, pp. 438–466.

Finkemeier, I., Kluge, C., Metwally, A., Georgi, M., Grotjohann, N. and Dietz, K.J. (2003a) 'Alterations in Cd induced gene expression under nitrogen deficiency', *Plant Cell Environment* **26**, 821–833.

Finkemeier, I., Lamkemeyer, P., Kandlbinder, A., Baier, M. and Dietz, K.J. (2003b) 'New insights into the antioxidant defence of mitochondria: the type II peroxiredoxin F (Prx IIF)', *Free Radical Research* **37**(Suppl.), 21–22.

Fomenko, D.E. and Gladyshev, V.N. (2003) 'Identity and functions of CxxC-derived motifs', *Biochemistry* **42**, 11214–11225.

Fratelli, M., Demol, H., Puype, M. *et al.* (2002) 'Identification by redox proteomics of glutathionylated proteins in oxidatively stressed human T lymphocytes', *Proceedings of the National Academy of Sciences USA* **99**, 3505–3510.

Freedman, R.B., Hirst, T.R. and Tuite, M.F. (1994) 'Protein disulphide isomerase: building bridges in protein folding', *Trends in Biochemical Science* **19**, 331–336.

Freedman, R.B., Klappa, P. and Ruddock, L.W. (2002) 'Protein disulfide isomerases exploit synergy between catalytic and specific binding domains', *EMBO Reports* **3**(2), 136–140.

Giovanelli, J., Mudd, S.H. and Datko, A.H. (1980) Sulfur amino acids in plants, in *The Biochemistry of Plants*, Vol. 5 (ed. B.J. Miflin), Academic Press, New York, pp. 453–505.

Grill, E., Löffler, S., Winnacker, E.L. and Zenk, M.H. (1989) 'Phytochelatins, the heavy metal binding peptides from plants, are synthesized from glutathione by a specific γ-glutamylcysteine dipeptidyl transpeptidase (phytochelatin synthase)', *Proceedings of the National Academy of Sciences USA* **86**, 6338–6342.

Harris, J.R., Schroder, E., Isupov, M.N. *et al.* (2001) 'Comparison of the decameric structure of peroxiredoxin-II by transmission electron microscopy and x-ray crystallography', *Biochimica et Biophysica Acta* **1547**(2), 221–234.

Haslekas, C., Grini, P.E., Nordgard, S.H. *et al.* (2003a) 'ABI3 mediates expression of the peroxiredoxin antioxidant AtPER1 gene and induction by oxidative stress', *Plant Molecular Biology* **53**(3), 313–326.

Haslekas, C., Viken, M.K., Grini, P.E. *et al.* (2003b) 'Seed 1-cysteine peroxiredoxin antioxidants are not involved in dormancy, but contribute to inhibition of germination during stress', *Plant Physiology* **133**(3), 1148–1157.

Hirota, K., Matsui, M., Iwata, S., Nishiyama, Y., Mori, K. and Yodoi, J. (1997) 'Ap-1 transcriptional activity is regulated by a direct association between thioredoxin and ref-1', *Proceedings of the National Academy of Sciences USA* **94**, 3633–3638.

Hofmann, B., Hecht, H.J. and Flohé, L. (2002) 'Peroxiredoxins', *Biological Chemistry* **383**, 347–364.

Horling, F., Baier, M. and Dietz, K.J. (2001) 'Redox-regulation of the expression of the peroxide-detoxifying chloroplast 2-Cys peroxiredoxin in the liverwort *Riccia fluitans*', *Planta* **214**, 304–313.

Horling, F., König, J. and Dietz, K.J. (2002) 'Type II peroxiredoxin C, a member of the peroxiredoxin family of *Arabidopsis thaliana*: its expression and activity in comparison with other peroxiredoxins', *Plant Physiology and Biochemistry* **40**(6–8), 491–499.

Horling, F., Lamkemeyer, P., König, J. *et al.* (2003) 'Divergent light-, ascorbate- and oxidative stress-dependent regulation of expression of the peroxiredoxin gene family in *Arabidopsis*', *Plant Physiology* **131**, 317–325.

Hung, C.Y., Lin, Y., Zhang, M., Pollock, S., Marks, M.D. and Schiefelbein, J. (1998) 'A common position-dependent mechanism controls cell-type patterning and GLABRA2 regulation in the root and hypocotyl epidermis of *Arabidopsis*', *Plant Physiology* **117**, 73–84.

Ito, H., Iwabuchi, M. and Ogawa, K. (2003) 'The sugar-metabolic enzymes aldolase and triosephosphate isomerase are targets of glutathionylation in *Arabidopsis thaliana*: detection using biotinylated glutathione', *Plant Cell Physiology* **44**(7), 655–660.

Jacquot, J.P., Lancelin, J.M. and Meyer, Y. (1997) 'Tansley review No. 94: Thioredoxins: structure and function in plant cells', *New Phytologist* **136**, 543–570.

Jacquot, J.P., Riveramadrid, R., Marinho, P. *et al.* (1994) '*Arabidopsis thaliana* NADPH thioredoxin reductase: cDNA characterization and expression of the recombinant protein in *Escherichia coli*', *Journal of Molecular Biology* **235**(4), 1357–1363.

Jeong, W., Cha, M.K. and Kim, I.H. (2000) 'Thioredoxin-dependent hydroperoxide peroxidase activity of bacterioferritin-comigratory protein (BCP) as a new member of the thiol-specific antioxidant protein (TSA)/alkyl hydroperoxide peroxidaseC (AhpC) family', *Journal of Biological Chemistry* **275**, 2924–2930.

Johnson, G.N. (2003) 'Thiol regulation of the thylakoid electron transport chain – a missing link in the regulation of photosynthesis', *Biochemistry* **42**, 3040–3044.

Kandlbinder, A., Finkemeier, I., Wormuth, D., Hanitzsch, M. and Dietz, K.J. (2004) 'The antioxidant status of photosynthesizing leaves under nutrient deficiency: redox regulation, gene expression and antioxidant activity in *Arabidopsis thaliana*', *Physiologia Plantarum* **120**, 63–73.

Kim, J. and Mayfield, S.P. (1997) 'Protein disulfide isomerase as a regulator of chloroplast translational activity', *Science* **278**, 1954–1957.

Kim, S.O., Merchant, K., Nudelman, R. *et al.* (2002) 'OxyR: a molecular code for redox regulated signalling', *Cell* **109**, 383–396.

Kong, W., Shiota, S., Shi, Y., Nakayama, H. and Nakayama, K. (2000) 'A novel peroxiredoxin of the plant Sedum limeare is a homologue of *Escherichia coli* bacterioferritin comigratory protein (Bcp)', *Biochemical Journal* **351**, 107–114.

König, J., Baier, M., Horling, F. *et al.* (2002) 'The plant-specific function of 2-Cys peroxiredoxin-mediated detoxification of peroxides in the redox-hierarchy of photosynthetic electron flux', *Proceedings of the National Academy of Sciences USA* **99**, 5738–5743.

König, J., Lotte, K., Plessow, R., Brockhinke, A., Baier, M. and Dietz, K.J. (2003) 'Reaction mechanism of plant 2-Cys peroxiredoxin: role of the C terminus and the quaternary structure', *Journal of Biological Chemistry* **278**, 24409–24420.

Koprivova, A., Meyer, A.J., Schween, G., Herschbach, C., Reski, R. and Kopriva, S. (2002) 'Functional knockout of the adenosine 5′-phosphosulfate reductase genes revives an old route of sulfate assimilation', *Journal of Biological Chemistry* **277**, 32195–32201.

Kruse, O. (2001) 'Light-induced short-term adaptation mechanisms under redox control in the PS II-LHCII supercomplex: LHC II state transitions and PS II repair cycle', *Naturwissenschaften* **88**(7), 284–292.

Laloi, C., Rayapuram, N., Chartier, Y., Grienenberg, J.M., Bonnard, G. and Meyer, Y. (2001) 'Identification and characterization of a mitochondrial thioredoxin system in plants', *Proceedings of the National Academy of Sciences USA* **98**, 14144–14149.

Lamkemeyer, P., Finkemeier, I., Kandlbinder, A., Baier, M. and Dietz, K.J. (2003) 'The role of peroxiredoxin Q in the antioxidant defence system of the chloroplast', *Free Radical Research* **37**(Suppl.), 40.

Lee, K., Lee, J., Kim, Y. *et al.* (2004) 'Defining the plant disulfide proteome', *Electrophoresis* **25**, 532–541.

Lemaire, S.D. (2004) 'The glutaredoxin family in oxygenic photosynthetic organisms', *Photosynthesis Research* **79**, 305–318.

Lillig, C.H., Prior, A., Schwenn, J.D. *et al.* (1999) 'New thioredoxins and glutaredoxins as electron donors of 3'-phosphoadenylylsulfate reductase', *Journal of Biological Chemistry* **274**, 7695–7698.

Link, G. (2003) 'Redox regulation of chloroplast transcription', *Antioxidants and Redox Signaling* **5**, 79–88.

Lu, P., Porat, R., Nadeau, J.A. and O'Neill, S.D. (1996) 'Identification of a meristem L1 layer-specific gene in *Arabidopsis* that is expressed during embryonic pattern formation and defines a new class of homeobox genes', *The Plant Cell* **8**, 2155–2168.

Marrs, K.A. (1996) 'The functions and regulation of glutathione-S-transferases in plants', *Annual Review of Plant Physiology and Plant Molecular Biology* **47**, 127–158.

Martin, J.L. (1995) 'Thioredoxin – a fold for all reasons', *Structure* **3**, 245–250.

Mestres-Ortega, D. and Meyer, Y. (1999) 'The *Arabidopsis thaliana* genome encodes at least four thioredoxin m and a new prokaryotic-like thioredoxin', *Gene* **240**, 307–316.

Meyer, Y., Verdoucq, L. and Vignols, F. (1999) 'Plant thioredoxins and glutaredoxins: identity and putative roles', *Trends in Plant Science* **4**, 388–394.

Minakuchi, K., Yabushita, T., Masumura, T., Ichihara, K. and Tanaka, K. (1994) 'Cloning and sequence analysis of a cDNA encoding rice glutaredoxin', *FEBS Letters* **337**, 157–160.

Morell, S., Follmann, H. and Häberlein, I. (1995) 'Identification and localization of the first glutaredoxin in leaves of a higher plant', *FEBS Letters* **369**, 149–152.

Motohashi, K., Kondoh, A., Stumpp, M.T. and Hisabori, T. (2001) 'A comprehensive survey of proteins targeted by chloroplast thioredoxins', *Proceedings of the National Academy of Sciences USA* **98**, 11224–11229.

Mouaheb, N., Thoma, D., Verdoucq, L., Monfort, P. and Meyer, Y. (1998) '*In vivo* functional discrimination between plant thioredoxins by heterologous expression in the yeast *Saccharomyces cerevisiae*', *Proceedings of the National Academy of Sciences USA* **95**, 3312–3317.

Mowda, S.B., Thomson, J.A, Farrant, J.M. and Mundree, S.G. (2002) 'A novel stress-inducible antioxidant enzyme identified in the resurrection plant *Xerophyta viscosa* Baker', *Planta* **215**, 716–726.

Noctor, G. and Foyer, C.H. (1998) 'Ascorbate and glutathione: keeping active oxygen under control', *Annual Review of Plant Physiology and Plant Molecular Biology* **49**, 249–279.

Paget, M.S.B. and Buttner, M.J. (2003) 'Thiol-based regulatory switches', *Annual Review in Genetics* **37**, 91–121.

Pastori, G.M. and Foyer, C.H. (2002) 'Common components, networks, and pathways of cross-tolerance to stress. The central role of "redox" and abscisic acid-mediated controls', *Plant Physiology* **129**, 460–468.

Pfannschmidt, T. (2003) 'Chloroplast redox signals: how photosynthesis controls its own genes', *Trends in Plant Science* **8**, 33–41.

Rigobello, M.P., Donella-Deana, A., Cesaro, L. and Bindoli, A. (2001) 'Distribution of protein disulphide isomerase in rat liver mitochondria', *Biochemical Journal* **356**, 567–570.

Romano, P,G.N., Horton, P. and Gray, J.E. (2004) 'The *Arabidopsis* cyclophilin gene family', *Plant Physiology* **134**, 1268–1282.

Rouhier, N., Gelhaye, E., Corbier, C. and Jacquot, J.P. (2004a) 'Active site mutagenesis and phospholipid hydroperoxide reductase activity of poplar type II peroxiredoxin', *Physiologia Plantarum* **120**, 57–62.

Rouhier, N., Gelhaye, E., Gualberto, J.M. *et al.* (2004b) 'Poplar peroxiredoxin Q: a thioredoxin-linked chloroplast antioxidant functional in pathogen defense', *Plant Physiology* **134**, 1027–1038.

Rouhier, N., Gelhaye, E., Villarejo, A. *et al.* (2004c) 'Identification of plant glutaredoxin targets', *Antioxidants and Redox Signaling*, in press.

Rouhier, N., Gelhaye, E., Sautiere, P.E. *et al.* (2001) 'Isolation and characterization of a new peroxiredoxin from poplar sieve tubes that uses either glutaredoxin or thioredoxin as a proton donor', *Plant Physiology* **127**, 1299–1309.

Rouhier, N. and Jacquot, J.P. (2002) 'Plant peroxiredoxins: alternative hydroperoxide scavenging enzymes', *Photosynthesis Research* **74**, 259–268.

Rouhier, N., Vlamis-Gardikas, A., Lillig, C.H. *et al.* (2003) 'Characterization of the redox properties of poplar glutaredoxin', *Antioxidants and Redox Signaling* **5**, 15–22.

Schafer, F.Q. and Buettner, G.R. (2001) 'Redox environment of the cell as viewed through the redox state of the glutathione disulfide/glutathione couple', *Free Radicals in Biology and Medicine* **30**, 1191–1212.

Scheibe, R. (1991) 'Redox modulation of chloroplast enzymes. A common principle for individual control', *Plant Physiology* **96**, 1–3.
Scheibe, R. (2004) 'Malate valves to balance cellular energy supply', *Physiologia Plantarum* **120**(1), 21–26.
Schürmann, P. and Jacquot, J.P. (2000) 'Plant thioredoxin system revisited', *Annual Review in Plant Physiology and Plant Molecular Biology* **51**, 371–400.
Segel, I.H. (1976) *Biochemical Calculations*, John Wiley, New York.
Shorrosh, B.S., Subramanian, J., Schubert, K.R. and Dixon, R.A. (1993) 'Expression and localization of plant protein disulfide isomerase', *Plant Physiology* **103**, 719–726.
Stacy, R.A.P., Nordeng, T.W., Culianez-Macia, F.A. and Aalen, R.B. (1999) 'The dormancy-related peroxiredoxin anti-oxidant, PER1, is localized to the nucleus of barley embryo and aleurone cells', *The Plant Journal* **19**(1), 1–8.
Storz, G., Tartaglia, L.A. and Ames, B.N. (1990) 'Transcriptional regulator of oxidative stress-inducible genes; direct activation by oxidation', *Science* **248**, 189–194.
Sweetlove, L.J., Heazlewood, J.L., Herald, V. *et al.* (2002) 'The impact of oxidative stress on *Arabidopsis* mitochondria', *The Plant Journal* **32**, 891–904.
Szederkenyi, J., Komor, E. and Schobert, C. (1997) 'Cloning of the cDNA for glutaredoxin, an abundant sieve-tube exudate protein from *Ricinus communis* L. and characterization of the glutathione-dependent thiol-reduction system in sieve tubes', *Planta* **202**, 349–356.
Tamarit, J., Belli, G., Cabiscol, E., Herrero, E. and Ros, J. (2003) 'Biochemical characterization of yeast mitochondrial GRX5 monothiol glutaredoxin', *Journal of Biological Chemistry* **278**, 25745–25751.
Theodoulou, F.L., Clark, I.M., He, X.L., Pallett, K.E., Cole, D.J. and Hallahan, D.L. (2003) 'Co-induction of glutathione-S-transferases and multidrug resistance associated protein by xenobiotics in wheat', *Pest Management Science* **59**, 202–214.
Trebitsh, T., Levitan, A., Sofer, A. and Danon, A. (2000) 'Translation of chloroplast psbA mRNA is modulated in the light by counteracting oxidizing and reducing activities', *Molecular Cell Biology* **20**, 1116–1123.
Trebitsh, T., Meire, E., Ostersetzer, O., Adam, Z. and Danon, A. (2001) 'The protein disulfide isomerase of *Chlamydomonas rheinhardtii* chloroplasts', *Journal of Biological Chemistry* **276**, 4564–4569.
Tron, A.E., Bertoncini, C.W., Chan, R.L. and Gonzalez, D.H. (2002) 'Redox regulation of plant homeodomain transcription factors', *Journal of Biological Chemistry* **277**, 34800–34807.
Walters, E.M. and Johnson, M.K. (2004) 'Ferredoxin : thioredoxin reductase: disulfide reduction catalyzed via novel site-specific [4Fe–4S]cluster chemistry', *Photosynthesis Research* **79**, 249–264.
Wheeler, G.L. and Grant, C.M. (2004) 'Regulation of redox homeostasis in the yeast *Saccharomyces cerevisiae*', *Physiologia Plantarum* **120**, 12–20.
Winter, H., Robinson, D.G. and Heldt, H.W. (1993) 'Subcellular volumes and metabolite concentrations in barley leaves', *Planta* **191**(2), 180–190.
Wood, Z.A., Poole, L.B., Hantgan, R.R. and Karplus, P.A. (2002) 'Dimers to doughnuts: redox-sensitive oligomerization of 2-cysteine peroxiredoxins', *Biochemistry* **41**, 5493–5504.
Wood, Z.A., Schröder, E., Harris, J.R. and Poole, L.B. (2003) 'Structure, mechanism and regulation of peroxiredoxins', *Trends in Biochemical Science* **28**, 32–40.
Yano, H., Wong, J.H., Lee, Y.M., Cho, M.J. and Buchanan, B. (2001) 'A strategy for the identification of proteins targeted by thioredoxin', *Proceedings of the National Academy of Sciences USA* **98**, 4794–4799.
Yao, Y., Zhou, Y. and Wang, C. (1997) 'Both isomerase and chaperone activities of protein disulfide isomerase are required for the reactivation of reduced and denatured acidic phospholipase A212', *EMBO Journal* **16**, 651–658.
Zheng, M., Åslund, F. and Storz, G. (1998) 'Activation of the OxyR transcription factor by reversible disulfide bond formation', *Science* **279**, 1718–1721.
Zheng, M., Wang, X., Templeton, L.J., Smulski, D.R., LaRossa, R.A. and Storz, G. (2001) 'DNA microarray-mediated transcriptional profiling of the *E. coli* response to hydrogen peroxide', *Journal of Bacteriology* **183**, 4562–4570.

3 Ascorbate, tocopherol and carotenoids: metabolism, pathway engineering and functions

Nicholas Smirnoff

3.1 Introduction

This chapter considers the biosynthesis and functions of ascorbate (vitamin C), tocopherols and carotenoids. Ascorbate is generally the most abundant small molecule antioxidant in plants – particularly in leaves, being about 5–10 times more concentrated than glutathione (GSH). Tocopherols and tocotrienols (collectively known as tocols or vitamin E), on the other hand, are membrane-localised antioxidants that are also particularly abundant in leaves and oil seeds. γ-tocopherol is associated with the triacylglycerol storage lipids in seeds. Ascorbate and tocopherols are included together in this chapter because they may interact in their functions: ascorbate can directly regenerate tocopheryl radicals back to the reduced tocopherol form. Finally, the antioxidant role of carotenoid pigments is also considered. Like tocopherols, these are lipophilic isoprenoids that are synthesised and localised in plastids. They are involved in the photosynthetic light-harvesting complex and also have photoprotective and antioxidant roles. Ascorbate, tocopherols and carotenoids are all required in the human diet, so there is considerable interest in the possibility of manipulating their synthesis in food plants.

3.2 Ascorbate

3.2.1 Distribution and subcellular localisation

Ascorbate occurs in almost all plant tissues, with the exception of dry seeds (DeGara *et al.*, 1997). It tends to be more concentrated in leaves than in roots and is also more concentrated in meristems. It is transported in the phloem from source leaves to sink tissues such as meristems (Franceschi & Tarlyn, 2002; Hancock *et al.*, 2003). High ascorbate concentration is popularly associated with fruit, particularly citrus. However, ascorbate concentration in fruit is not always higher than in leaves, and may be considerably lower in many species (Davey *et al.*, 2000). Ascorbate occurs in all subcellular compartments including the cell wall, but it generally has a low concentration in vacuoles (Davey *et al.*, 2000). The structures of ascorbic acid and its oxidation products are shown in Figure 3.1. At physiological pH, it is

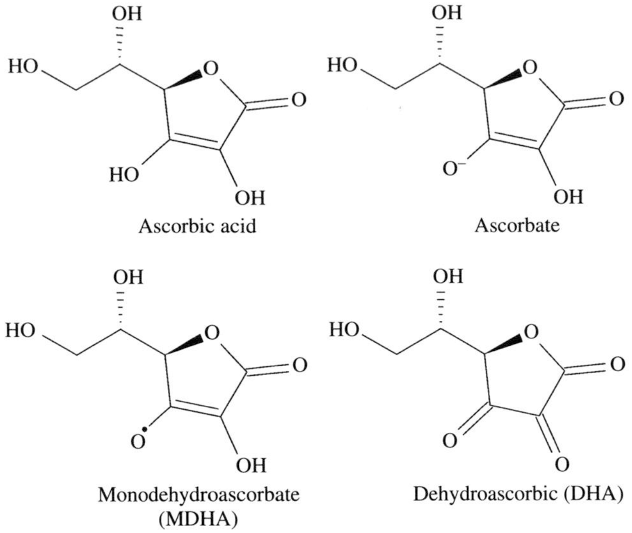

Figure 3.1   The structure of ascorbic acid (vitamin C) and its oxidation products.

predominantly present as the ascorbate anion. This readily loses an electron from its *ene*-diol group to produce the monodehydroascorbate (MDHA) radical. Further oxidation results in dehydroascorbate (DHA), which is uncharged and more properly referred to as dehydroascorbic. The ability of ascorbate to donate an electron and the relatively low reactivity of the resulting MDHA radical is the basis of its biologically useful antioxidant and free radical scavenging activity (Buettner & Schafer, 2004).

3.2.2   Ascorbate biosynthesis

The proposed pathways of ascorbate biosynthesis from hexose sugars are shown in Figure 3.2. ^{14}C-mannose, L-galactose (L-Gal) and L-galactonolactone (L-GalL) are all incorporated into ascorbate, the latter two compounds causing large increases in ascorbate pool size. The pathway of ascorbate biosynthesis differs between plants and animals. The plant pathway *via* mannose and L-Gal was proposed relatively recently (Wheeler *et al.*, 1998). The evidence for this route is provided by radio-tracer studies (Wheeler *et al.*, 1998) and from mutants and transgenic plants. The *Arabidopsis thaliana vtc1* mutant, which has low guanosine diphosphate (GDP)-Man pyrophosphorylase activity and transgenic potatoes with reduced GDP-Man pyrophosphorylase activity, have low ascorbate content (Conklin *et al.*, 1999; Keller *et al.*, 1999). GDP-mannose-3′,5′-epimerase, which converts GDP-Man to GDP-L-Gal, has been purified and cloned (Wolucka *et al.*, 2001). It is inhibited by ascorbate, suggesting that it could be controlled by feedback inhibition. It also co-purifies with a heat shock protein (Wolucka & Van Montagu, 2003) but the significance of this has not been established. GDP-L-Gal is converted to L-Gal by two enzymes,

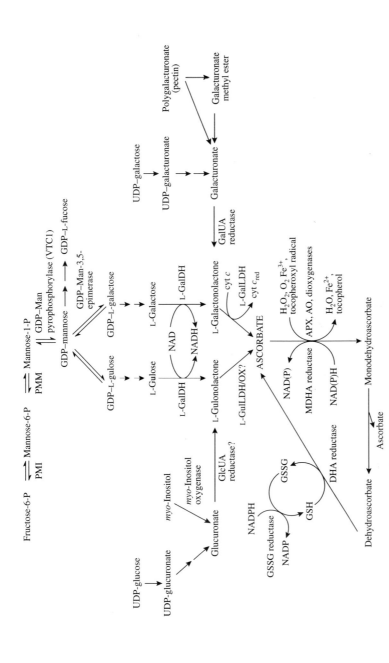

Figure 3.2 The pathways of ascorbate biosynthesis and its redox reactions in plants. Question marks indicate that the predicted enzymes have not been characterised but there is evidence for the reactions. All sugars are in the D-configuration unless indicated.

Abbreviations: AO, ascorbate oxidase; APX, ascorbate peroxidase; cyt c, cytochrome c; DHA, dehydroascorbate; Gal, galactose; L-GalL, L-galactono-1,4-lactone; GSH, glutathione; GSSG, oxidised glutathione; L-Gul, L-gulose; L-GulL, L-gulono-1,4-lactone; L-GulLO, L-GulL oxidase; PMI, phosphomannose isomerase; PMM, phosphomannose mutase.

which sequentially form L-Gal-1-P and L-Gal. These two enzymes have been purified but the genes that encode them have not been identified yet (S. Gatzek & N. Smirnoff, unpublished results). Antisense suppression of L-galactose dehydrogenase also reduces ascorbate content (Gatzek *et al.*, 2002). The final enzyme in the pathway, L-galactonolactone dehydrogenase (L-GalLDH), was identified in the 1950s (Mapson & Breslow, 1958). It is localised in mitochondria, apparently associated with complex I of the electron transport chain. It uses cytochrome *c* as the electron acceptor and its activity in isolated mitochondria is stimulated by electron flow through complex I (Millar *et al.*, 2003). Antisense reduction of L-GalLDH in tobacco BY2 cells reduces their ascorbate content (Tabata *et al.*, 2001). Overall, there is strong evidence for ascorbate synthesis *via* mannose and L-Gal. Green (chlorophyte) algae also synthesise ascorbate *via* the mannose/L-Gal pathway (Running *et al.*, 2003).

Two other potential routes to ascorbate *via* D-galacturonic acid (D-GalUA) and D-glucuronic acid (D-GlcUA)/L-gulose (L-Gul) have been proposed. First, it was suggested some time ago that ascorbate could be formed from D-GalUA acid (Loewus, 1999) (Figure 3.2). The methyl ester of D-GalUA can be absorbed and converted to ascorbate by *A. thaliana* cell cultures (Isherwood *et al.*, 1954; Davey *et al.*, 1999). The recent identification of an aldo-ketose reductase from ripening strawberry fruits whose expression correlates with ascorbate accumulation suggests a pathway that could incorporate GalUA into ascorbate via L-GalL (Agius *et al.*, 2003). This enzyme has high specificity for D-GalUA and NADPH, reducing it to L-GalL. Expression of the gene in *A. thaliana* resulted in a two- to three-fold increase in ascorbate. This result shows that there is a potential for this pathway to operate because a supply of D-GalUA must be available in *A. thaliana* leaves. It needs to be established whether this is derived directly either from UDP-GalUA or from breakdown of pectin (Figure 3.2). So far, the extent to which this pathway operates in unmodified *A. thaliana* leaves is unknown but the earlier labelling studies of Loewus (1999) suggest that it may only provide a minor route. Another alternative pathway resembles the mammalian route (Smirnoff, 2001). In this case, the immediate precursor of ascorbate is L-gulonolactone (L-GulL). This is formed by reduction of D-GlcUA (Figure 3.2). In the mammalian pathway, D-GlcUA is formed from UDP-Glc. Plant tissues can convert exogenous GulL and glucuronolactone (which is easily hydrolysed to GlcUA) to ascorbate, however, the enzyme involved has not been identified (Davey *et al.*, 1999). The enzyme that converts GulL to ascorbate has also not been characterised (Davey *et al.*, 1999): it is not oxidised by purified L-GalLDH (Ostergaard *et al.*, 1997). In mammals, GuL is oxidised to ascorbate by an oxidase (GulLO) that uses both L-GulL and L-GalL as substrate (Smirnoff, 2001). Transgenic plants expressing rat L-GulL oxidase (L-GulLO) contain more ascorbate (Jain & Nessler, 2000; Radzio *et al.*, 2003), suggesting that there is a potential for the L-GulL pathway. Since rat L-GulLO can utilise both L-GulL *and* L-GalDH as substrate, it is necessary to determine if overexpression is increasing oxidation of either or both of these compounds. Since L-GulLO expression can rescue ascorbate concentration to above the WT level in the *vtc1* mutant (Radzio *et al.*, 2003), which is affected in L-GalL formation via L-Gal (Figure 3.2), it is likely that the enzyme is acting on endogenous L-GulL. The potential sources of L-GulL are shown in Figure 3.2. Oxidation of uridine diphosphate (UDP)-Glc to UDP-GlcUA, followed by release of free GlcUA (with or without cycling through wall polysaccharides) and its reduction to L-GulL is one possible route. Davey *et al.* (1999a) reported an uncharacterised NADPH-dependent enzyme activity from *A. thaliana* cell suspension cultures that reduces GlcUA to L-GulL. More recently, in an exciting development, Lorence *et al.* (2004), have shown that the overexpression of *myo*-inositol oxygenase (*miox4*) in *A. thaliana* increases leaf ascorbate content. This enzyme cleaves the ring of *myo*-inositol, a cyclic polyol present in *A. thaliana* leaves, to produce D-GlcUA. This experiment shows that the leaves contain all the machinery needed to convert this to ascorbate. Therefore, there is potential for this pathway to occur but its quantitative importance in un-manipulated plants is not yet established. *Miox4* is expressed at very low level in leaves but at much higher level in reproductive tissue and pollen (Lorence *et al.*, 2004). Another source of L-GuL for this route could be from GDP-Man. L-GuL is the product of the 5'-epimerisation of GDP-Man by GDP-Man-3,5-epimerase and is formed by the enzyme *in vitro*, additionally to L-Gal (Wolucka & Van Montagu, 2003). L-Gulose that is oxidised at C1 by L-galactose dehydrogenase (Wheeler *et al.*, 1998), could form L-GulL, which can be oxidised to ascorbate by plants (Davey *et al.*, 1999b).

To assess the role of the uronic acid pathways *versus* the mannose/L-Gal pathway, it will be necessary to produce mutations or knockouts of the pathway-specific genes. It is an exciting possibility that multiple pathways exist: possibly their relative importance and control over flux could differ between tissues or environmental conditions. Further understanding of ascorbate metabolism will come from characterisation of mutants with altered ascorbate content. Four low ascorbate *A. thaliana* (*vtc*) mutants have been isolated (Conklin *et al.*, 2000). VTC1 (GDP-Man pyrophosphorylase) is a biosynthetic enzyme, as described above. VTC2 has been sequenced, but the role of the predicted protein has not been established (Jander *et al.*, 2002). *VTC3* and *VTC4* have not yet been identified.

Most studies of ascorbate metabolism have focused on leaves. The ascorbate content of non-photosynthetic tissues tends to be lower than that of leaves (although some fruits such as blackcurrants and rosehips are famous for their exceptionally high ascorbate content). Recently, ascorbate has been detected in the phloem sap of *A. thaliana* (Franceschi & Tarlyn, 2002) and a range of other species (Hancock *et al.*, 2004). In *Cucurbita* species, concentrations of ~5 mM were recorded (Hancock *et al.*, 2004). This raises the possibility that ascorbate could be transported from source to sink tissues in the phloem. Whole plant autoradiography, after feeding ^{14}C-ascorbate to source leaves, shows that the label is transported to sink tissues and to meristems (Franceschi & Tarlyn, 2002). Furthermore, feeding source leaves with L-GalL to elevate their ascorbate content also increased the ascorbate concentration of remote sink tissues (Franceschi & Tarlyn, 2002). Together, these results demonstrate source to sink ascorbate transport in the phloem.

In *Cucurbita*, from which it is possible to obtain phloem sap-rich exudate in relatively large quantities, Hancock *et al.* (2004) were able to detect all the known enzymes of the mannose/L-galactose pathway, with the exception of L-GalLDH. Lack of L-GalLDH could be because the sieve cells contain few mitochondria, so this final step may take place in the companion cells. Support for this was provided by the conversion of mannose, L-Gal and L-GalL to ascorbate by isolated phloem strands (containing sieve cells and companion cells) isolated from celery (*Apium graveolens*). Therefore, ascorbate is both synthesised and transported in the phloem. The ascorbate content of economically important sink organs such as fruits and tubers may therefore depend on both import and *in situ* synthesis.

3.2.3 Ascorbate recycling

Ascorbate is oxidised when acting as an oxidant or as a reductant for dioxygenase enzymes. The first product is the MDHA radical (sometimes referred to as ascorbyl radical or semidehydroascorbate). DHA is produced by disproportionation of two MDHA radicals to produce ascorbate and DHA (Buettner & Schafer, 2004). The ascorbate pool in healthy leaves is usually about 90% reduced, with most of the rest present as DHA (Noctor & Foyer, 1998). It tends to be more oxidised in cell walls (Pignocchi *et al.*, 2003; Sanmartin *et al.*, 2003). MDHA does not accumulate to high levels. It can however be detected *in vivo* by electron paramagnetic resonance spectroscopy in leaves exposed to severe oxidative stresses (Heber *et al.*, 1996; Hideg *et al.*, 1997). In normal circumstances, the ascorbate pool is kept reduced by a regeneration system that involves a number of players. These include monodehydroascorbate reductases (MDHARs), dehydroascorbate reductases (DHARs) and GSH as part of the ascorbate–glutathione cycle (Figure 3.2; see Chapter 1) and reduced ferredoxin in photosystem I (PSI) (Miyake & Asada, 1992; Asada, 1999). MDHARs are localised in the cytosol, chloroplasts, peroxisomes, mitochondria and plasma membrane (Jimenez *et al.*, 1997; Berczi & Moller, 1998; Noctor & Foyer, 1998). They are FAD-containing enzymes that catalyse NAD(P)H-dependent reduction of MDHA to ascorbate (Hossain & Asada, 1985). About 25 MDHAR sequences from 11 species have been identified. The *A. thaliana* genome has five genes with predicted MDHAR activity (Chew *et al.*, 2003). One *A. thaliana* MDHAR (*Arabidopsis* Genome Initiative locus number At1g63940) is targeted to both chloroplasts and mitochondria. The mitochondrial location was confirmed by transient expression of a green fluorescent protein (GFP) fusion protein in soybean cells (Chew *et al.*, 2003). Obara *et al.* (2002), using GFP fusions, also identified dual targeting of MDHAR, but in contrast, they suggested that the gene is transcribed differentially into longer and shorter transcripts, the former being mitochondrial and the latter chloroplastic. Prediction of the subcellular localisation of the other *A. thaliana* MDHAR genes using various algorithms (ARAMEMNON http://aramemnon. botanik.uni-koeln.de, PSORT http://psort.ims.u-tokyo.ac.jp/form.html, TMHMM http://www.cbs.dtu.dk/services/TMHMM/ and SOSUI http://sosui.proteome.bio. tuat.ac.jp/sosuiframe0E.html) suggests the following localisations. At3g27820 has a predicted secretory, ER membrane or plasma membrane targeting sequence and could represent the activity associated with plasma membranes. It has one predicted transmembrane domain (459–481) and a hydrophobic N-terminus that may be a signal peptide. At3g52880 has a microbody/peroxisomal or chloroplast location predicted by PSORT. The other two (At5g03630 and At3g09940) are predicted to be soluble proteins (TMHMM and SOSUI) but have potential ER membrane target peptides (PSORT). MDHAR can also reduce phenoxyl radicals (e.g. the quercetin radical; see Chapter 6) in an NADH-dependent manner and inhibits oxidation of ferulic, coniferyl and chlorogenic acids by peroxidase/hydrogen peroxide (Sakihama *et al.*, 2000). It also catalyses ferredoxin-dependent oxygen photoreduction by PSI, along with other flavin containing enzymes such as glutathione reductase (Miyake *et al.*, 1998). DHA is readily reduced non-enzymatically to ascorbate by thiols such as GSH and by a number of SH containing proteins. These include GSH-dependent DHAR, glutaredoxins, glutathione-S-transferases, protein disulfide isomerase, thioredoxin reductase and Kunitz-type trypsin inhibitors (Trumper *et al.*, 1994; Hou & Lin, 1997; Hou *et al.*, 1999, 2000; May & Asard, 2004). Exogenous DHA is taken up by plant tissues and reduced to ascorbate. Interestingly, this process does not deplete or oxidise the GSH pool in tobacco cell cultures and is not inhibited in cells depleted of GSH by diethylmaleate, suggesting that reductants other than GSH can reduce DHA *in vivo* (Potters *et al.*, 2004). Despite some initial controversy about the existence of specific DHAR activity in plants (Foyer & Mullineaux, 1998; Morell *et al.*, 1998), GSH-dependent DHARs have been purified, sequenced and cloned from a number of species (Hossain & Asada, 1984; Dipierro & Borraccino, 1991; Kato *et al.*, 1997; Foyer & Mullineaux, 1998; Shimaoka *et al.*, 2000; Urano *et al.*, 2000). The structure and reaction mechanism of spinach DHAR has been described (Mizohata *et al.*, 2001; Shimaoka *et al.*, 2003). GSH-dependent DHAR activity occurs in chloroplasts, mitochondria and peroxisomes (Jimenez *et al.*, 1997; Noctor & Foyer, 1998). There are five DHAR genes in *A. thaliana* (Chew *et al.*, 2003). They have some sequence homology to glutathione-S-transferases. One of them (At5g16710) is predicted to be a chloroplast or mitochondrial protein with a 42 amino acid targeting sequence (Target P, PSORT). However, this localisation could not be confirmed by *in vitro* protein chloroplast and mitochondria import experiments (Chew *et al.*, 2003). Chen *et al.* (2003) cloned a predicted GSH-dependent DHAR from wheat. The recombinant enzyme expressed in *Escherichia coli* had GSH-dependent DHAR activity and has high sequence similarity to *A. thaliana*, rice, maize, tobacco and tomato proteins. Overexpression of wheat DHAR in maize and tobacco increased the ascorbate/dehydroascorbate pool size and increased the proportion present as ascorbate in leaves. This suggests that the capacity of DHAR (and MDHAR) is limiting for ascorbate regeneration and that the instability and breakdown of DHA may be responsible for loss of ascorbate measured in leaves. However, increased ascorbate over the normal amount present may not necessarily be beneficial and could perturb processes that depend on the 'redox state' (see Section 3.2.8). The concentration and reduction state of the GSH pool was also increased, suggesting some cross-talk with ascorbate (Chen *et al.*, 2003). Plasma membrane and tonoplast-localised cytochrome b_{561} can accept electrons from ascorbate and donate electrons to MDHA (Horemans *et al.*, 1994, 2000; Griesen *et al.*, 2004; May & Asard, 2004). These cytochromes could be involved in transmembrane electron transport and in reducing apoplastic and vacuolar MDHA (Villalba *et al.*, 2004).

3.2.4 · Ascorbate and dehydroascorbate transport across membranes

Ascorbate and DHA can be taken up by plant cells and transported from the cytosol into organelles. Saturable carriers for ascorbate and DHA have been identified although their molecular identity has not been established yet (Horemans *et al.*, 2000; May & Asard, 2004). Ascorbate is taken up by chloroplasts by facilitated diffusion (Foyer & Lelandais, 1996). Uptake of ascorbate into vacuoles does not have saturation kinetics, indicating lack of a specific transporter (Rautenkranz *et al.*, 1994). The plasma membrane appears to have transporters for ascorbate and DHA. DHA uptake is dependent on the proton electrochemical gradient because it is inhibited by the uncoupler carbonylcyanide-3-chlorophenylhydrazone (CCCP; Horemans *et al.*, 2000) and by the proton-ATPase inhibitor erythrosin B (Kollist *et al.*, 2001). Evidence from DHA uptake by plasma membrane vesicles suggests that DHA uptake occurs by exchange with ascorbate (Horemans *et al.*, 1998). This would provide a mechanism for regenerating apoplastic DHA after transport into the cytosol while maintaining the apoplastic ascorbate pool.

3.2.5   Enzymes involved in ascorbate oxidation

Plants have two enzymes that catalyse ascorbate oxidation. Ascorbate peroxidase (APX) has members in all subcellular compartments. APX catalyses the ascorbate-dependent reduction of hydrogen peroxide to water. It is central in the antioxidant function of ascorbate in plants and is covered in Chapter 4. Ascorbate oxidase (AO), a member of the blue copper oxidase family (Messerschmidt & Huber, 1990), is a glycoprotein that is exported to the apoplast and its activity can be detected in isolated extracellular fluid (Esaka *et al.*, 1989). AO activity is particularly high in expanding cucurbit fruits and in other rapidly expanding tissues (Diallinas *et al.*, 1997; Al-Madhoun *et al.*, 2003). Its function is obscure but it has been suggested that a low wall ascorbate concentration is needed for cell expansion (see Chapter 9). Overexpression and antisense suppression of activity in tobacco has produced plants with increased or decreased activity in the apoplast, and these have, respectively, a less or more oxidised apoplastic ascorbate pool (Pignocchi & Foyer, 2003; Pignocchi *et al.*, 2003; Sanmartin *et al.*, 2003). Therefore, it is clear that AO can control the redox state of ascorbate in the cell wall but the effect on growth and development needs further investigation. AO expression in tobacco cell cultures appears to increase the rate of cell expansion in response to auxin (Kato & Esaka, 2000), supporting a role in cell expansion. Because ascorbate is the major redox active compound in the apoplast, Pignocchi and Foyer (2003) have proposed that AO has a role in controlling cell wall redox state. AO activity is high and ascorbate concentration is low in the quiescent centre (QC) (zone of non-dividing cells) in the maize root meristem (Kerk & Feldman, 1995).

The final class of enzymes associated with ascorbate oxidation are 2-oxoglutarate-Fe(II)-dioxygenases. They require ascorbate and non-haem iron as cofactors and ascorbate, because it can readily reduce Fe(III), may have a role in maintaining the iron in the Fe(II) state needed for activity. Ascorbate may have a specific binding site on the enzyme near the Fe binding site (Lukacin & Britsch, 1997). Dioxygenases have diverse metabolic roles in plants and animals (Arrigoni & De Tullio, 2002). The most famous example is hydroxylation of proline to form hydroxylproline in collagen. The symptoms of scurvy disease caused by ascorbate deficiency in animals that cannot synthesise ascorbate (e.g. humans and other primates and guinea pigs) are largely caused by collagen deficiency (Davey *et al.*, 2000; De Tullio, 2004). Plants also synthesise hydroxyproline-containing extracellular proteins (e.g. extensin) essential for cell wall function (De Tullio *et al.*, 1999). Ascorbate-dependent oxygenases are also involved in hormone (ethylene, GA, ABA), anthocyanin and alkaloid synthesis. The *A. thaliana* genome contains about 60 genes with predicted 2-oxoglutarate-Fe(II)-dependent dioxygenase activity but the ascorbate requirement has only been directly demonstrated in a few plant dioxygenases (De Tullio *et al.*, 1999; Arrigoni & De Tullio, 2002). Violaxanthin de-epoxidase is also an ascorbate-dependent enzyme (Muller-Moule *et al.*, 2002; see Chapter 10) but it has little sequence homology with the 2-oxoglutarate-dependent dioxygenases.

3.2.6 Ascorbate catabolism

Ascorbate is converted to oxalate, tartrate and threonate (Saito & Loewus, 1989; Keates *et al.*, 2000). The enzymes involved in cleavage of the ascorbate C-skeleton to produce these organic acids have not been identified. The extent and mechanism of catabolism varies between species and has been reviewed by Loewus (1999) and Banhegyi and Loewus (2004).

3.2.7 Control of ascorbate synthesis and metabolic engineering

Very little is known about the control of ascorbate synthesis, e.g. in relation to leaf age and light. Changes in L-GalLDH activity are associated with changes in ascorbate pool size during leaf senescence and changes in light intensity (Smirnoff, 2000). The rate of L-Gal synthesis is also increased by high light intensity suggesting that early steps of the L-Gal pathway have increased flux (Gatzek *et al.*, 2002). There is evidence that ascorbate synthesis could be regulated by feedback inhibition (Pallanca & Smirnoff, 2000). Because of its importance in human nutrition, and its potential roles in stress tolerance, ascorbate is a target for metabolic engineering or directed breeding to increase its concentration. In general, overexpression of enzymes in the mannose/L-Gal pathway has not resulted in increased ascorbate concentration (N. Smirnoff, S. Gatzek & J. Dowdle, unpublished results) although potentially important enzymes converting GDP-L-Gal to L-Gal (Figure 3.2) have not been cloned yet. However, expression of genes that encode enzymes of the alternative uronic acid pathways (Figure 3.2) D-galacturonate reductase (Agius *et al.*, 2003), *miox4* (Lorence *et al.*, 2004) and L-gulonolactone oxidase (Jain & Nessler, 2000; Radzio *et al.*, 2003) have provided increases in ascorbate of around at least two-fold in *A. thaliana*, tobacco and lettuce. Antisense reduction of AO in tobacco had no effect on total leaf ascorbate but decreased the proportion present as DHA in the apoplast (Pignocchi *et al.*, 2003). The overexpression of recycling enzymes also has some effect. Overexpression of wheat DHAR in tobacco and maize increased the total ascorbate pool and decreased the proportion of DHA (Chen *et al.*, 2003) while human DHAR increased reduction state but not pool size (Kwon *et al.*, 2003). Similarly, overexpression of glutathione reductase in *Populus tremula* increased ascorbate pool size (Foyer *et al.*, 1995). In addition to the transgenic approach, natural variation, QTL analysis and activation tagging also have promise to identify genes that control ascorbate concentration.

3.2.8   The functions of ascorbate

Recently, progress in understanding the functions of ascorbate has accelerated. Clearly, it would be predicted to have a role as an antioxidant and as an enzyme cofactor. Some of the *vtc* mutants are more sensitive to UV radiation and ozone (Conklin *et al.*, 1996, 2000). APX has been overexpressed in transgenic plants and increases in tolerance to oxidative and photo-oxidative stress have been reported (Kwon *et al.*, 2002; Kornyeyev *et al.*, 2003; Yan *et al.*, 2003; Chapter 4). Similarly APX expressed in tobacco increased the tolerance of leaf discs to hydrogen peroxide and superoxide generated by methyl viologen treatment (Kwon *et al.*, 2003). A recent study of *vtc1* and mutants with reduced flavonoid (see Chapter 6 for a discussion of phenolic compounds as antioxidants) content shows that they have increased incidence of DNA damage as indicated by expression of a GUS reporter gene that is activated by homologous recombination (Filkowski *et al.*, 2004). This illustrates the importance of antioxidants in genome stability. It has been implicated in the control of cell expansion (Tabata *et al.*, 2001) and cell division (Potters *et al.*, 2002). Partly this may be associated with synthesis of hormones and extracellular structural proteins containing hydroxyproline (De Tullio *et al.*, 1999). The redox state of apoplastic ascorbate is controlled by AO (Pignocchi *et al.*, 2003; Sanmartin *et al.*, 2003) and this may affect growth and ozone resistance. Ascorbate can potentially act as a pro-oxidant by reducing iron, copper and manganese. The reduced forms of these transition metals can generate hydroxyl radicals in the presence of hydrogen peroxide (Fenton reaction; see Chapter 9). Mixtures of transition metals and ascorbate can therefore cause oxidative damage to DNA (Poulsen *et al.*, 2004). Apoplastic ascorbate has been proposed to interact with copper to produce hydroxyl radicals that can degrade wall polysaccharides (Fry, 1998). This could be involved in wall softening in ripening fruit (Fry *et al.*, 2001; see Chapter 9). In mammalian systems, pro-oxidant effects of ascorbate administration has been reported but are not consistent (Suh *et al.*, 2003).

The *A. thaliana vtc1* (Veljovic-Jovanovic *et al.*, 2001) and *vtc2* mutants (Conklin *et al.*, 2000) and low ascorbate potato plants (Keller *et al.*, 1999) are smaller than wild type plants. However, because GDP-Man pyrophosphorylase is required for synthesis of mannose containing polysaccharides and glycoproteins as well as ascorbate, the smaller size cannot yet be unambiguously attributed to low ascorbate. However, it is interesting to note that *vtc1* has higher abscisic acid (ABA) content than wild type plants (Pastori *et al.*, 2003). This hormone generally inhibits growth. Another possibility to consider is that peroxidative cross-linking of wall polysaccharides and proteins and increased lignification, could occur in low ascorbate plants, leading to reduced wall extensibility. These peroxidative reactions are inhibited by ascorbate through hydrogen peroxide removal and direct inhibition of peroxidases. This has not been investigated in detail, but its importance may be small, because transgenic tobacco plants with increased apoplastic AO activity, and consequently lower apoplastic ascorbate, are not significantly smaller than wild type (WT) plants (Pignocchi *et al.*, 2003; Sanmartin *et al.*, 2003) although cultured tobacco cells overexpressing pumpkin AO expanded more rapidly (Kato & Esaka, 2000). The growth rate of tobacco cell cultures in which ascorbate content was reduced by antisense suppression of GalLDH was decreased (Tabata *et al.*, 2001). Manipulation of ascorbate and DHA in tobacco cell cultures suggests that DHA inhibits cell division while ascorbate stimulates it (Potters *et al.*, 2000, 2002, 2004). The QC of maize roots, a zone of the meristem with non-dividing cells has low ascorbate content, caused by high AO activity and low GSH content (Kerk & Feldman, 1995; Jiang *et al.*, 2003). This corresponds with higher superoxide and hydrogen peroxide generation in this part of the meristem. Removing the root cap increased the concentration of reduced ascorbate and GSH and activated cell division in the QC (Jiang *et al.*, 2003), providing evidence that cell division is arrested by the oxidised ascorbate and GSH pools. Low ascorbate antisense potato plants also have accelerated leaf senescence (Keller *et al.*, 1999). Similarly, the expression of several senescence-associated genes (SAG13, 15 and 27) is increased in *vtc1* leaves and detached *vtc1* leaves have faster dark-induced senescence. Ascorbate concentration decreases in older leaves as they approach senescence, as does the activity of GalLDH (Bartoli *et al.*, 2000). These observations all suggest an association between ascorbate and leaf ageing. It is tempting to speculate that ascorbate changes associated with senescence require a more 'oxidising' environment that is facilitated by a lower ascorbate concentration. It should be noted that not all antioxidants decrease in senescing leaves, and some such as tocopherol increase (Molinatorres & Martinez, 1991). Response to pathogens also requires ROS (Grant & Loake, 2000). A comparison of the transcriptome of *vtc1* with wild type leaves shows an increase in expression of PR genes, e.g. PR1 and PR5 (Pastori *et al.*, 2003; Barth *et al.*, 2004). Furthermore Barth *et al.* (2004) also showed that salicylic acid (SA) and SA glucoside are higher in *vtc1*, particularly after inoculation with virulent strains of bacterial (*Pseudomonas syringae*) and fungal (*Peronospora parasitica*) pathogens. Significantly, the growth of the pathogens and disease symptoms were reduced in the mutant. It is, therefore, apparent that ascorbate can influence the expression of pathogen defence genes. As many of these genes are induced through a signalling system involving SA, it may be that this has a significant role. SA and SA-responsive genes also increase in expression during leaf senescence (Morris *et al.*, 2000; Robatzek & Somssich, 2001). This may be tied in with the decrease of the ascorbate pool during leaf senescence. A further analysis of the *vtc1* transcriptome data produced by Pastori *et al.* (2003) shows a considerable similarity with between the set of genes induced in *vtc1* and those that increase in senescing leaves (Smirnoff, unpublished analysis). About 32% of these show an SA-dependent increase in expression in older senescence because they do not increase in the NahG transgenic line of *A. thaliana*, in which SA is inactivated by hydroxylation (Delaney *et al.*, 1994). This new information suggests that ascorbate could be closely related to pathogen response and leaf senescence in a causal manner. Possibly the 'redox state' of cells is influenced by ascorbate because of its very high concentration. This may affect the levels of hydrogen peroxide (through the action of APX), other ROS and products of oxidative reactions (e.g. lipid oxidation) that could act as signals of redox state (op den Camp *et al.*, 2003; Laloi *et al.*, 2004). It is interesting that ascorbate deficiency does not lead to a marked increase in expression of a large number of antioxidant genes (Pastori *et al.*, 2003). Indeed in another study, the expression of three antioxidant genes encoding catalase, and Fe superoxide dismutase and a chloroplastic glutathione peroxidase was lower in *vtc1* than wild type, although induction on exposure to UV-C was bigger and faster in *vtc1* (Filkowski *et al.*, 2004). Possibly higher expression of antioxidant genes in non-stressed plants would defeat the ability of ascorbate concentration to modulate defence responses. Improved understanding of how ascorbate synthesis and redox state are controlled becomes more important in view of this model. Interestingly, infection of *A. thaliana* leaves with the necrotrophic pathogen *Botrytis cinerea* causes extensive oxidation of ascorbate, not only at the site of the lesions (as might be expected in dead cells), but also at a distance where the fungus is not present in the tissue and there is no visible damage (Muckenschnabel *et al.*, 2002). This supports the model proposed above and suggests control over ascorbate content and redox state in relation to pathogen infection.

Ascorbate, particularly in the apoplast, has been implicated in defence against ozone (Luwe *et al.*, 1993; Plochl *et al.*, 2000; Maddison *et al.*, 2002) (see Chapter 11 for a detailed discussion). This has been clearly illustrated in transgenic tobacco plants expressing AO in the apoplast. They have less ascorbate and more DHA in their apoplastic fluid and are also more sensitive to ozone (Sanmartin *et al.*, 2003). *vtc1* and *vtc2-1* are ozone hypersensitive (Conklin *et al.*, 1996, 2000) but, oddly, the *vtc2-2* and *2-3* alleles are not, despite similarly low ascorbate concentrations. *vtc1* is also UV-B (Conklin *et al.*, 1996) and UV-C sensitive (Filkowski *et al.*, 2004). Considering the connections between ozone and UV-B responses and pathogen responses (see Chapter 11), it is paradoxical that *vtc1* is ozone sensitive but is more resistant to virulent pathogens. Of the 79 genes with high expression in *vtc1* (Pastori *et al.*, 2003), 41% are induced by ozone and 38% by UV-B (Smirnoff, unpublished analysis of gene expression data from public domain Affymetrix GeneChips, http://arabidopsis.info/prototype). Barth *et al.* (2004) suggest that this can be explained as follows: ozone treatment induces SA production in *vtc1* adding to that already present. This then triggers programmed cell death that results in the lesions induced by ozone. The difference in ozone sensitivity of different *vtc* mutants could then be explained by differences in SA content. In the pathogen response, it is possible that the SA-induced defences reduce pathogen spread. The role of ROS in pathogen responses and programmed cell death are discussed in more detail in Chapters 8 and 11.

ROS (e.g. hydrogen peroxide and singlet oxygen) have been implicated as signalling molecules in a number of processes (see Chapters 7 and 8). These include cell expansion growth, e.g. in root hairs (Foreman *et al.*, 2003; Rentel *et al.*, 2004) and ABA signalling in guard cells. Ascorbate and other antioxidants disrupt root hair growth, causing bursting at the tip (see Chapter 8). Clearly, ascorbate concentration in the root hair apoplast must be low. ABA-induced stomatal closure requires superoxide/hydrogen peroxide formation via an NADPH oxidase (see Chapters 6 and 8). Tobacco plants overexpressing DHAR, which have higher ascorbate with a smaller proportion of DHA, have impaired drought and ABA-induced stomatal closure and wilt more readily. The hydrogen peroxide concentration in guard cells shown by fluorescence microscopy was higher. Antisense reduction of DHAR activity had the opposite effect (Chen & Gallie, 2004). This paper is highly significant because it illustrates the subtle but pervasive effect that ascorbate redox state may have: the balance between ascorbate and ROS in guard cells is a factor in environmental control of stomatal behaviour that may explain some phenomena, e.g. mid-day stomatal closure. In engineering increased ascorbate, there is a possibility that some processes could be compromised. There may be a difference between engineering ascorbate by improving recycling, as in this case, or by increasing the rate of biosynthesis. In the case of increased biosynthesis, disruption to redox sensitive processes may be less if redox state is not altered.

As has been noted above, ascorbate concentration in leaves is light-dependent in many species, and chloroplast concentrations can be very high (Smirnoff, 2000). This suggests an important role in photosynthesis and photoprotection. This aspect of ascorbate has been well reviewed (Noctor & Foyer, 1998; Noctor *et al.*, 2000) and is briefly outlined. Two distinct roles for ascorbate in chloroplasts can be identified. First, its role in superoxide/hydrogen peroxide scavenging via APX, which includes a thylakoid membrane-associated form (see Chapter 10) and its regeneration by reduced ferredoxin and the ascorbate–glutathione cycle (Asada, 1999; Noctor *et al.*, 2000). Superoxide and hydrogen peroxide are formed by PSI oxygen photoreduction (Mehler reaction) and in PSII (Chapter 10). Singlet oxygen is formed by photosensitisation in PSII (op den Camp *et al.*, 2003). Ascorbate will react with any singlet oxygen that escapes from the thylakoid membrane and may also directly regenerate tocopheryl radicals formed by oxidation of tocopherol (see Section 3.3.3). The role of ascorbate in protection against excess light is illustrated by the higher sensitivity of the *vtc* mutants to photo-bleaching caused by exposure to high light intensity and high salinity (Smirnoff, 2000). However, under low light intensity laboratory conditions, photosynthesis is not impaired in *vtc1* (Veljovic-Jovanovic *et al.*, 2001). APX activity, which catalyses the removal of hydrogen peroxide by ascorbate is higher in high light acclimated leaves. Also a specific cytosolic APX (APX2) is rapidly induced by sudden exposure to high light (Karpinski *et al.*, 1997; Fryer *et al.*, 2003). Recently it has been shown that it is localised in the bundle sheath cells of leaf vascular bundles, where it may have a role in controlling hydrogen peroxide movement from the leaf (Fryer *et al.*, 2003). As photosynthesis produces ROS in chloroplasts and hydrogen peroxide in peroxisomes from photorespiration, it is clear that ROS homeostasis is required to maintain optimal poise for response to abiotic stresses and pathogens. The light regulation of ascorbate pool size and APX activity is a key component of this system.

The role of GSH, the other soluble small molecule antioxidant in chloroplasts, is discussed in Chapter 1. It is possible that the redox state of GSH is to some extent independent of ascorbate (Noctor *et al.*, 2000). Because of the relative redox potentials of ascorbate and GSH, oxidised GSSG can exist in the presence of the high ascorbate concentration in the chloroplast stroma. Therefore, the high stromal ascorbate concentration should not interfere with the redox control of the thioredoxin system and other thiol–disulphide interconversions (see Chapters 1 and 2). On the other hand, ascorbate, because of abundant APX and its ability to react with superoxide and singlet oxygen, could act as a redox buffer. If oxidation of ascorbate increases substantially under severe stress, then the ascorbate–glutathione cycle will be activated resulting in increased GSSG formation. APX is discussed in more detail in Chapter 4 and the role of ROS in photosynthesis is discussed in more detail in Chapter 10. The second involvement of ascorbate in photosynthesis is in the xanthophyll cycle. In this case, development of NPQ and zeaxanthin formation is retarded in *vtc* mutants (Smirnoff, 2000; Muller-Moule *et al.*, 2002). When exposed to very high light *vtc2* shows bleaching of older leaves, while wild type plants are unaffected. A double mutant of *vtc2* and *npq1* (which is deficient in violaxanthin de-epoxidase) shows greatly increased damage (Muller-Moule *et al.*, 2004).

Finally, a word of caution is needed in the use of mutants, particularly *vtc1*, in elucidating the functions of ascorbate. *Vtc1* is not specifically affected in ascorbate content and second, it is an EMS mutagenised line, even after several back-crosses to wild type, it may still harbour other mutations. Therefore, more experiments are needed with multiple alleles of the same gene or with different genes (e.g. *vtc2, 3* and *4*). Alternatively, as was done in the case of the *vtc1* transcriptome (Pastori *et al.*, 2003), it is necessary to show that restoring the ascorbate content of the mutant by feeding ascorbate or a precursor restores the wild type phenotype. Although recent results from the mutants is revealing a consistent picture of the functions of ascorbate, more work is needed, particularly to identify the functions of VTC2, 3 and 4.

3.3   Vitamin E: tocopherols and tocotrienols

3.3.1   Isoprenoid antioxidants

Plants synthesise a variety of isoprenoid compounds in their plastids (Figure 3.3).
A number of these, e.g. plastoquinone, have roles in electron transport and sensing
of the redox state of the chloroplast electron transport chain (Pfannschmidt *et al.*,
1999). They also have roles in photoprotection (e.g. carotenoids) and as antioxidants (e.g. tocopherols, tocotrienols and carotenoids). Having long isoprenoid
chains, these molecules are hydrophobic and associated with membranes. They not
only have a key role in photosynthesis and in antioxidant defence, but they are also
important in the human diet. The tocopherols and tocotrienols, in particular
α-tocopherol, are required in the diet and are collectively known as vitamin E (or
tocols). α-tocopherol deficiency leads to a number of disorders and there is strong
evidence that vitamin E supplementation is beneficial to health (Bramley *et al.*,

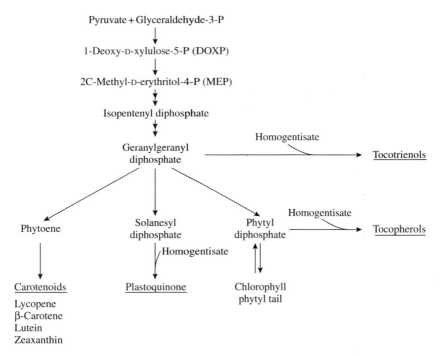

Figure 3.3   An outline of the synthesis of isoprenoid compounds in plastids. The precursor
geranylgeranyl diphosphate (GGDP) is synthesised by 2C-methyl-D-erythritol 4-phosphate
(MEP) or non-mevalonate pathway. This provides the isoprenoid chain of carotenoids, the
phytyl chain of chlorophyll, and the isoprenoid component of plastoquinone (involved in
photosynthetic electron transport and redox sensing) and tocols (tocopherols and
tocotrienols). The homogentisate needed for plastoquinone and tocol synthesis is derived
from the shikimic acid pathway.

2000). Carotenoids are also important dietary components. β-carotene (provitamin A) is needed for synthesis of retinol (vitamin A), which is used for synthesis of the light absorbing rhodopsin pigment in the eye. Vitamin A deficiency is a major cause of blindness. Rice with β-carotene synthesis engineered into its endosperm ('Golden rice') has been proposed, with some controversy, as a means of reducing blindness caused by vitamin A deficiency in developing countries (Ye et al., 2000; Potrykus, 2001). Carotenoids derived from the diet may also have a photoprotective role in the eye, as they do in the chloroplast. Indeed it is striking that both the eye and chloroplasts are rich in ascorbate, tocopherol and carotenoids (Halliwell & Gutteridge, 1999). A variety of plant-derived carotenoids (e.g. lycopene and zea-xanthin) have been proposed to have beneficial antioxidant effects in humans (Bramley, 2000; van den Berg et al., 2000). Given the potential health benefits, there is great interest in understanding and manipulating the synthesis of vitamin E and carotenoids in plants. This has met with considerable success and is described below.

3.3.2   Structure and antioxidant activity of tocopherols and tocotrienols

Vitamin E comprises a number of related compounds (tocopherols and tocotrienols) characterised by a hydrophobic isoprenoid tail and a hydrophobic chromanol head (Figure 3.4). The tail is buried in the lipid bilayer of membranes and the more hydrophilic head is held at the membrane surface. Tocopherols have a saturated tail derived from phytyl diphosphate, while tocotrienols have an unsaturated tail derived from geranylgeranyl diphosphate. In each case, there are four forms differing in degree and position of methylation of the chromanol ring (Figure 3.4).

Tocopherols and tocotrienols are made in higher plant plastids and by cyanobacteria. α-tocopherol is the main form in leaves while γ-tocopherol is the major form in seeds. δ-tocopherol and β-tocopherol are much less abundant. The tocotrienols are found only in the seeds of monocot species (e.g. cereals) and a few other dicot species such as tobacco (Cahoon et al., 2003; Falk et al., 2003). In seeds, γ-tocopherol is associated with triacylglycerols in the oil bodies and possibly in glyoxysomes where they are oxidised during germination (Sattler et al., 2004). Tocopherol must presumably be transported from the plastids to these locations. Animals have a tocopherol transfer protein (TTP) (Azzi et al., 2002), but it is not known if plants have similar proteins. The occurrence and function of tocopherols in plants have been reviewed recently (Fryer, 1992; Bramley et al., 2000; Munne-Bosch & Falk, 2004) and the reader should refer to these for additional details.

The chromanol ring can donate single electrons resulting in production of a resonance-stabilised tocopheroxyl (tocopheryl or chromanoxyl) radical (Figure 3.4). This is the basis of its ability to prevent lipid peroxidation chain reactions. Lipid peroxyl radicals are converted to hydroperoxides by reaction with tocopherol. For tocopherols, the order of effectiveness in protection against lipid peroxidation *in vitro* is $\delta > \beta = \gamma > \alpha$, however α-tocopherol has the highest vitamin E activity

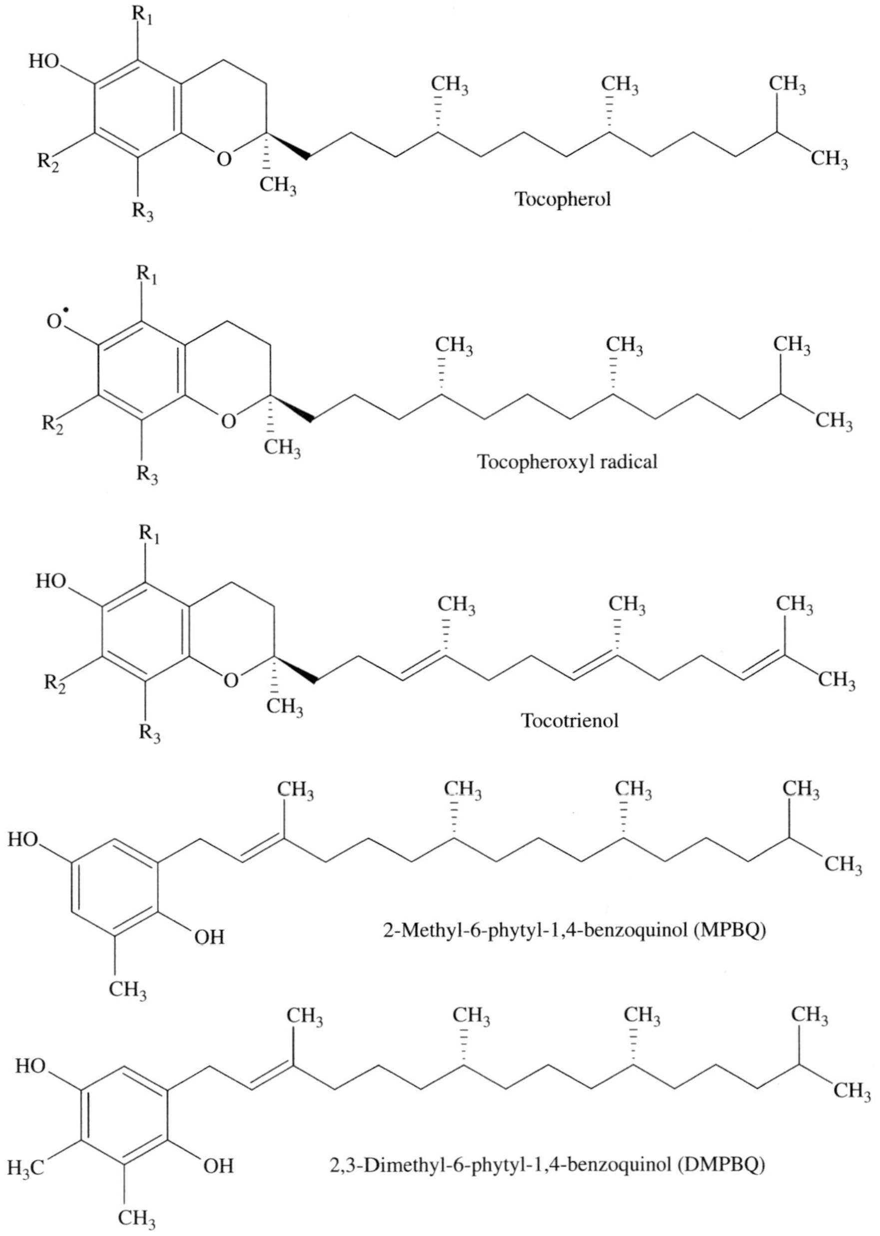

Figure 3.4 The structure of tocols and related compounds. δ-tocopherol/tocotrienol: R_1 and $R_2 = H$; $R_3 = CH_3$. β-tocopherol/tocotrienol: $R_2 = H$; R_1 and $R_3 = CH_3$. γ-tocopherol/tocotrienol: $R_1 = H$; R_2 and $R_3 = CH_3$. α-tocopherol/tocotrienol: R_1, R_2 and $R_3 = CH_3$.

in the human diet. This appears to be because TTP has high specificity for α-tocopherol (Azzi et al., 2002). Moreover, chemically synthesised α-tocopherol also has less vitamin E action than natural α-tocopherol because it is a mixture of eight stereoisomers: TTP has higher affinity for the natural isomer (Grusak & DellaPenna, 1999). Tocotrienols have equal or higher antioxidant activity (Rippert et al., 2004) but lower vitamin E activity than tocopherols.

Tocopherol and tocotrienols can react with singlet oxygen, produced by photo-sensitised pigments such as chlorophyll (see Chapter 10) in two ways. First, they can quench it by absorbing excitation energy, causing the singlet oxygen to return to ground state oxygen. Second, they can react with singlet oxygen to form an oxidised product: tocopherylquinone (Figure 3.4). 2-methyl-6-phytyl-1,4-benzo-quinone (MPBQ) and 2,3-dimethyl-5-phytyl-1,4-benzoquinone (DMPBQ), the quinol precursors of tocopherols (Figure 3.4), cannot carry out this reaction because the full chromanol ring structure is needed. The reaction with superoxide is relatively slow but as expected, very fast with hydroxyl radicals (Halliwell & Gutteridge, 1999). Little is known about the extent of tocopherylquinone produc-tion or its fate in plants. In animals, it is conjugated to glucuronate and excreted or degraded to tocopheronic acid in kidneys and excreted (Brigelius-Flohé & Traber, 1999). It is generally considered that ascorbate is the reductant that regenerates tocopherol from the tocopheroxyl radical. This reaction can easily be demonstrated in vitro and a large amount of evidence in both plants and animals suggests that these antioxidants can act synergistically in protecting against oxidative stress (Buettner & Schafer, 2004). Interestingly, dry seeds contain little or no ascorbate (DeGara et al., 1997; Pallanca & Smirnoff, 2000), which suggests that the specific role of tocopherol in maintaining seed viability (Sattler et al., 2004) does not require tocopherol regeneration by ascorbate, possibly because this reaction would be slow in a low water environment. It is likely that the potential for interaction between ascorbate and tocopherol will become clearer when double mutant plants low in ascorbate and tocopherol are produced, e.g. by crossing the vtc and vte (see Section 3.3.3) mutants.

3.3.3 Functions of tocopherol

Tocopherol concentration in leaves generally increases on exposure to high light intensity and expression of genes encoding some of the biosynthetic enzymes is upregulated in high light (Collakova & DellaPenna, 2003). Tocopherol also accu-mulates in response to drought stress in a number of species (Munne-Bosch et al., 1999). Both high light and reduced CO_2 assimilation during drought can cause photo-oxidative stress (see Chapter 10) suggesting that tocopherol may be involved in defence. An increase in tocopherol during leaf senescence is a distinctive feature (Rise et al., 1989; Molinatorres & Martinez, 1991). Accumulation during senes-cence may be because phytol, which is released during chlorophyll degradation, can be utilised in tocopherol synthesis (Figure 3.5). Various tocopherol biosynthesis genes, particularly 4-hydroxy phenolpyruvate dioxygenase (HPPD) (Figure 3.5)

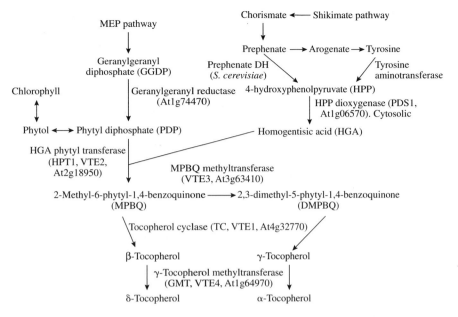

Figure 3.5 The biosynthetic pathway of tocopherols in plastids. The abbreviated enzyme names refer to genes or mutants identified in *A. thaliana* and the *Arabidopsis* Genome Initiative locus numbers indicate the genes that encode each enzyme.

are upregulated in senescing leaves (Falk *et al.*, 2002), which suggests that toco-pherol accumulation may have a specific role. Lipid catabolism by β-oxidation in glyoxysomes is a major metabolic pathway in senescing leaves and this produces hydrogen peroxide. Since recent work (Sattler *et al.*, 2004) suggests tocopherol may have a key role in protection against oxidative stress caused by β-oxidation, it could be that tocopherol also has this role during senescence.

Tocopherol is very abundant in the thylakoid membranes that contain polyunsat-urated fatty acids and are in close proximity to ROS produced during photosynthesis (Chapter 10; Fryer, 1992) and circumstantial and correlative evidence strongly sug-gests an antioxidant role for tocopherol (Munne-Bosch & Alegre, 2002). However, the ability to critically assess the physiological roles of tocopherol has only recently become available with characterisation of plant and cyanobacterial mutants affected in its biosynthesis and transgenic plants with increased tocopherol and tocotrienol content (see Section 3.3.4). Surprisingly, the first studies of low toco-pherol plants indicated that they have no strong phenotype, and particularly that they are only marginally more sensitive to photooxidative stress (Porfirova *et al.*, 2002; Havaux *et al.*, 2003) but they do show increased lipid peroxidation under high light stress (Havaux *et al.*, 2003). These tobacco plants were affected in geranylgeranyl reductase (Figure 3.5). A recent study of seed viability and germi-nation of *A. thaliana* mutants in homogentisate phytyltransferase (HPT, VTE2) and tocopherol cyclase (TC, VTE1) has provided new information on tocopherol function in plants. *vte1* lacks tocopherols but accumulates DMPBQ (Figure 3.5), while *vte2* has reduced HPT activity, which specifically reduces tocopherol synthesis without accumulation of the intermediates MPBQ and DMPBQ. MPBQ and DMPBQ have antioxidant activity by donation of two electrons and, unlike the tocopheroxyl radical, would need enzymatic reduction (Liebler & Burr, 2000). *vte1* and *vte2* seeds lost viability more rapidly than wild type during storage and artificial ageing at elevated temperature and humidity, demonstrating a requirement for tocopherol in seed preservation as had been suggested by many previous studies (Sattler *et al.*, 2004). In contrast, in *vte2* the root growth was reduced and cotyledon development was defective but *vte1* was unaffected. This was associated with massive lipid peroxidation in *vte2*. This shows two key points. First, there appears to be a high oxidative load during germination, perhaps associated with utilisation of the storage triacylglyerols by β-oxidation. This process generates hydrogen peroxide in the glyoxysomes, and tocopherol is required for protection against lipid peroxidation. Second, since *vte1* is unaffected but has no tocopherol, it suggests that DMPBQ can substitute in this case. It is intriguing that protection of the dry seed requires tocopherol itself and further work is needed to understand the different roles of tocopherol and DMPBQ.

Tocopherols may have non-antioxidant functions. They can stabilise, for example, membrane structure by interacting with polyunsaturated fatty acyl chains (Sattler *et al.*, 2003). In animals there is strong evidence for roles in modulating signal transduction, e.g. protein kinase C (Clement *et al.*, 2002) and eicanosoid synthesis (Sattler *et al.*, 2003). Signal transduction and gene expression may be modulated by tocopherol-associated proteins (TAPs). Human TAPs, for example, bind to phosphatidylinositol 3-kinase (PI3-K), a key enzyme in phosphoinositide signalling. This inhibits its activity in the absence of α-tocopherol but stimulates it five-fold in the presence of α-tocopherol (Kempna *et al.*, 2003). TAPs also affect transcription in a tocopherol-dependent manner (Yamauchi *et al.*, 2001). It is not known if plants have homologues of TAPs. It will be interesting to determine the extent to which tocopherol can affect signals generated in chloroplasts by singlet oxygen or the lipid peroxidation product malondialdehyde, both of which affect the expression of specific sets of genes (Almeras *et al.*, 2003; op den Camp *et al.*, 2003; Weber *et al.*, 2004). A maize mutant (*sxd1*) is unable to export sucrose from its leaves because the plasmodesmata between the bundle sheath cells and the phloem are blocked by callose deposition (Provencher *et al.*, 2001; Hofius & Sonnewald, 2003). SXD1 is 62% similar to TC (VTE1) suggesting a link between this phenotype and either MPBQ/DMPBQ accumulation or tocopherol deficiency. Reduction of SDX1 activity in tobacco by RNAi decreases tocopherol and also induces the same blockage of plasmodesmata and sucrose export (Hofius *et al.*, 2004) suggesting a general pleiotropic effect. However, *A. thaliana* TC mutants do not accumulate sucrose in source leaves (Sattler *et al.*, 2003), which suggests that this phenotype does not occur in all species.

3.3.4 Biosynthesis of tocopherols and tocotrienols

The overall pathway of tocopherol synthesis in plants (Figure 3.5) was determined by classical biochemical techniques (Soll, 1987). A large part of tocopherol synthesis is localised in plastids and the enzymes involved are associated with the inner envelope. The membrane localisation of the enzymes proved to be a barrier to their purification and sequencing, so the genes encoding tocopherol biosynthesis enzymes have only been identified more recently using comparative genomics approaches (DellaPenna, 2001) and mutants (Porfirova *et al.*, 2002; Van Eenennaam *et al.*, 2003). Tocols are derived from two biosynthetic pathways: the isoprenoid tail from the plastid isoprenoid pathway and the hydrophilic chromanol head from the shikimic acid pathway (Grusak & DellaPenna, 1999; Ajjawi & Shintani, 2004). The hydrophobic isoprenoid tail, derived from GGDP, is synthesised in the chloroplast by the non-mevalonic acid isoprenoid biosynthesis pathway. GGDP is reduced to phytyl diphosphate by GDDP reductase. Phytyl diphosphate then forms the tocopherol tail. It also provides the hydrophobic tail of chlorophyll. The tail of tocotrienols is derived directly from GGDP. The chromanol head group of tocopherols and tocotrienols is derived from homogentisic (HGA). This is synthesised from 4-hydroxy-phenylpyruvate (HPP) by a cytosolic ascorbate-dependent Fe(II) dioxygenase HPP dioxygenase (PDS1; Tsegaye *et al.*, 2002). It is inhibited by sulcotrione, an herbicide that causes severe bleaching due to inhibition of PQ and carotenoid synthesis (PQ is required as a cofactor for carotenoid synthesis) (Tsegaye *et al.*, 2002). HGA is condensed with phytyl diphosphate to produce MPBQ in a reaction catalysed by homogentisate prenyltransferase (VTE2, HPT1). MPBQ is methylated by 2-methyl-6-phytyl-1,4-benzoquinone methyltransferase (MPBQMT, VTE3) to produce DMPBQ. MPBQ and DMPBQ do not have the ability to donate one electron, but are redox active quinones, and so could possibly carry out some of the functions of tocopherol (Liebler & Burr, 2000). The key enzyme for converting both of these quinones to tocopherols is TC (VTE1; Porfirova *et al.*, 2002). MPBQ gives rise to δ-tocopherol and DMPBQ gives rise to γ-tocopherol. δ-tocopherol and γ-tocopherol are then further methylated to β-tocopherol and α-tocopherol respectively by γ-tocopherol methyltransferase (VTE4) (Shintani & DellaPenna, 1998).

The synthesis of tocotrienols in seeds has received little attention. The enzyme required for condensing HGA with GGDP geranylgeranyl transferase (HGGT) has been characterised and cloned from barley, wheat and rice (Cahoon *et al.*, 2003). It is specific for geranylgeranyl diphosphate and does not utilise the phytyl diphosphate used by HPT. It is likely that all the other enzymes used for tocotrienol synthesis are the same as those used for tocopherol synthesis, since spinach chloroplasts can convert the product 2-methyl-6-geranylgeranylbenzoquinol into tocotrienols (Soll, 1987) and expression of HGGT in *A. thaliana* causes tocotrienol accumulation (Cahoon *et al.*, 2003).

3.3.5 Control and engineering of tocopherol and tocotrienol biosynthesis

The cloning of tocopherol biosynthesis genes described above has provided the opportunity to overexpress them in transgenic plants with the aim of altering the amount and composition of tocols in plants. Achieving this goal provides information on the steps that exert control over tocol synthesis and provides the basis for producing crop plants with increased tocol content and vitamin E activity. Compared with the modest progress that has generally been made in metabolic pathway engineering in plants, and with ascorbate in particular, engineering tocol synthesis has proved to be successful. Two aspects of tocol metabolism can be manipulated. First, flux through the pathway can be increased to boost the total tocol content and second, the balance between the $\alpha/\beta/\delta/\gamma$ forms and tocopherol/tocotrienols can be altered. Ajjawi and Shintani (2004) provide reviews of tocopherol engineering. The first attempt at engineering tocopherol composition was with *A. thaliana* seeds. These oil-storing seeds contain 95% γ-tocopherol and less than 1% α-tocopherol and, because the former has much lower vitamin E activity than the latter, it would be useful to convert one to the other. This was achieved by overexpression of γ-tocopherol methyltransferase (GMT, VTE4) under a seed-specific promoter, converting the tocopherol pool to 90% α-tocopherol (Shintani & DellaPenna, 1998). A novel approach to metabolic engineering has also been used in *A. thaliana* seeds in which a synthetic zinc finger transcription factor, designed to upregulate the endogenous GMT, was expressed in a seed-specific manner. This increased the GMT activity and the proportion of α-tocopherol in the oil (Van Eenennaam *et al.*, 2004). A similar change in soybean seed oil was also achieved, but in this case δ- as well as γ-tocopherol is present in wild type seeds. Therefore, the greatest increase in α-tocopherol was obtained by expressing both GMT (VTE4) and MPBQMT (VTE3) in seeds (Van Eenennaam *et al.*, 2003). These results show that it is feasible to increase the proportion of α-tocopherol in seed oil but the consequences of this for seed function have not been explored: given that γ-tocopherol may have a specific role in seeds (Sattler *et al.*, 2004). To increase flux through the pathway, and therefore increase total tocopherol, earlier pathway enzymes have been manipulated. HPT (VTE2) overexpression in *A. thaliana* increases leaf tocopherol four-fold and by 40% in seeds and crossing these plants to plants overexpressing GMT increased the proportion of α-tocopherol (Savidge *et al.*, 2002). Increasing the formation of PDP by engineering GGDP reductase has not been attempted, but increasing an expression of deoxyxylulose phosphate synthase, an early enzyme in 2C-methyl-D-erythritol 4-phosphate (MEP) or non-mevalonate pathway, caused a 40% increase in leaf tocopherol (Lichtenthaler, 1999; Estevez *et al.*, 2001).

The effect of altering HGA synthesis has been investigated by overexpressing HPPD in tobacco (Falk *et al.*, 2003) and *A. thaliana* (Tsegaye *et al.*, 2002). This modestly increased total tocopherol in seeds and *A. thaliana* leaves but the effect was only evident in senescent tobacco leaves. Another successful strategy was to increase the synthesis of HPP by diverting the shikimic acid pathway intermediate, prephenate, to HPP and away from the phenylalanine branch that leads to phenyl-propanoids. This was achieved by expressing a yeast prephenate dehydrogenase (PDH) that converts prephenate directly to HPP in tobacco (Rippert et al., 2004). Unexpectedly, this resulted in a small accumulation of α-tocotrienol in the trans-genic plants and, when PDH was expressed in plants already overexpressing HPPD, there was a massive accumulation of tocotrienols in the leaves. Although tobacco seeds can synthesise tocotrienols (Falk et al., 2003), this result cannot be explained. It reveals a potentially complex regulation of tocopherol/tocotrienol syn-thesis. A deliberate attempt to engineer tocotrienol synthesis in to A. thaliana leaves was made by overexpressing barley HGGT, the enzyme required for tocotrienol synthesis in seeds. This resulted in a massive increase in tocotrienol, 10–15-fold over the original tocopherol content (Cahoon et al., 2003).

3.4 Carotenoids

The structures of some of the commonly occurring plant carotenoids are shown in Figure 3.6. The carotenoids have essential roles in photosynthesis. They contribute to light harvesting, the xanthophylls being localised in the light-harvesting complex and are also associated with the photosystem II (PSII) reaction centre (Robert et al., 2004). It is also well established that they contribute to photoprotection by quench-ing triplet chlorophyll and singlet oxygen (Havaux & Niyogi, 1999). Inhibition of carotenoid synthesis by various inhibitors/herbicides, or inhibition of electron transfer from PSII by DCMU, results in rapid photo-oxidative bleaching of chloroplasts. Bleaching reflects loss of double bonds or destruction of the carotenoids and occurs in a specific order according to their reactivity (Barry et al., 1990). It is noteworthy that the effect of carotenoid deficiency appears to be much more severe than toco-pherol deficiency, which was discussed above. This might suggest that carotenoids are more effective than tocopherols as chloroplast antioxidants, or they can substi-tute effectively in tocopherol-deficient plants. The overlapping roles of tocopherols and carotenoids are discussed below. It is also possible that their localisation in thy-lakoid membranes and the photosystems differs. The specific role of xanthophyll cycle pigments (zeaxanthin, violaxanthin and antheroxanthin) and their interconver-sion in the xanthophyll cycle is discussed in Chapter 10. Zeaxanthin is involved in non-photochemical quenching of excitation energy in PSII in which excess energy in PSII is transferred to zeaxanthin and re-radiated as heat.

3.4.1 Carotenoids as antioxidants

In relation to their antioxidant activity, the following reactions of carotenoids have been demonstrated. Carotenoids can quench singlet oxygen in a similar manner to tocopherols. They are also able to react directly with superoxide and other free

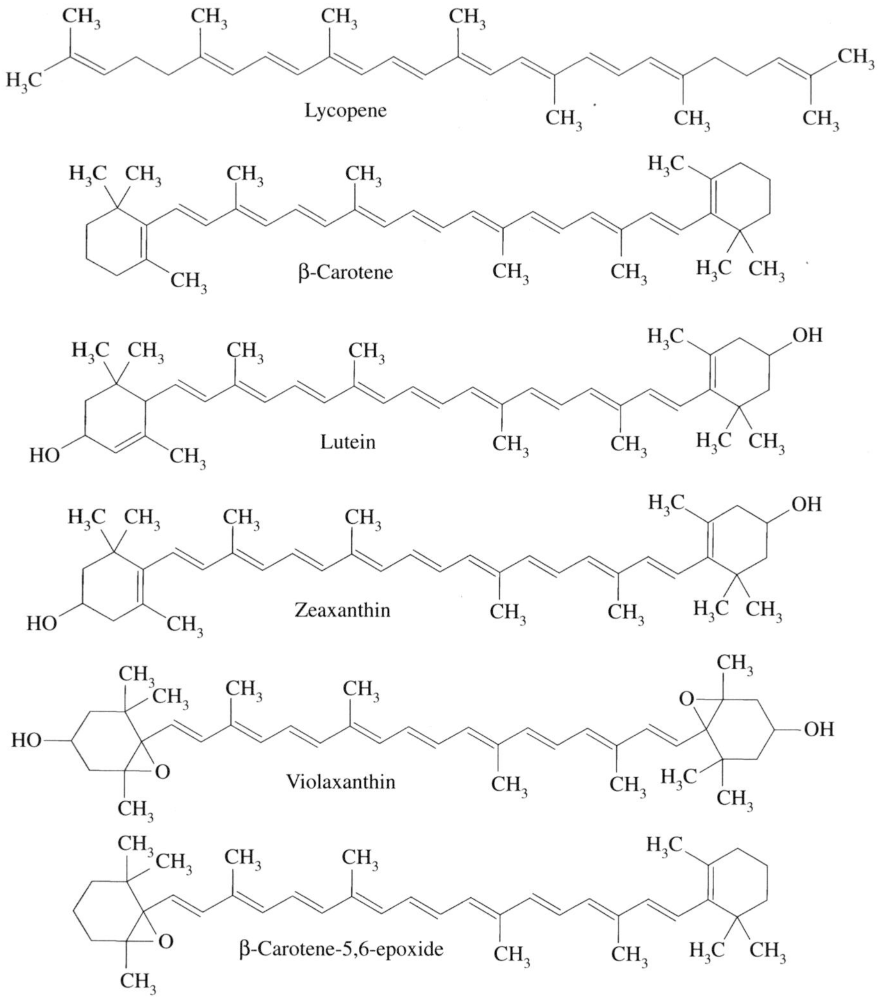

Figure 3.6 The structure of plant carotenoids.

radicals (reviewed by Young & Lowe, 2001; Krinsky & Yeum, 2003). Carotenoids (CAR) can form resonance-stabilised carbon-centred radicals, e.g. by reaction with lipid peroxyl radicals (ROO˙):

$$CAR + ROO^{\cdot} \rightarrow ROO{-}CAR^{\cdot}$$

This could form a new peroxyl radical in the presence of oxygen which would be the basis of pro-oxidant effect of carotenoids although this may not occur *in vivo* (Young & Lowe, 2001). They can transfer electrons forming a radical cation (CAR˙⁺):

$$CAR + R^{\cdot} \rightarrow CAR^{\cdot+} + R$$

Alternatively, they can accept electrons, e.g. from superoxide, forming a radical anion, in the case of lycopene:

$$\text{Lycopene} + O_2^{-\bullet} \rightarrow \text{Lycopene}^{-\bullet} + O_2$$

Allylic hydrogen abstraction, e.g. from lipid peroxyl radicals, could also occur:

$$\text{CAR} + \text{ROO}^\bullet \rightarrow \text{CAR}^\bullet + \text{ROOH}$$

The reactions with lipid peroxyl radicals suggest that carotenoids could prevent lipid peroxidation chain reactions. This has been shown in model systems such as liposomes (Stahl *et al.*, 1998) and *in vivo* in animal cell cultures (Wrona *et al.*, 2004). In liposomes the relative effectiveness was: lycopene>α-tocopherol>α-carotene>β-cryptoxanthin>zeaxanthin=β-carotene>lutein and there was a synergistic effect between lycopene and lutein (Stahl *et al.*, 1998). The susceptibility to photobleaching of carotenoids in leaves or isolated chloroplasts exposed to high light or DCMU (3-[3,4-dichlorophenyl]-1,1-dimethylurea) that blocks electron transfer from PSII, is β-carotene>neoxanthin>violaxanthin=zeaxanthin> lutein, and all these bleach before chlorophyll (Young & Britton, 1990). Evidence from a zeaxanthin-deficient *A. thaliana* mutant (*npq1*; Havaux & Niyogi, 1999) and *lut2npq2* double mutants in which all the xanthophyll pigments are replaced by zeaxanthin (Havaux *et al.*, 2004) show that zeaxanthin, in addition to its role in non-photochemical dissipation of excess excitation in PSII (Chapter 10), has a direct protective role against lipid peroxidation in thylakoid membranes (Davison *et al.*, 2002). Zeaxanthin may also protect thylakoid membranes by decreasing membrane fluidity (Havaux, 1998).

It has also been proposed that carotenoids can interact with tocopherols and ascorbate (Krinsky & Yeum, 2003; Wrona *et al.*, 2004). Carotenoids could react with lipid peroxyl radicals to produce lipid hydroperoxides and carotenoid radicals. The carotenoid radical could then be regenerated by tocopherol. Both carotenoid radicals and tocopheryl radicals have the potential to be reduced by ascorbate (Figure 3.7). Synergistic interactions between carotenoids and tocopherol in protecting against lipid peroxidation *in vivo* and *in vitro* have been observed (Mortensen *et al.*, 1998; Yeum *et al.*, 2000, 2004). The extent to which carotenoids and tocopherol interact will depend on the ability of the chromanol head group on the membrane surface to interact with carotenoid radicals. It should also be noted that carotenoids can potentially act as pro-oxidants in the presence of high oxygen concentrations, although this may not occur *in vivo* (Young & Lowe, 2001; Krinsky & Yeum, 2003).

Very little is known about the fate, or possible biological functions of oxidised carotenoids and their breakdown products. In animal systems, some of them have an anti-proliferative effect on cancer cells, or induce apoptosis (Krinsky & Yeum, 2003). β-carotene-5,6-epoxide (Figure 3.6) can be detected in plants, but only under conditions of severe photo-oxidative stress, and could therefore be considered as an indicator of damage (Young & Britton, 1990).

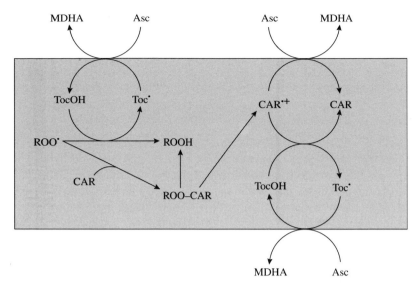

Figure 3.7   The proposed interactions between tocopherol, carotenoids and ascorbate in scavenging lipid peroxyl radicals in membranes. Both carotenoids (CAR) and tocopherol (TocOH) can react with lipid peroxyl radicals (ROO˙), which are produced as a result of lipid peroxidation. The tocopheryl radical (Toc˙) is regenerated by ascorbate (Asc) with the production of monodehydroascorbate (MDHA), which is regenerated to ascorbate by MDHA reductase. The carotenoid radical ($CAR^{˙+}$) can potentially be converted back to CAR by either tocopherol or ascorbate. These reactions may explain the synergistic interactions sometimes observed between these antioxidants in protecting against lipid peroxidation chain reactions.

3.4.2   Carotenoid biosynthesis and metabolic engineering

The proposed health benefit of carotenoids in the human diet has provided the driving force for understanding carotenoid biosynthesis and for engineering the production of desirable and novel carotenoids in plants and bacteria (reviewed by Fraser & Bramley, 2004). It has been possible, e.g. to increase the concentration of β-carotene in tomato fruits at the expense of lycopene (producing yellow fruit), to increase zeaxanthin content of potatoes and to introduce β-carotene synthesis into rice endosperm (Rosati *et al.*, 2000; Ye *et al.*, 2000; Romer *et al.*, 2002). The engineering of carotenoid biosynthesis is reviewed by Fraser and Bramley (2004), Giuliano *et al.* (2000) and van den Berg *et al.* (2000).

References

Agius, F., Gonzalez-Lamothe, R., Caballero, J.L., Munoz-Blanco, J., Botella, M.A. and Valpuesta, V. (2003) 'Engineering increased vitamin C levels in plants by overexpression of a D-galacturonic acid reductase', *Nature Biotechnology* **21**, 177–181.

Ajjawi, I. and Shintani, D. (2004) 'Engineered plants with elevated vitamin E: a nutraceutical success story', *Trends in Biotechnology* **22**, 104–107.

Al-Madhoun, A.S., Sanmartin, M. and Kanellis, A.K. (2003) 'Expression of ascorbate oxidase isoenzymes in cucurbits and during development and ripening of melon fruit', *Postharvest Biology and Technology* **27**, 137–146.

Almeras, E., Stolz, S., Vollenweider, S., Reymond, P., Mene-Saffrane, L. and Farmer, E.E. (2003) 'Reactive electrophile species activate defense gene expression in *Arabidopsis*', *The Plant Journal* **34**, 202–216.

Arrigoni, O. and De Tullio, M.C. (2002) 'Ascorbic acid: much more than just an antioxidant', *Biochimica et Biophysica Acta – General Subjects* **1569**, 1–9.

Asada, K. (1999) 'The water–water cycle in chloroplasts: scavenging of active oxygens and dissipation of excess photons', *Annual Review of Plant Physiology and Plant Molecular Biology* **50**, 601–639.

Azzi, A., Ricciarelli, R. and Zingg, J.M. (2002) 'Non-antioxidant molecular functions of α-tocopherol (vitamin E)', *FEBS Letters* **519**, 8–10.

Banhegyi, G. and Loewus, F.A. (2004) Ascorbic acid catabolism: breakdown pathways in animals and plants, in *Vitamin C. Function and Biochemistry in Animals and Plants* (ed. H. Asard, J.M. May and N. Smirnoff), Bios Scientific Publishers, Oxford, pp. 31–48.

Barry, P., Young, A.J. and Britton, G. (1990) 'Photodestruction of pigments in higher plants by herbicide action. 1. The effect of DCMU (Diuron) on isolated chloroplasts', *Journal of Experimental Botany* **41**, 123–129.

Barth, C., Moeder, W., Klessig, D.F. and Conklin, P.L. (2004) 'The timing of senescence and response to pathogens is altered in the ascorbate-deficient *Arabidopsis* mutant vitamin *vtc1*', *Plant Physiology* **134**, 1784–1792.

Bartoli, C.G., Pastori, G.M. and Foyer, C.H. (2000) 'Ascorbate biosynthesis in mitochondria is linked to the electron transport chain between complexes III and IV', *Plant Physiology* **123**, 335–343.

Berczi, A. and Moller, I.M. (1998) 'NADH-monodehydroascorbate oxidoreductase is one of the redox enzymes in spinach leaf plasma membranes', *Plant Physiology* **116**, 1029–1036.

Bramley, P.M. (2000) 'Is lycopene beneficial to human health?', *Phytochemistry* **54**, 233–236.

Bramley, P.M., Elmadfa, I., Kafatos, A. *et al.* (2000) 'Vitamin E', *Journal of the Science of Food and Agriculture* **80**, 913–938.

Brigelues-Flohe, R. and Traber, M.G. (1999) 'Vitamin E: function and metabolism', *FASEB Journal* **13**, 1145–1155.

Buettner, G.R. and Schafer, F.Q. (2004) Ascorbate as an antioxidant, in *Vitamin C. Function and Biochemistry in Animals and Plants* (ed. H. Asard, J.M. May and N. Smirnoff), Bios Scientific Publishers, Oxford, pp.173–188.

Cahoon, E.B., Hall, S.E., Ripp, K.G., Ganzke, T.S., Hitz, W.D. and Coughlan, S.J. (2003) 'Metabolic redesign of vitamin E biosynthesis in plants for tocotrienol production and increased antioxidant content', *Nature Biotechnology* **21**, 1082–1087.

Chen, Z. and Gallie, D.R. (2004) 'The ascorbic acid redox state controls guard cell signaling and stomatal movement', *Plant Cell* **16**, 1143–1162.

Chen, Z., Young, T.E., Ling, J., Chang, S.C. and Gallie, D.R. (2003) 'Increasing vitamin C content of plants through enhanced ascorbate recycling', *Proceedings of the National Academy of Sciences USA* **100**, 3525–3530.

Chew, O., Whelan, J. and Millar, A.H. (2003) 'Molecular definition of the ascorbate–glutathione cycle in *Arabidopsis* mitochondria reveals dual targeting of antioxidant defenses in plants', *Journal of Biological Chemistry* **278**, 46869–46877.

Clement, S.A., Tan, C.C., Guo, J.L., Kitta, K. and Suzuki, Y.J. (2002) 'Roles of protein kinase C and α-tocopherol in regulation of signal transduction for GATA-4 phosphorylation in HL-1 cardiac muscle cells', *Free Radical Biology and Medicine* **32**, 341–349.

Collakova, E. and DellaPenna, D. (2003) 'The role of homogentisate phytyltransferase and other tocopherol pathway enzymes in the regulation of tocopherol synthesis during abiotic stress', *Plant Physiology* **133**, 930–940.

Conklin, P.L., Norris, S.R., Wheeler, G.L., Williams, E.H., Smirnoff, N. and Last, R.L. (1999) 'Genetic evidence for the role of GDP-mannose in plant ascorbic acid (vitamin C) biosynthesis', *Proceedings of the National Academy of Sciences USA* **96**, 4198–4203.

Conklin, P.L., Saracco, S.A., Norris, S.R. and Last, R.L. (2000) 'Identification of ascorbic acid-deficient *Arabidopsis thaliana* mutants', *Genetics* **154**, 847–856.

Conklin, P.L., Williams, E.H. and Last, R.L. (1996) 'Environmental stress sensitivity of an ascorbic acid-deficient *Arabidopsis* mutant', *Proceedings of the National Academy of Sciences USA* **93**, 9970–9974.

Davey, M.W., Gilot, C., Persiau, G. *et al.* (1999) 'Ascorbate biosynthesis in *Arabidopsis* cell suspension culture', *Plant Physiology* **121**, 535–543.

Davey, M.W., Van Montagu, M., Inze, D. *et al.* (2000) 'Plant L-ascorbic acid: chemistry, function, metabolism, bioavailability and effects of processing', *Journal of the Science of Food and Agriculture* **80**, 825–860.

Davison, P.A., Hunter, C.N. and Horton, P. (2002) 'Overexpression of β-carotene hydroxylase enhances stress tolerance in *Arabidopsis*', *Nature* **418**, 203–206.

de Tullio, M. (2004) How does ascorbate prevent scurvy? A survey of the nonantioxidant functions of vitamin C, in *Vitamin C. Function and Biochemistry in Animals and Plants* (ed. H. Asard, J.M. May and N. Smirnoff), Bios Scientific Publishers, Oxford, pp. 159–171.

De Tullio, M.C., Paciolla, C., la Veechia, F. *et al.* (1999) 'Changes in onion root development induced by the inhibition of peptidyl-prolyl hydroxylase and influence of the ascorbate system on cell division and elongation', *Planta* **209**, 424–434.

DeGara, L., dePinto, M.C. and Arrigoni, O. (1997) 'Ascorbate synthesis and ascorbate peroxidase activity during the early stage of wheat germination', *Physiologia Plantarum* **100**, 894–900.

Delaney, T.P., Uknes, S., Vernooij, B. *et al.* (1994) 'A central role of salicylic acid in plant disease resistance', *Science* **266**, 1247–1250.

DellaPenna, D. (2001) 'Plant metabolic engineering', *Plant Physiology* **125**, 160–163.

Diallinas, G., Pateraki, I., Sanmartin, M. *et al.* (1997) 'Melon ascorbate oxidase: cloning of a multigene family, induction during fruit development and repression by wounding', *Plant Molecular Biology* **34**, 759–770.

Dipierro, S. and Borraccino, G. (1991) 'Dehydroascorbate reductase from potato tubers', *Phytochemistry* **30**, 427–429.

Esaka, M., Fukui, H., Suzuki, K. and Kubota, K. (1989) 'Secretion of ascorbate oxidase by suspension-cultured pumpkin cells', *Phytochemistry* **28**, 117–119.

Estevez, J.M., Cantero, A., Reindl, A., Reichler, S. and Leon, P. (2001) '1-Deoxy-D-xylulose-5-phosphate synthase, a limiting enzyme for plastidic isoprenoid biosynthesis in plants', *Journal of Biological Chemistry* **276**, 22901–22909.

Falk, J., Andersen, G., Kernebeck, B. and Krupinska, K. (2003) 'Constitutive overexpression of barley 4-hydroxyphenylpyruvate dioxygenase in tobacco results in elevation of the vitamin E content in seeds but not in leaves', *FEBS Letters* **540**, 35–40.

Falk, J., Krauss, N., Dahnhardt, D. and Krupinska, K. (2002) 'The senescence associated gene of barley encoding 4-hydroxyphenylpyruvate dioxygenase is expressed during oxidative stress', *Journal of Plant Physiology* **159**, 1245–1253.

Filkowski, J., Kovalchuk, O. and Kovalchuk, I. (2004) 'Genome stability of *vtc1*, *tt4*, and *tt5 Arabidopsis thaliana* mutants impaired in protection against oxidative stress', *The Plant Journal* **38**, 60–69.

Foreman, J., Demidchik, V., Bothwell, J.H.F. *et al.* (2003) 'Reactive oxygen species produced by NADPH oxidase regulate plant cell growth', *Nature* **422**, 442–446.

Foyer, C.H. and Lelandais, M. (1996) 'A comparison of the relative rates of transport of ascorbate and glucose across the thylakoid, chloroplast and plasmalemma membranes of pea leaf mesophyll cells', *Journal of Plant Physiology* **148**, 391–398.

Foyer, C.H. and Mullineaux, P.M. (1998) 'The presence of dehydroascorbate and dehydroascorbate reductase in plant tissues', *FEBS Letters* **425**, 528–529.

Foyer, C.H., Souriau, N., Perret, S. *et al.* (1995) 'Overexpression of glutathione reductase but not glutathione synthetase leads to increases in antioxidant capacity and resistance to photoinhibition in poplar trees', *Plant Physiology* **109**, 1047–1057.

Franceschi, V.R. and Tarlyn, N.M. (2002) 'L-Ascorbic acid is accumulated in source leaf phloem and transported to sink tissues in plants', *Plant Physiology* **130**, 649–656.

Fraser, P.D. and Bramley, P.M. (2004) 'The biosynthesis and nutritional uses of carotenoids', *Progress in Lipid Research* **43**, 228–265.

Fry, S.C. (1998) 'Oxidative scission of plant cell wall polysaccharides by ascorbate-induced hydroxyl radicals', *Biochemical Journal* **332**, 507–515.

Fry, S.C., Dumville, J.C. and Miller, J.G. (2001) 'Fingerprinting of polysaccharides attacked by hydroxyl radicals *in vitro* and in the cell walls of ripening pear fruit', *Biochemical Journal* **357**, 729–737.

Fryer, M.J. (1992) 'The antioxidant effects of thylakoid vitamin E (α-tocopherol)', *Plant Cell and Environment* **15**, 381–392.

Fryer, M.J., Ball, L., Oxborough, K., Karpinski, S., Mullineaux, P.M. and Baker, N.R. (2003) 'Control of Ascorbate Peroxidase 2 expression by hydrogen peroxide and leaf water status during excess light stress reveals a functional organisation of *Arabidopsis* leaves', *The Plant Journal* **33**, 691–705.

Gatzek, S., Wheeler, G.L. and Smirnoff, N. (2002) 'Antisense suppression of L-galactose dehydrogenase in *Arabidopsis thaliana* provides evidence for its role in ascorbate synthesis and reveals light modulated L-galactose synthesis', *The Plant Journal* **30**, 541–553.

Giuliano, G., Aquilani, R. and Dharmapuri, S. (2000) 'Metabolic engineering of plant carotenoids', *Trends in Plant Science* **5**, 406–409.

Grant, J.J. and Loake, G.J. (2000) 'Role of reactive oxygen intermediates and cognate redox signaling in disease resistance', *Plant Physiology* **124**, 21–29.

Griesen, D., Su, D., Berczi, A. and Asard, H. (2004) 'Localization of an ascorbate-reducible cytochrome b_{561} in the plant tonoplast', *Plant Physiology* **134**, 726–734.

Grusak, M.A. and DellaPenna, D. (1999) 'Improving the nutrient composition of plants to enhance human nutrition and health', *Annual Review of Plant Physiology and Plant Molecular Biology* **50**, 133–161.

Halliwell, B. and Gutteridge, J.M.C. (1999) *Free Radicals in Biology and Medicine*, Oxford University Press, Oxford.

Hancock, R.D., McRae, D., Haupt, S. and Viola, R. (2003) 'Synthesis of L-ascorbic acid in the phloem', *BMC Plant Biology* **3**, 7.

Havaux, M. (1998) 'Carotenoids as membrane stabilizers in chloroplasts', *Trends in Plant Science* **3**, 147–151.

Havaux, M. and Niyogi, K.K. (1999) 'The violaxanthin cycle protects plants from photooxidative damage by more than one mechanism', *Proceedings of the National Academy of Sciences USA* **96**, 8762–8767.

Havaux, M., Dall'Osto, L., Cuine, S., Giuliano, G. and Bassi, R. (2004) 'The effect of zeaxanthin as the only xanthophyll on the structure and function of the photosynthetic apparatus in *Arabidopsis thaliana*', *Journal of Biological Chemistry* **279**, 13878–13888.

Havaux, M., Lutz, C. and Grimm, B. (2003) 'Chloroplast membrane photostability in *chlP* transgenic tobacco plants deficient in tocopherols', *Plant Physiology* **132**, 300–310.

Heber, U., Miyake, C., Mano, J., Ohno, C. and Asada, K. (1996) 'Monodehydroascorbate radical detected by electron paramagnetic resonance spectrometry is a sensitive probe of oxidative stress in intact leaves', *Plant and Cell Physiology* **37**, 1066–1072.

Hideg, E., Mano, J., Ohno, C. and Asada, K. (1997) 'Increased levels of monodehydroascorbate radical in UV-B-irradiated broad bean leaves', *Plant and Cell Physiology* **38**, 684–690.

Hofius, D. and Sonnewald, U. (2003) 'Vitamin E biosynthesis: biochemistry meets cell biology', *Trends in Plant Science* **8**, 6–8.

Hofius, D., Hajirezaei, M.R., Geiger, M., Tschiersch, H., Melzer, M. and Sonnewald, U. (2004) 'RNAi-mediated tocopherol deficiency impairs photoassimilate export in transgenic potato plants', *Plant Physiology* **135**, 1256–1268.

Horemans, N., Asard, H. and Cauberg, R.J. (1994) 'The role of ascorbate free-radical as an electron-acceptor to cytochrome *b*-mediated trans-plasma membrane electron transport in higher plants', *Plant Physiology* **104**, 1455–1458.

Horemans, N., Asard, H. and Cauberg, R.J. (1998) 'Carrier mediated uptake of dehydroascorbate into higher plant plasma membrane vesicles shows trans-stimulation', *FEBS Letters* **421**, 41–44.

Horemans, N., Foyer, C.H. and Asard, H. (2000) 'Transport and action of ascorbate at the plant plasma membrane', *Trends in Plant Science* **5**, 263–267.

Hossain, M.A. and Asada, K. (1984) 'Purification of dehydroascorbate reductase from spinach and its characterization as a thiol enzyme', *Plant and Cell Physiology* **25**, 85–92.

Hossain, M.A. and Asada, K. (1985) 'Monodehydroascorbate reductase from cucumber is a flavin adenine dinucleotide enzyme', *Journal of Biological Chemistry* **260**, 2920–2926.

Hou, W.C. and Lin, Y.H. (1997) 'Dehydroascorbate reductase and monodehydroascorbate reductase activities of trypsin inhibitors, the major sweet potato (*Ipomoea batatas* [L.] Lam) root storage protein', *Plant Science* **128**, 151–158.

Hou, W.C., Chen, H.J. and Lin, Y.H. (1999) 'Dioscorins, the major tuber storage proteins of yam (*Dioscorea batatas* Decne), with dehydroascorbate reductase and monodehydroascorbate reductase activities', *Plant Science* **149**, 151–156.

Hou, W.C., Wang, Y.T., Lin, Y.H. *et al.* (2000) 'A complex containing both trypsin inhibitor and dehydroascorbate reductase activities isolated from mitochondria of etiolated mung bean (*Vigna radiata* L. (Wilczek) cv. Tainan No. 5) seedlings', *Journal of Experimental Botany* **51**, 713–719.

Isherwood, F.A., Chen, Y.T. and Mapson, L.W. (1954) 'Synthesis of L-ascorbic acid in plants and animals', *Biochemical Journal* **56**, 1–15.

Jain, A.K. and Nessler, C.L. (2000) 'Metabolic engineering of an alternative pathway for ascorbic acid biosynthesis in plants', *Molecular Breeding* **6**, 73–78.

Jander, G., Norris, S.R., Rounsley, S.D., Bush, D.F., Levin, I.M. and Last, R.L. (2002) '*Arabidopsis* map-based cloning in the post-genome era', *Plant Physiology* **129**, 440–450.

Jiang, K., Meng, Y.L. and Feldman, L.J. (2003) 'Quiescent center formation in maize roots is associated with an auxin-regulated oxidizing environment', *Development* **130**, 1429–1438.

Jimenez, A., Hernandez, J.A., delRio, L.A. and Sevilla, F. (1997) 'Evidence for the presence of the ascorbate-glutathione cycle in mitochondria and peroxisomes of pea leaves', *Plant Physiology* **114**, 275–284.

Karpinski, S., Escobar, C., Karpinska, B., Creissen, G. and Mullineaux, P.M. (1997) 'Photosynthetic electron transport regulates the expression of cytosolic ascorbate peroxidase genes in *Arabidopsis* during excess light stress', *The Plant Cell* **9**, 627–640.

Kato, N. and Esaka, M. (2000) 'Expansion of transgenic tobacco protoplasts expressing pumpkin ascorbate oxidase is more rapid than that of wild-type protoplasts', *Planta* **210**, 1018–1022.

Kato, Y., Urano, J., Maki, Y. and Ushimaru, T. (1997) 'Purification and characterization of dehydroascorbate reductase from rice', *Plant and Cell Physiology* **38**, 173–178.

Keates, S.E., Tarlyn, N.M., Loewus, F.A. and Franceschi, V.R. (2000) 'L-Ascorbic acid and L-galactose are sources for oxalic acid and calcium oxalate in *Pistia stratiotes*', *Phytochemistry* **53**, 433–440.

Keller, R., Springer, F., Renz, A. and Kossmann, J. (1999) 'Antisense inhibition of the GDP-mannose pyrophosphorylase reduces the ascorbate content in transgenic plants leading to developmental changes during senescence', *The Plant Journal* **19**, 131–141.

Kempna, P., Zingg, J.M., Ricciarelli, R., Hierl, M., Saxena, S. and Azzi, A. (2003) 'Cloning of novel human SEC14p-like proteins: ligand binding and functional properties', *Free Radical Biology and Medicine* **34**, 1458–1472.

Kerk, N.M. and Feldman, L.J. (1995) 'A biochemical model for the initiation and maintenance of the quiescent center – implications for organization of root meristems', *Development* **121**, 2825–2833.

Kollist, H., Moldau, H., Oksanen, E. and Vapaavuori, E. (2001) 'Ascorbate transport from the apoplast to the symplast in intact leaves', *Physiologia Plantarum* **113**, 377–383.

Kornyeyev, D., Logan, B.A., Allen, R.D. and Holaday, A.S. (2003) 'Effect of chloroplastic overproduction of ascorbate peroxidase on photosynthesis and photoprotection in cotton leaves subjected to low temperature photoinhibition', *Plant Science* **165**, 1033–1041.

Krinsky, N.I. and Yeum, K.J. (2003) 'Carotenoid-radical interactions', *Biochemical and Biophysical Research Communications* **305**, 754–760.

Kwon, S.Y., Choi, S.M., Ahn, Y.O. *et al.* (2003) 'Enhanced stress-tolerance of transgenic tobacco plants expressing a human dehydroascorbate reductase gene', *Journal of Plant Physiology* **160**, 347–353.

Kwon, S.Y., Jeong, Y.J., Lee, H.S. *et al.* (2002) 'Enhanced tolerances of transgenic tobacco plants expressing both superoxide dismutase and ascorbate peroxidase in chloroplasts against methyl viologen-mediated oxidative stress', *Plant Cell and Environment* **25**, 873–882.

Laloi, C., Apel, K. and Danon, A. (2004) 'Reactive oxygen signalling: the latest news', *Current Opinion in Plant Biology* **7**, 323–328.

Lichtenthaler, H.K. (1999) 'The 1-deoxy-D-xylulose-5-phosphate pathway of isoprenoid biosynthesis in plants', *Annual Review of Plant Physiology and Plant Molecular Biology* **50**, 47–65.

Liebler, D.C. and Burr, J.A. (2000) 'Antioxidant reactions of α-tocopherolhydroquinone', *Lipids* **35**, 1045–1047.

Loewus, F.A. (1999) 'Biosynthesis and metabolism of ascorbic acid in plants and of analogs of ascorbic acid in fungi', *Phytochemistry* **52**, 193–210.

Lorence, A., Chevone, B.I., Mendes, P. and Nessler, C.L. (2004) '*myo*-Inositol oxygenase offers a possible entry point into plant ascorbate biosynthesis', *Plant Physiology* **134**, 1200–1205.

Lukacin, R. and Britsch, L. (1997) 'Identification of strictly conserved histidine and arginine residues as part of the active site in *Petunia hybrida* flavanone 3 beta-hydroxylase', *European Journal of Biochemistry* **249**, 748–757.

Luwe, M.W.F., Takahama, U. and Heber, U. (1993) 'Role of ascorbate in detoxifying ozone in the apoplast of spinach (*Spinacia oleracea* L.) leaves', *Plant Physiology* **101**, 969–976.

Maddison, J., Lyons, T., Plochl, M. and Barnes, J. (2002) 'Hydroponically cultivated radish fed L-galactono-1,4-lactone exhibit increased tolerance to ozone', *Planta* **214**, 383–391.

Mapson, L.W. and Breslow, E. (1958) 'Biological synthesis of ascorbic acid: L-galactono-1, 4-lactone dehydrogenase', *Biochemical Journal* **68**, 395–406.

May, J.M. and Asard, H. (2004) Ascorbate recycling, in *Vitamin C. Function and Biochemistry in Animals and Plants* (ed. H. Asard, J.M. May and N. Smirnoff), Bios Scientific Publishers, Oxford, pp. 139–157.

Messerschmidt, A. and Huber, R. (1990) 'The blue oxidases, ascorbate oxidase, laccase and ceruloplasmin – modeling and structural relationships', *European Journal of Biochemistry* **187**, 341–352.

Millar, A.H., Mittova, V., Kiddle, G. *et al.* (2003) 'Control of ascorbate synthesis by respiration and its implications for stress responses', *Plant Physiology* **133**, 443–447.

Miyake, C. and Asada, K. (1992) 'Thylakoid bound ascorbate peroxidase scavenges hydrogen peroxide – photoreduction of monodehydroascorbate radical', *Photosynthesis Research* **34**, 156.

Miyake, C., Schreiber, U., Hormann, H., Sano, S. and Asada, K. (1998) 'The FAD-enzyme monodehydroascorbate radical reductase mediates photoproduction of superoxide radicals in spinach thylakoid membranes', *Plant and Cell Physiology* **39**, 821–829.

Mizohata, E., Kumei, M., Matsumura, H. *et al.* (2001) 'Crystallization and preliminary x-ray diffraction analysis of glutathione-dependent dehydroascorbate reductase from spinach chloroplasts', *Acta Crystallographica Section D – Biological Crystallography* **57**, 1726–1728.

Molinatorres, J. and Martinez, M.L. (1991) 'Tocopherols and leaf age in *Xanthium Strumarium* L', *New Phytologist* **118**, 95–99.

Morell, S., Follmann, H., de Tullio, M. and Haberlein, I. (1998) 'Dehydroascorbate reduction: the phantom remaining', *FEBS Letters* **425**, 530–531.

Morris, K., Mackerness, S.A.H., Page, T. *et al.* (2000) 'Salicylic acid has a role in regulating gene expression during leaf senescence', *The Plant Journal* **23**, 677–685.

Mortensen, A., Skibsted, L.H., Willnow, A. and Everett, S.A. (1998) 'Re-appraisal of the tocopheroxyl radical reaction with β-carotene: evidence for oxidation of vitamin E by the β-carotene radical cation', *Free Radical Research* **28**, 69–80.

Muckenschnabel, I., Goodman, B.A., Williamson, B., Lyon, G.D. and Deighton, N. (2002) 'Infection of leaves of *Arabidopsis thaliana* by *Botrytis cinerea*: changes in ascorbic acid, free radicals and lipid peroxidation products', *Journal of Experimental Botany* **53**, 207–214.

Muller-Moule, P., Conklin, P.L. and Niyogi, K.K. (2002) 'Ascorbate deficiency can limit violaxanthin de-epoxidase activity *in vivo*', *Plant Physiology* **128**, 970–977.

Muller-Moule, P., Golan, T. and Niyogi, K.K. (2004) 'Ascorbate-deficient mutants of *Arabidopsis* grow in high light despite chronic photooxidative stress', *Plant Physiology* **134**, 1163–1172.

Munne-Bosch, S. and Alegre, L. (2002) 'The function of tocopherols and tocotrienols in plants', *Critical Reviews in Plant Sciences* **21**, 31–57.

Munne-Bosch, S. and Falk, J. (2004) 'New insights into the function of tocopherols in plants', *Planta* **218**, 323–326.

Munne-Bosch, S., Schwarz, K. and Alegre, L. (1999) 'Enhanced formation of α-tocopherol and highly oxidized abietane diterpenes in water-stressed rosemary plants', *Plant Physiology* **121**, 1047–1052.

Noctor, G. and Foyer, C.H. (1998) 'Ascorbate and glutathione: keeping active oxygen under control', *Annual Review of Plant Physiology and Plant Molecular Biology* **49**, 249–279.

Noctor, G., Veljovic-Jovanovic, S. and Foyer, C.H. (2000) 'Peroxide processing in photosynthesis: antioxidant coupling and redox signalling', *Philosophical Transactions of the Royal Society of London Series B – Biological Sciences* **355**, 1465–1475.

Obara, K., Sumi, K. and Fukuda, H. (2002) 'The use of multiple transcription starts causes the dual targeting of *Arabidopsis* putative monodehydroascorbate reductase to both mitochondria and chloroplasts', *Plant and Cell Physiology* **43**, 697–705.

op den Camp, R.G.L., Przybyla, D., Ochsenbein, C. *et al.* (2003) 'Rapid induction of distinct stress responses after the release of singlet oxygen in *Arabidopsis*', *The Plant Cell* **15**, 2320–2332.

Ostergaard, J., Persiau, G., Davey, M.W., Bauw, G. and Vanmontagu, M. (1997) 'Isolation of a cDNA coding for L-galactono-γ-lactone dehydrogenase, an enzyme involved in the biosynthesis of ascorbic acid in plants – purification, characterization, cDNA cloning, and expression in yeast', *Journal of Biological Chemistry* **272**, 30009–30016.

Pallanca, J.E. and Smirnoff, N. (2000) 'The control of ascorbic acid synthesis and turnover in pea seedlings', *Journal of Experimental Botany* **51**, 669–674.

Pastori, G.M., Kiddle, G., Antoniw, J. *et al.* (2003) 'Leaf vitamin C contents modulate plant defense transcripts and regulate genes that control development through hormone signaling', *The Plant Cell* **15**, 939–951.

Pfannschmidt, T., Nilsson, A. and Allen, J.F. (1999) 'Photosynthetic control of chloroplast gene expression', *Nature* **397**, 625–628.

Pignocchi, C. and Foyer, C.H. (2003) 'Apoplastic ascorbate metabolism and its role in the regulation of cell signalling', *Current Opinion in Plant Biology* **6**, 379–389.

Pignocchi, C., Fletcher, J.M., Wilkinson, J.E., Barnes, J.D. and Foyer, C.H. (2003) 'The function of ascorbate oxidase in tobacco', *Plant Physiology* **132**, 1631–1641.

Plochl, M., Lyons, T., Ollerenshaw, J. and Barnes, J. (2000) 'Simulating ozone detoxification in the leaf apoplast through the direct reaction with ascorbate', *Planta* **210**, 454–467.

Porfirova, S., Bergmuller, E., Tropf, S., Lemke, R. and Dormann, P. (2002) 'Isolation of an *Arabidopsis* mutant lacking vitamin E and identification of a cyclase essential for all tocopherol biosynthesis', *Proceedings of the National Academy of Sciences USA* **99**, 12495–12500.

Potrykus, I. (2001) 'Golden rice and beyond', *Plant Physiology* **125**, 1157–1161.

Potters, G., De Gara, L., Asard, H. and Horemans, N. (2002) 'Ascorbate and glutathione: guardians of the cell cycle, partners in crime?', *Plant Physiology and Biochemistry* **40**, 537–548.

Potters, G., Horemans, N., Bellone, S. *et al.* (2004) 'Dehydroascorbate influences the plant cell cycle through a glutathione-independent reduction mechanism', *Plant Physiology* **134**, 1479–1487.

Potters, G., Horemans, N., Caubergs, R.J. and Asard, H. (2000) 'Ascorbate and dehydroascorbate influence cell cycle progression in a tobacco cell suspension', *Plant Physiology* **124**, 17–20.

Poulsen, H.E., Moller, P., Lykkesfeldt, J., Weiman, A. and Loft, S. (2004) Vitamin C and oxidative damage to DNA, in *Vitamin C. Function and Biochemistry in Animals and Plants* (ed. H. Assard, J.M. May and N. Smirnoff), Bios Scientific Publishers, Oxford, pp. 189–202.

Provencher, L.M., Miao, L., Sinha, N. and Lucas, W.J. (2001) 'Sucrose export defective1 encodes a novel protein implicated in chloroplast-to-nucleus signaling', *The Plant Cell* **13**, 1127–1141.

Radzio, J.A., Lorence, A., Chevone, B.I. and Nessler, C.L. (2003) 'L-Gulono-1,4-lactone oxidase expression rescues vitamin C deficient *Arabidopsis* (*vtc*) mutants', *Plant Molecular Biology* **53**, 837–844.

Rautenkranz, A.A.F., Li, L.J., Machler, F., Martinoia, E. and Oertli, J.J. (1994) 'Transport of ascorbic and dehydroascorbic acids across protoplast and vacuole membranes isolated from barley (*Hordeum vulgare* L. cv. Gerbel) leaves', *Plant Physiology* **106**, 187–193.

Rentel, M.C., Lecourieux, D., Ouaked, F. *et al.* (2004) 'OXI1 kinase is necessary for oxidative burst-mediated signalling in *Arabidopsis*', *Nature* **427**, 858–861.

Rippert, P., Scimemi, C., Dubald, M. and Matringe, M. (2004) 'Engineering plant shikimate pathway for production of tocotrienol and improving herbicide resistance', *Plant Physiology* **134**, 92–100.

Rise, M., Cojocaru, M., Gottlieb, H.E. and Goldschmidt, E.E. (1989) 'Accumulation of α-tocopherol in senescing organs as related to chlorophyll degradation', *Plant Physiology* **89**, 1028–1030.

Robatzek, S. and Somssich, I.E. (2001) 'A new member of the Arabidopsis WRKY transcription factor family, AtWRKY6, is associated with both senescence- and defence-related processes', *The Plant Journal* **28**, 123–133.

Robert, B., Horton, P., Pascal, A.A. and Ruban, A.V. (2004) 'Insights into the molecular dynamics of plant light-harvesting proteins *in vivo*', *Trends in Plant Science* **9**, 385–390.

Romer, S., Lubeck, J., Kauder, F., Steiger, S., Adomat, C. and Sandmann, G. (2002) 'Genetic engineering of a zeaxanthin-rich potato by antisense inactivation and co-suppression of carotenoid epoxidation', *Metabolic Engineering* **4**, 263–272.

Rosati, C., Aquilani, R., Dharmapuri, S. *et al.* (2000) 'Metabolic engineering of β-carotene and lycopene content in tomato fruit', *The Plant Journal* **24**, 413–419.

Running, J.A., Burlingame, R.P. and Berry, A. (2003) 'The pathway of L-ascorbic acid biosynthesis in the colourless microalga *Prototheca moriformis*', *Journal of Experimental Botany* **54**, 1841–1849.

Saito, K. and Loewus, F.A. (1989) 'Formation of tartaric acid in vitaceous plants – relative contributions of L-ascorbic acid inclusive and non-inclusive pathways', *Plant and Cell Physiology* **30**, 905–910.

Sakihama, Y., Mano, J., Sano, S., Asada, K. and Yamasaki, H. (2000) 'Reduction of phenoxyl radicals mediated by monodehydroascorbate reductase', *Biochemical and Biophysical Research Communications* **279**, 949–954.

Sanmartin, M., Drogoudi, P.D., Lyons, T., Pateraki, I., Barnes, J. and Kanellis, A.K. (2003) 'Over-expression of ascorbate oxidase in the apoplast of transgenic tobacco results in altered ascorbate and glutathione redox states and increased sensitivity to ozone', *Planta* **216**, 918–928.

Sattler, S.E., Cahoon, E.B., Coughlan, S.J. and DellaPenna, D. (2003) 'Characterization of tocopherol cyclases from higher plants and cyanobacteria. Evolutionary implications for tocopherol synthesis and function', *Plant Physiology* **132**, 2184–2195.

Sattler, S.E., Gilliland, L.U., Magallanes-Lundback, M., Pollard, M. and DellaPenna, D. (2004) 'Vitamin E is essential for seed longevity, and for preventing lipid peroxidation during germination', *The Plant Cell* **16**, 1419–1432.

Savidge, B., Weiss, J.D., Wong, Y.H.H. *et al.* (2002) 'Isolation and characterization of homogentisate phytyltransferase genes from *Synechocystis* sp. PCC 6803 and *Arabidopsis*', *Plant Physiology* **129**, 321–332.

Shimaoka, T., Miyake, C. and Yokota, A. (2003) 'Mechanism of the reaction catalyzed by dehydroascorbate reductase from spinach chloroplasts', *European Journal of Biochemistry* **270**, 921–928.

Shimaoka, T., Yokota, A. and Miyake, C. (2000) 'Purification and characterization of chloroplast dehydroascorbate reductase from spinach leaves', *Plant and Cell Physiology* **41**, 1110–1118.

Shintani, D. and DellaPenna, D. (1998) 'Elevating the vitamin E content of plants through metabolic engineering', *Science* **282**, 2098–2100.

Smirnoff, N. (2000) 'Ascorbate biosynthesis and function in photoprotection', *Philosophical Transactions of the Royal Society of London Series B – Biological Sciences* **355**, 1455–1464.

Smirnoff, N. (2001) 'L-Ascorbic acid biosynthesis', *Vitamins and Hormones – Advances in Research and Applications* **61**, 241–266.

Soll, J. (1987) 'α-Tocopherol and plastoquinone synthesis in chloroplast membranes', *Methods in Enzymology* **148**, 383–392.

Stahl, W., Junghans, A., de Boer, B., Driomina, E.S., Briviba, K. and Sies, H. (1998) 'Carotenoid mixtures protect multilamellar liposomes against oxidative damage: synergistic effects of lycopene and lutein', *FEBS Letters* **427**, 305–308.

Suh, J., Zhu, B.Z. and Frei, B. (2003) 'Ascorbate does not act as a pro-oxidant towards lipids and proteins in human plasma exposed to redox-active transition metal ions and hydrogen peroxide', *Free Radical Biology and Medicine* **34**, 1306–1314.

Tabata, K., Oba, K., Suzuki, K. and Esaka, M. (2001) 'Generation and properties of ascorbic acid-deficient transgenic tobacco cells expressing antisense RNA for L-galactono-1,4-lactone dehydrogenase', *The Plant Journal* **27**, 139–148.

Trumper, S., Follmann, H. and Haberlein, I. (1994) 'A novel dehydroascorbate reductase from spinach chloroplasts homologous to plant trypsin inhibitor', *FEBS Letters* **352**, 159–162.

Tsegaye, Y., Shintani, D.K. and DellaPenna, D. (2002) 'Overexpression of the enzyme p-hydroxyphenolpyruvate dioxygenase in *Arabidopsis* and its relation to tocopherol biosynthesis', *Plant Physiology and Biochemistry* **40**, 913–920.

Urano, J., Nakagawa, T., Maki, Y. *et al.* (2000) 'Molecular cloning and characterization of a rice dehydroascorbate reductase', *FEBS Letters* **466**, 107–111.

van den Berg, H., Faulks, R., Granado, H.F. *et al.* (2000) 'The potential for the improvement of carotenoid levels in foods and the likely systemic effects', *Journal of the Science of Food and Agriculture* **80**, 880–912.

Van Eenennaam, A.L., Li, G.F., Venkatramesh, M. *et al.* (2004) 'Elevation of seed α-tocopherol levels using plant-based transcription factors targeted to an endogenous locus', *Metabolic Engineering* **6**, 101–108.

Van Eenennaam, A.L., Lincoln, K., Durrett, T.P. *et al.* (2003) 'Engineering vitamin E content: from *Arabidopsis* mutant to soy oil', *The Plant Cell* **15**, 3007–3019.

Veljovic-Jovanovic, S.D., Pignocchi, C., Noctor, G. and Foyer, C.H. (2001) 'Low ascorbic acid in the *vtc1* mutant of *Arabidopsis* is associated with decreased growth and intracellular redistribution of the antioxidant system', *Plant Physiology* **127**, 426–435.

Villalba, J.M., Cordoba-Pedregosa, M.C. and Gonzalez-Reyes, J.A. (2004) Membrane redox proteins involved in ascorbate-mediated reactions, in *Vitamin C. Function and Biochemistry in Animals and Plants* (ed. H. Asard, J.M. May and N. Smirnoff), Bios Scientific Publishers, Oxford, pp. 119–137.

Weber, H., Chetelat, A., Reymond, P. and Farmer, E.E. (2004) 'Selective and powerful stress gene expression in *Arabidopsis* in response to malondialdehyde', *The Plant Journal* **37**, 877–888.

Wheeler, G.L., Jones, M.A. and Smirnoff, N. (1998) 'The biosynthetic pathway of vitamin C in higher plants', *Nature* **393**, 365–369.

Wolucka, B.A. and Van Montagu, M. (2003) 'GDP-mannose 3′,5′-epimerase forms GDP-L-gulose, a putative intermediate for the de novo biosynthesis of vitamin C in plants', *Journal of Biological Chemistry* **278**, 47483–47490.

Wolucka, B.A., Persiau, G., Van Doorsselaere, J. *et al.* (2001) 'Partial purification and identification of GDP-mannose 3′,5′-epimerase of *Arabidopsis thaliana*, a key enzyme of the plant vitamin C pathway', *Proceedings of the National Academy of Sciences USA* **98**, 14843–14848.

Wrona, M., Manowska, M. and Sarna, T. (2004) 'Zeaxanthin in combination with ascorbic acid or α-tocopherol protects ARPE-19 cells against photosensitized peroxidation of lipids', *Free Radical Biology and Medicine* **36**, 1094–1101.

Yamauchi, J., Iwamoto, T., Kida, S., Masushige, S., Yamada, K. and Esashi, T. (2001) 'Tocopherol-associated protein is a ligand-dependent transcriptional activator', *Biochemical and Biophysical Research Communications* **285**, 295–299.

Yan, J.Q., Wang, J., Tissue, D., Holaday, A.S., Allen, R. and Zhang, H. (2003) 'Photosynthesis and seed production under water-deficit conditions in transgenic tobacco plants that overexpress an *Arabidopsis* ascorbate peroxidase gene', *Crop Science* **43**, 1477–1483.

Ye, X.D., Al-Babili, S., Kloti, A. *et al.* (2000) 'Engineering the provitamin A (β-carotene) biosynthetic pathway into (carotenoid-free) rice endosperm', *Science* **287**, 303–305.

Yeum, K.J., Ferreira, A.L.D., Smith, D., Krinsky, N.I. and Russell, R.M. (2000) 'The effect of alpha-tocopherol on the oxidative cleavage of β-carotene', *Free Radical Biology and Medicine* **29**, 105–114.

Yeum, K.J., Russell, R.M., Krinsky, N.I., Aldini, G. and Mayer, J. (2004) 'Synergistic interactions of β-carotene, α-tocopherol and ascorbic acid in delipidized human serum enriched with phosphatidyl choline micelles', *FASEB Journal* **18**, A480.

Young, A. and Britton, G. (1990) Carotenoids, in *Stress Responses in Plants: Adaptation and Acclimation* (ed. R.G. Alscher and J.R. Cumming), Wiley-Liss, New York, pp. 87–112.

Young, A.J. and Lowe, G.M. (2001) 'Antioxidant and prooxidant properties of carotenoids', *Archives of Biochemistry and Biophysics* **385**, 20–27.

4   Ascorbate peroxidase

Ron Mittler and Thomas L. Poulos

4.1   Enzymatic removal of hydrogen peroxide in plants

Detoxification of hydrogen peroxide (H_2O_2) in plants is essential for cell protection and cell signaling (Dat *et al.*, 2000; Mittler, 2002; Neill *et al.*, 2002; Apel and Hirt, 2004). Plants contain at least five different enzymes capable of rapid and efficient H_2O_2 removal. These are encoded by five distinct gene families: ascorbate peroxidases (APX, nine genes in *Arabidopsis*; Mittler *et al.*, 2004), catalases (CAT, three genes in *Arabidopsis*; Vandenabeele *et al.*, 2004), glutathione peroxidases (GPX, eight genes in *Arabidopsis*; Rodriguez *et al.*, 2003), peroxiredoxins (PrxR, eleven genes in *Arabidopsis*; Dietz, 2003) and type III peroxidases (Prx, seventy-three genes in *Arabidopsis*; Tognolli *et al.*, 2002). Together with the antioxidants ascorbic acid and glutathione (Noctor & Foyer, 1998), these enzymes provide cells with highly efficient machinery for detoxifying H_2O_2.

The coordinated function of the five different H_2O_2-removal enzymes, together with the activity of the superoxide (O_2^-)-scavenging enzyme superoxide dismutase (SOD, eight genes in *Arabidopsis*; Kliebenstein *et al.*, 1998), is especially important to prevent the formation of the highly toxic hydroxyl radical (HO˙). Only by maintaining a delicate balance between the steady-state levels of O_2^- and H_2O_2, and by sequestering free metal ions such as iron and copper (by ferritin and other metal-binding proteins), can the formation of HO˙ via the Haber–Weiss or the Fenton reactions be prevented (Halliwell & Gutteridge, 1989; Bowler *et al.*, 1991). The coordinated function of all reactive oxygen scavenging enzymes is also critical for controlling signal transduction pathways that utilize H_2O_2, O_2^- and O_2^1 in plants. These pathways are involved in the control and regulation of processes such as pathogen defense, stomatal signaling, stress response, growth, development and hormonal signaling (Kovtun *et al.*, 2000; Pei *et al.*, 2000; Mullineaux & Karpinski, 2002; Torres *et al.*, 2002; Foreman *et al.*, 2003; Kwak *et al.*, 2003; Overmyer *et al.*, 2003; Mittler *et al.*, 2004).

In contrast to CAT, GPX, PrxR and class III peroxidases, APX (EC. 1.11.1.11) appears to be a unique enzyme found mainly in plants and algae (Asada & Takahashi, 1987; Shigeoka *et al.*, 2002; Raven, 2003). It has a high preference for ascorbic acid as a reducing substrate and it catalyzes the reaction:

$$2\text{Ascorbate} + H_2O_2 \rightarrow 2\text{Monodehydroascorbate} + 2H_2O$$

Ascorbate peroxidase is found in almost every compartment of the plant cell and it participates in the removal of H_2O_2 as part of the ascorbate–glutathione or

Asada–Halliwell–Foyer pathway (Figure 4.1; Foyer & Halliwell, 1976; Asada & Takahashi, 1987; Chapters 1 and 3). As shown in Figure 4.1, the function of APX is dependent upon the availability of a pool of reduced ascorbic acid, and, in some cases, reduced glutathione (depending on the relative activities of monodehydroascorbate reductase and dehydroascorbate; Polle, 2001). The cellular pools of these antioxidants are maintained in their reduced state by a set of enzymes capable of using NAD(P)H to regenerate oxidized glutathione or ascorbic acid (e.g. monodehydroascorbate reductase, dehydroascorbate reductase, glutaredoxin and glutathione reductase; Figure 4.1). Although dehydroascorbate reductases and glutaredoxins are indicated in Figure 4.1 as capable of reducing dehydroascorbic acid, many other enzymes in plants can catalyze this reaction with different efficiencies (Chew et al., 2003).

The initial identification of APX was in isolated intact chloroplasts and algae (Gordon & Beck, 1979; Shigeoka et al., 1980; Nakano & Asada, 1981). In chloroplasts, APX is found in at least three different forms: thylakoid APX that is bound to the thylakoid membrane, and stromal and lumen APXs that are present as soluble isoforms (Asada, 1999; Shigeoka et al., 2002; Mittler et al., 2004). In some plants (e.g. spinach, pumpkin and tobacco), the thylakoid and stromal isoforms are generated from the same gene by alternative splicing, whereas in others (e.g. Arabidopsis), they are encoded by two distinct genes (Shigeoka et al., 2002). Interestingly, in Arabidopsis, stromal APX is dually targeted to the stroma and the IMS space of the mitochondria (Chew et al., 2003). In addition to stromal APX, two

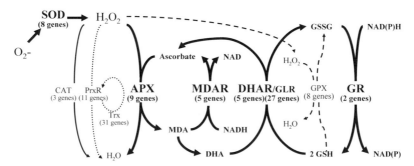

Figure 4.1 The main cellular pathways for H_2O_2 removal in plants. The superoxide dismutase–ascorbate–glutathione pathway is shown in bold. Enzymes representing this pathway are found in the chloroplast, mitochondria, cytosol, peroxisomes, microsomes and apoplast. The number of genes in Arabidopsis encoding each of the different enzymes is also shown. Due to the large number of possible substrates utilized by class III peroxidases, they are not included in the figure. Abbreviations: APX, ascorbate peroxidase; DHA, dehydroascorbate; DHAR, dehydroascorbate reductase; MDA, monodehydroascorbate; MDAR, MDA reductase; GLR, glutaredoxin; GPX, glutathione peroxidase; GR, glutathione reductase; GSH and GSSG, reduced and oxidized glutathione; PrxR, peroxiredoxin; SOD, superoxide dismutase; Trx, thioredoxin.

other stromal enzymes of the ascorbate–glutathione cycle (Figure 4.1) are dually targeted to the chloroplast and mitochondria. These stromal enzymes are monodehydroascorbate reductase and glutathione reductase (Creissen et al., 1995; Obara et al., 2002; Chew et al., 2003). In addition to the chloroplast and mitochondria, enzymes of the ascorbate–glutathione cycle are also present in peroxisomes, glycosomes and the apoplast, suggesting that the ascorbate–glutathione pathway is a key H_2O_2-removal pathway in plants (Corpas et al., 2001; Pignocchi & Foyer, 2003).

Substrate affinity is an important parameter in assessing the relative contribution of the different enzymes shown in Figure 4.1 to H_2O_2 detoxification (Mittler, 2002). Thus, enzymes such as APX and GPX, have a very high affinity to H_2O_2 (at the micromolar and submicromolar range), whereas enzymes such as CAT and PrxR have a relatively low affinity (millimolar range). The effective removal of H_2O_2 by CAT and PrxR is usually achieved by a high concentration of these enzymes in specific cellular compartments.

Table 4.1 provides a gene and protein annotation to the APX gene family in Arabidopsis (Mittler et al., 2004). Of the nine different genes encoding APX in Arabidopsis, four are thought to encode chloroplastic isozymes. These include the thylakoid and stromal APXs, APX4 that is thought to be targeted to the lumen and APX6 that is one of the two small APX proteins of Arabidopsis. Two different APXs are thought to be targeted to peroxisomes/glyoxysomes, i.e. APX3 and APX5 (both containing a putative membrane anchoring domain), and two different APXs are thought to be located in the cytosol (APX1 and APX2). APX7 and the dually targeted stromal APX are thought to be localized to the mitochondria. The two microsomal APXs (APX3 and APX5) might bind to the external surface

Table 4.1 Gene and protein annotation for the ascorbate peroxidase (APX) gene family in *Arabidopsis*

Protein name	Gene identifier	Localization	MW	Isoelectric point	Membrane domains
APX1	At1g07890.1	Cytosol	27 561	5.72	0
APX2	At3g09640.1	Cytosol	27 415	6.01	0
APX3	At4g35000.1	Microsomes	31 571	6.47	1
APX4	At4g09010.1	Lumen	37 934	8.59	0
APX5	At4g35970.1	Microsomes	30 895	8.8	1
APX6	At4g32320.1	Chloroplast	18 214	5.92	0
APX7	At1g33660.1	Mitochondria	11 721	10.15	0
Stromal APX	At4g08390.2	Stroma/ mitochondria	40 407	8.31	0
Thylakoid APX	At1g77490.1	Thylakoid	46 092	6.81	1

Note: Localization, predicted molecular weight (MW), isoelectric point and membrane-domain data were obtained as described by Mittler et al. (2004).

of glyoxysomes or be transported into peroxisomes (Shigeoka et al., 2002). Interestingly, two of the APX genes shown in Table 4.1, i.e. APX6 and APX7, have a relatively small predicted molecular weight. APX enzymes with such a small size were not previously characterized biochemically and it would be interesting to find how they differ from the previously characterized cytosolic and chloroplastic isozymes that have a much larger molecular weight (Table 4.1).

Gene expression studies in plants subjected to biotic and abiotic stress revealed that cytosolic APX isozymes are the most stress-responsive among the different members of the APX gene family. APX1 has a relatively high level of expression in the absence of stress; however, its expression is dramatically enhanced in response to almost all biotic and abiotic stresses studied (summarized in Mittler, 2002 and Shigeoka et al., 2002; see also Mittler & Zilinskas, 1992; Yoshimura et al., 2000). In contrast, APX2 is not expressed under non-stressful controlled conditions and its expression is significantly elevated in response to very high light stress or heat shock (Mullineaux & Karpinski, 2002; Panchuk et al., 2002). The expression of APX1 is also controlled post-transcriptionally during pathogen attack and recovery from drought stress (Mittler & Zilinskas, 1994; Mittler et al., 1998).

4.2   Functional analysis of APX

The function of some of the APX isozymes in tobacco and Arabidopsis was studied using two different reverse-genetics approaches, i.e. antisense and gene-knockouts, and using transgenic plants that over-express APX. Initial studies in tobacco using antisense constructs for cytosolic APX1 revealed that, compared to wild-type or control plants, plants with suppressed expression of APX1 are less tolerant to high light stress, ozone fumigation and oxidative stress (Orvar & Ellis, 1997; Mittler et al., 1999). Suppression of APX1 in tobacco was also found to affect the sensitivity of plants to pathogen attack. Thus, plants with suppressed expression of APX1 activated their hypersensitive-programmed cell death-response upon infection with a very low titer of bacterial pathogens – a titer that did not activate a cell death response in control or wild-type plants (Mittler et al., 1999). This finding linked between the post-transcriptional suppression of APX1 during pathogen attack (Mittler et al., 1998), and the suggested role of H_2O_2 as a signaling molecule that triggers the hypersensitive response (Neill et al., 2002). Thus, to activate the hypersensitive response that is dependent upon the accumulation of H_2O_2, the expression and activity of APX (and CAT) needs to be suppressed (Mittler, 2002).

Attempts to suppress the expression of thylakoid APX by antisense expression in tobacco were unsuccessful suggesting that this enzyme is absolutely required for plant survival (Yabuta et al., 2002). A wheat mutant with 40% decrease in thylakoid APX activity was, however, recently characterized (Danna et al., 2003). This mutant shows impaired electron transport and reduced growth and photosynthetic activity when grown under controlled conditions. Overexpression of APX1 or thylakoid APX resulted in enhanced tolerance to oxidative stress, supporting the reverse-genetic and mutant studies (Wang *et al.*, 1999; Yabuta *et al.*, 2002).

The availability of gene-knockouts in *Arabidopsis* opened the way for an extensive analysis of the APX gene family. Analysis of knockout plants for APX1 revealed that plants lacking APX1 have a delayed growth and development phenotype. In addition, the guard cells of knockout-APX1 plants were impaired in their response to light, and a moderate level of light stress resulted in the augmented induction of catalase and heat shock proteins in knockout plants (Pnueli *et al.*, 2003). The abnormality in stomatal responses in knockout APX1 could be explained by the role H_2O_2 was suggested to play in abscisic-acid-induced stomatal closure (Kwak *et al.*, 2003). Thus, in the absence of APX1, H_2O_2 levels accumulate and induce abnormal closure of stomata (Pnueli *et al.*, 2003). The augmented induction of heat shock proteins during light stress in knockout APX1 plants can also suggest that H_2O_2 plays an essential role as a signaling molecule during abiotic stress (Mittler, 2002; Neill *et al.*, 2002).

Microarray analysis of plants lacking APX1 identified a number of signal transduction transcripts that might be involved in the response of plants to H_2O_2 stress (Pnueli *et al.*, 2003; Mittler *et al.*, unpublished). Of these, the zinc-finger protein Zat12 was found to be required for the enhanced expression of APX1 during H_2O_2 stress (Rizhsky *et al.*, 2004). Interestingly, the expression of the other APX genes shown in Table 4.1 was not elevated in knockout APX1 plants grown under controlled conditions or subjected to a moderate level of light stress (Mittler *et al.*, unpublished), suggesting that APX1 could not be replaced by any of the other APX genes. The expression of catalase was, however, elevated in knockout APX1 plants in response to light stress (Pnueli *et al.*, 2003). Additional studies using knockouts for other APX genes, and using double and triple knockouts, should reveal the function of the different APX genes in *Arabidopsis*. These studies should be performed with plants grown under controlled conditions and with plants subjected to different biotic or abiotic stress treatments.

Recent studies in *Arabidopsis* have suggested that the mode of coordination between the different H_2O_2-removal enzymes of plants is complex (Mittler *et al.*, 2004). For example, the application of light stress to *Arabidopsis* results in the induction of cytosolic and not chloroplastic APXs (Karpinski *et al.*, 1997, 1999; Pnueli *et al.*, 2003). This finding is in spite of the fact that most H_2O_2 produced during light stress is thought to be generated in chloroplasts or peroxisomes (Asada, 1999). As indicated above, APX and CAT appear to compensate for each other at least to a certain degree (Pnueli *et al.*, 2003; Vandenabeele *et al.*, 2004). However, in a study that compared the sensitivity of antisense APX plants, antisense CAT plants and double antisense APX/CAT plants to light stress, it was found that in the absence of APX *and* CAT, plants activated a redundant/alternative pathway(s) that involves the suppression of photosynthesis and the induction of chloroplastic alternative oxidase and other defense enzymes (Rizhsky *et al.*, 2002). Identifying the exact components of the alternative pathway is currently underway in *Arabidopsis*.

4.3 APX structure

4.3.1 Overall structure

A summary of structure–function relationships in APX has been published in a recent review (Raven, 2003). The first APX crystal structure to be solved was pea cytosolic APX (Patterson & Poulos, 1995) whereas, more recently, the tobacco chloroplastic APX structure was determined (Wada *et al.*, 2003). Since the structures are very similar (rms deviation of 232 common Cα = 0.86 Å), the following discussion will focus on pea cytosolic APX. APX exhibits the overall peroxidase fold that has been observed in a number of class I, II and III peroxidases (Patterson & Poulos, 1995). At 33% sequence identity, the closest homolog to APX is CCP, another class I intra-cellular peroxidase. On comparing CCP and APX, the average rms difference for 137 Cα carbon atoms in a helical conformation is 0.9 Å whereas that for 249 topologically equivalent Cα atoms is 1.3 Å. A schematic diagram of APX is shown in Figure 4.2. The domain structure of the monomer unit is nearly identical to CCP consisting of Domain I (N- and C-termini, helices A–D) and Domain II (helices F–J). The domains are connected via the E helix. The J helix, which begins in Domain II, extends into Domain I where the C-terminus ends within 4 Å of the N-terminus. APX differs from CCP in that two elements of secondary structure found in CCP are missing in APX. APX has a truncated C-terminus as compared to CCP and the CCP J′ helix is missing. The two β-strands consisting of residues 210–231 in CCP are also missing in APX.

The access channel connecting the molecular surface to the distal heme pocket is very much the same in size and shape in both peroxidases (indicated by the arrow in Figure 4.2a). The heme also is anchored in place by similar sets of interactions

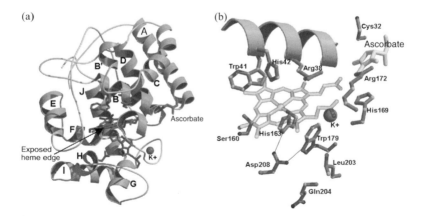

Figure 4.2   (a) A schematic diagram of APX. The helices are labeled according to Patterson and Poulos (1995). The arrow indicates the exposed heme edge where traditional phenolic peroxidase substrates are thought to bind. (b) A close-up view of the active site. The thin lines indicate Asp208–His163 and Asp208–Trp179 H-bonds. Note that ascorbate directly interacts with Arg172 near one of the heme propionates.

although with some variations. Like CCP, APX has a His residue (His169 in APX and His181 in CCP, Figure 4.2b) H-bonds with one heme propionate. In CCP, however, His181 also interacts with Asp37 which is part of an unusual left-handed helical turn. This interaction is not present in APX since APX is missing the left-handed helical turn containing Asp37 found in CCP. Where CCP has Asn184 directly interacting with a heme propionate, APX has Arg172.

A major difference between APX and CCP is that APX forms a homodimer. The dimer interface of the APX homodimer is mediated primarily by electrostatic interactions among Asp, Glu, Arg and Lys side chains. Site-directed mutagenesis of residues at the dimer interface demonstrated that the monomer is still fully active although calorimetric studies showed that the dimer contributes substantially to stability (Mandelman *et al.*, 1998b). There also is a major structural difference between the cytosolic and chloroplastic APXs. The chloroplastic APX has an extra loop, residues 176–193, near the heme propionates (Wada *et al.*, 2003). In addition, the chloroplastic enzyme has no direct salt bridges or H-bonds between the heme propionates and amino side chains found in the cytosolic enzyme and CCP. The functional significance of these differences is unclear although the chloroplastic enzyme is substantially less stable than the cytosolic enzyme in the absence of ascorbate (Nakano & Asada, 1987).

4.3.2   Active site structure

APX exhibits the same active site structure as all other class I–III peroxidases. The proximal His heme ligand H-bonds with a buried and conserved Asp (Figure 4.2). In the distal pocket, the conserved distal His42 and Arg38 surround the peroxide-binding pocket and are key components in the formation of Compound I (Poulos & Kraut, 1980). Only class I peroxidases have Trp at these positions whereas most other peroxidases have Phe. However, only in CCP does the proximal Trp191 (Trp179 in APX) have functional relevance. Trp191 in CCP forms a stable cationic radical in Compound I (Sivaraja *et al.*, 1989) that is critical for enzyme activity (Mauro *et al.*, 1987). However, APX forms the more traditional porphyrin π cation radical (Patterson *et al.*, 1995) and mutation of Trp179 to Phe has no effect on activity (Pappa *et al.*, 1996).

Unraveling why CCP forms a Trp radical whereas APX forms the traditional porphyrin radical has required a combination of crystallography, spectroscopy and computational methods. The current view is that differences in electrostatic stabilization surrounding the proximal side Trp residue controls where oxidizing equivalents are stored. One important structural difference contributing to such stabilization is the cation site present in APX but absent in CCP (Figure 4.2). All peroxidases, except CCP, have a cation at this position. This cation site is about 8 Å from Trp179 and, hence, formation of a cationic Trp179 radical may not be as favorable in APX owing to the nearby K^+. Indeed, engineering the APX cation site into CCP greatly diminishes the stability of the Trp radical (Bonagura *et al.*, 1996, 1999a,b; Bhaskar *et al.*, 2000, 2002) although the cation site alone cannot account for all the stability of the Trp radical in CCP (Jensen *et al.*, 1998). Further mutagenesis work has shown that three Met residues in the proximal pocket of CCP are also crucial in stabilization of the Trp191 radical (Barrows *et al.*, 2004). The corresponding residues in APX are Ser160, Gln204 and Leu203 (Figure 4.2). Therefore, the absence of the three electronegative Met sulfur atoms and the presence of the K^+ site are why APX is unable to stabilize a cationic radical centered on Trp179.

4.3.3   Substrate binding

The favored site for binding peroxidase substrates is at the exposed heme edge (Figure 4.2). This provides the closest approach of the substrate to the heme so that reduction of the porphyrin radical can proceed as a direct electron transfer process. However, very early chemical modification results indicated that APX may have an alternate binding site. Chemical modification of the single Cys32 (Figure 4.2) in APX leads to the loss of ascorbate peroxidase activity (Chen & Asada, 1989). Further work on APX was hampered by the lack of sufficient material but the cloning and expression of APX (Patterson & Poulos, 1994) enabled more detailed studies to proceed. Subsequent chemical modification and mutagenesis work led to the conclusion that blocking Cys32 with a bulky thiol reagent prevents ascorbate from binding (Mandelman *et al.*, 1999a). However, the ability to oxidize traditional phenolic peroxidase substrates like guaiacol remains unaltered (Mandelman *et al.*, 1999a). One of the key amino acids predicted to be involved in ascorbate binding is Arg172 (Mandelman *et al.*, 1999a) which mutagenesis (Bursey & Poulos, 2000) showed is, indeed, critical for ascorbate peroxidase activity but not for guaiacol peroxidase activity (Bursey & Poulos, 2000). The crystal structure of the APX–ascorbate complex (Figure 4.2) (Sharp *et al.*, 2003) supported these conclusions and showed that Arg172 is involved with ascorbate binding. More recently, the structure of APX complexed with an aromatic inhibitor, salicylhydroxamic acid, was determined (Sharp *et al.*, 2004) and, as expected, the salicylhydroxamic acid binds at the exposed heme edge. It should be cautioned, however, that hydroxamic acids are not quite the same as true phenolic substrates since the hydroxamic acids have H-bonding groups that interact with active site side chains. Phenolic substrates do not have this capability and are unlikely to bind in quite the same way as observed in peroxidase–hydroxamic acid complexes. Nevertheless, it does appear that APX has at least two binding sites for substrate. The physiologically important substrate, ascorbate, binds near a heme propionate and Arg172 whereas phenolic substrates very likely bind near the exposed heme edge (Figure 4.2).

4.4   Evolution of APXs

The initial cloning and sequencing of APX1 from pea (Mittler & Zilinskas, 1991) revealed that APX is more similar to yeast CCP and bacterial catalase–peroxidases (both class I peroxidases) than to typical plant peroxidases (class III peroxidases) or fungal peroxidases (class II peroxidases). To date, an APX protein has not been isolated from bacteria or cyanobacteria, although a complex between an iron atom and a low molecular weight peptide with a distinct APX activity was isolated and characterized from cyanobacteria (Rozen *et al.*, 1991). In contrast, eukaryotic algae were found to contain APX (Shigeoka *et al.*, 1980, 2002).

The earliest event in APX evolution probably resulted in the formation of two separate groups of APX: cytosolic and chloroplastic (Jespersen *et al.*, 1997; Zamocky *et al.*, 2000; Shigeoka *et al.*, 2002). Chloroplastic APXs have apparently diverged only recently into stromal and thylakoid APX via gene duplication in *Arabidopsis*, or alternative splicing in pumpkin and spinach (Jespersen *et al.*, 1997; Shigeoka *et al.*, 2002). The formation of a separate cytosolic APX class (e.g. Spinach APX2) might have also only recently occurred in a species-specific manner from the original cytosolic APX1 branch (Jespersen *et al.*, 1997; Shigeoka *et al.*, 2002). In contrast, in all species, microsomal APX might have originated from the APX ancestor or the cytosolic APX1 branch via gene duplication (Jespersen *et al.*, 1997; Shigeoka *et al.*, 2002).

A phylogenetic tree for all *Arabidopsis* APXs (Table 4.1) is shown in Figure 4.3. As shown in Figure 4.3, APX1 and APX2, the two cytosolic APXs, APX5 and

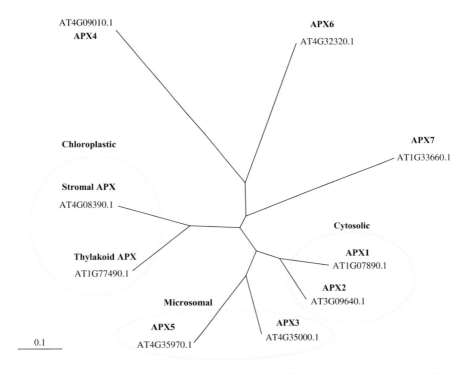

Figure 4.3   A phylogenetic tree for *Arabidopsis* APXs (Table 4.1). Neighbor-joining alignment was performed with ClustalW using entire ORFs (deduced amino acid sequences). Unrooted tree was drawn by TreeView.

APX3, the two microsomal APXs, and stromal and thylakoid APXs form three distinct groups labeled: cytosolic, microsomal and chloroplastic. In contrast, APX4 and APX6 that are also thought to be chloroplastic are very distinct from the stromal and thylakoid APXs, and APX7 that is thought to be mitochondrial appears also to be distinct from the other groups. In agreement with the analysis performed with APX sequences from other plants (Jespersen et al., 1997; Shigeoka et al., 2002), at least three different groups of APXs were initially generated from the ancient APX gene. These were cytosolic, chloroplastic (i.e. stromal and thylakoid APX) and microsomal (Figure 4.3).

4.5 Summary

Ascorbate peroxidase is a unique class I peroxidase found mainly in photosynthetic algae and plants. It has a high structural homology to yeast CCP and uses ascorbic acid as its preferred reducing substrate. Ascorbate peroxidase is encoded by a small gene family and can be found in almost every compartment of the plant cell. Reverse genetics studies, as well as overexpression analysis in transgenic plants, have demonstrated that APX is a key defense enzyme involved in the removal of H_2O_2 in plants. The involvement of APX in the control and regulation of H_2O_2 signaling in plants has recently emerged as an additional function (Mittler, 2002). The high affinity of APX to H_2O_2, its presence in every compartment of the plant cell and its highly regulated expression (transcriptional and post-transcriptional) makes it a suitable candidate to participate in the removal of H_2O_2 for signaling purposes. The finding that knockout and transgenic plants with suppressed expression of APX1 are impaired in their physiological and molecular responses to biotic and abiotic stress can be seen as genetic evidence for the involvement of APX in H_2O_2 signaling (Mittler et al., 1999; Pnueli et al., 2003). Ascorbate peroxidase might therefore function as part of the H_2O_2-removal network of plants that controls the steady-state level of H_2O_2 used for cellular signaling. Thus, together with other H_2O_2-removal enzymes of the cell, it balances the different cellular systems that produce H_2O_2 in plants (e.g. NADPH oxidase coupled with SOD), and controls, in a cellular- or even a subcellular-specific manner, the level of H_2O_2 used for signaling during biotic or abiotic stress. Further studies are required to elucidate how the function of the different APX genes/proteins of plants is coordinated and whether or not APX have additional functions in plants (Mittler et al., 2004).

Acknowledgments

Research in the laboratory of R. Mittler is supported by funding from The National Science Foundation (NSF-0431327) and Nevada NSF-EPSCoR.

References

Apel, K. and Hirt, H. (2004) 'Reactive oxygen species: metabolism, oxidative stress, and signal transduction', *Annual Review of Plant Biology* **55**, 373–399.

Asada, K. (1999) 'The water–water cycle in chloroplasts: scavenging of active oxygen and dissipation of excess photons', *Annual Review of Plant Physiology and Plant Molecular Biology* **50**, 601–639.

Asada, K. and Takahashi, M. (1987) Production and scavenging of active oxygen in photosynthesis, in *Photoinhibition* (ed. D.J. Kyle, C.B. Osmond and C.J. Arntzen), Elsevier, Amsterdam, pp. 227–287.

Barrows, T.P., Bhaskar, B. and Poulos, T.L. (2004) 'Formation and stability of the tryptophan cation radical in cytochrome c peroxidase: electrostatic control', *Biochemistry* **43**, 8826–8834.

Bhaskar, B., Bonagura, C.A., Jamal, J. and Poulos, T.L. (2000) 'Loop stability in the engineered potassium binding site of cytochrome c peroxidase', *Tetrahedron* **56**, 9471–9475.

Bhaskar, B., Bonagura, C.A., Li, H. and Poulos, T.L. (2002) 'Cation-induced stabilization of the engineered cation-binding loop in cytochrome c peroxidase', *Biochemistry* **41**, 2684–2693.

Bonagura, C.A., Bhaskar, B., Sundaramoorthy, M. and Poulos, T.L. (1999a) 'Conversion of an engineered potassium-binding site into a calcium-selective site in cytochrome c peroxidase', *Biochemistry* **274**, 37827–37833.

Bonagura, C.A., Sundaramoorthy, M., Bhaskar, B. and Poulos, T.L. (1999b) 'The effects of an engineered cation site on the structure, activity, and EPR properties of cytochrome c peroxidase', *Biochemistry* **38**, 5528–5545.

Bonagura, C.A., Sundaramoorthy, M., Pappa, H.S., Patterson, W.R. and Poulos, T.L. (1996) 'An engineered cation site in cytochrome c peroxidase alters the reactivity of the redox active tryptophan', *Biochemistry* **35**, 6107–6115.

Bowler, C., Slooten, L., Vandenbranden, S. *et al.* (1991) 'Manganese superoxide dismutase can reduce cellular damage mediated by oxygen radicals in transgenic plants', *EMBO Journal* **10**, 1723–1732.

Bursey, E.H. and Poulos, T.L. (2000) 'Two substrate binding sites in ascorbate peroxidase: the role of arginine 172', *Biochemistry* **39**, 7374–7379.

Chen, G.X. and Asada, K. (1989) 'Ascorbate peroxidase in tea leaves: occurrence of two isozymes and the difference in their enzymatic and molecular properties', *Plant Cell Physiology* **30**, 987–998.

Chew, O., Whelan, J. and Millar, A.H. (2003) 'Molecular definition of the ascorbate–glutathione cycle in *Arabidopsis* mitochondria reveals dual targeting of antioxidant defenses in plants', *The Journal of Biological Chemistry* **278**, 46869–46877.

Corpas, F.J., Barroso, J.B. and del Rio, L.A. (2001) 'Peroxisomes as a source of reactive oxygen species and nitric oxide signal molecules in plant cells', *Trends in Plant Science* **6**, 145–150.

Creissen, G., Reynolds, H., Xue, Y. and Mullineaux, P. (1995) 'Simultaneous targeting of pea glutathione reductase and of a bacterial fusion protein to chloroplasts and mitochondria in transgenic tobacco', *The Plant Journal* **8**, 167–175.

Danna, C.H., Bartoli, C.G., Sacco, F. *et al.* (2003) 'Thylakoid-bound ascorbate peroxidase mutant exhibits impaired electron transport and photosynthetic activity', *Plant Physiology* **132**, 2116–2125.

Dat, J., Vandenabeele, S., Vranova, E., Van Montagu, M., Inze, D. and Van Breusegem, F. (2000) 'Dual action of the active oxygen species during plant stress responses', *Cellular and Molecular Life Sciences* **57**, 779–795.

Dietz, K.J. (2003) 'Plant peroxiredoxins', *Annual Review of Plant Biology* **54**, 93–107.

Foreman, J., Demidchik, V., Bothwell, J.H. *et al.* (2003) 'Reactive oxygen species produced by NADPH oxidase regulate plant cell growth', *Nature* **422**, 442–446.

Foyer, C.H. and Halliwell, B. (1976) 'The presence of glutathione and glutathione reductase in chloroplasts: a proposed role in ascorbic acid metabolism', *Planta* **133**, 21–25.

Groden, D. and Beck, E. (1979) 'H_2O_2 destruction by ascorbate-dependent systems from chloroplasts', *Biochimica et Biophysica Acta* **546**, 426–435.

Halliwell, B. and Gutteridge, J.M.C. (1989) *Free Radicals in Biology and Medicine*, Clarendon, Oxford.

Jensen, G.M., Bunte, S.W., Warshel, A. and Goodin, D.B. (1998) 'Energetics of cation radical formation at the proximal active site tryptophan of cytochrome c peroxidase and ascorbate peroxidase', *The Journal of Physical Chemistry* B **102**, 8221–8228.

Jespersen, H.M., Kjaersgard, I.V., Ostergaard, L. and Welinder, K.G. (1997) 'From sequence analysis of three novel ascorbate peroxidases from *Arabidopsis thaliana* to structure, function and evolution of seven types of ascorbate peroxidase', *The Biochemical Journal* **326**, 305–310.

Karpinski, S., Escobar, C., Karpinska, B., Creissen, G. and Mullineaux, P.M. (1997) 'Photosynthetic electron transport regulates the expression of cytosolic ascorbate peroxidase genes in *Arabidopsis* during excess light stress', *The Plant Cell* **9**, 627–640.

Karpinski, S., Reynolds, H., Karpinska, B., Wingsle, G., Creissen, G. and Mullineaux, P. (1999) 'Systemic signaling and acclimation in response to excess excitation energy in *Arabidopsis*', *Science* **284**, 654–657.

Kliebenstein, D.J., Monde, R.A. and Last, R.L. (1998) 'Superoxide dismutase in *Arabidopsis*: an eclectic enzyme family with disparate regulation and protein localization', *Plant Physiology* **118**, 637–650.

Kovtun, Y., Chiu, W.L., Tena, G. and Sheen, J. (2000) 'Functional analysis of oxidative stress-activated mitogen-activated protein kinase cascade in plants', *Proceedings of the National Academy of Sciences USA* **97**, 2940–2945.

Kwak, J.M., Mori, I.C., Pei, Z.M. *et al.* (2003) 'NADPH oxidase AtrbohD and AtrbohF genes function in ROS-dependent ABA signaling in *Arabidopsis*', *EMBO Journal* **22**, 2623–2633.

Mandelman, D., Jamal, J. and Poulos, T.L. (1998a) 'Identification of two electron transfer sites in ascorbate peroxidase', *Biochemistry* **37**, 17610–17617.

Mandelman, D., Sshwarz, F.P., Li, H. and Poulos, T.L. (1998b) 'The role of quaternary interactions on the stability and activity of ascorbate peroxidase', *Protein Science* **7**, 2089–2098.

Mauro, J.M., Fishel, L.A., Hazzard, J.T. *et al.* (1987) 'Tryptophan-191→phenylalanine, a proximal-side mutation in yeast cytochrome c peroxidase that strongly affects the kinetics of ferrocytochrome c oxidation', *Biochemistry* **27**, 6243–6256.

Mittler, R. (2002) 'Oxidative stress, antioxidants and stress tolerance', *Trends in Plant Science* **9**, 405–410.

Mittler, R. and Zilinskas, B. (1991) 'Molecular cloning and nucleotide sequence analysis of a cDNA encoding pea cytosolic ascorbate peroxidase', *FEBS Letters* **289**, 257–259.

Mittler, R. and Zilinskas, B. (1992) 'Molecular cloning and characterization of a gene encoding pea cytosolic ascorbate peroxidase', *The Journal of Biological Chemistry* **267**, 21802–21807.

Mittler, R. and Zilinskas, B. (1994) 'Regulation of pea cytosolic ascorbate peroxidase and other antioxidant enzymes during the progression of drought stress and following recovery from drought', *The Plant Journal* **5**, 397–406.

Mittler, R., Feng, X. and Cohen, M. (1998) 'Post-transcriptional suppression of cytosolic ascorbate peroxidase expression during pathogen-induced programmed cell death in tobacco', *The Plant Cell* **10**, 461–474.

Mittler, R., Hallak-Herr, E., Orvar, B.L. *et al.* (1999) 'Transgenic tobacco plants with reduced capability to detoxify reactive oxygen intermediates are hyper-responsive to pathogen infection', *Proceedings of the National Academy of Sciences USA* **96**, 14165–14170.

Mittler, R., Vanderauwera, S., Gollery, M. and Van Breusegem, F. (2004) 'The reactive oxygen gene network of plants', *Trends in Plant Science* **9**, 490–498.

Mullineaux, P. and Karpinski, S. (2002) 'Signal transduction in response to excess light: getting out of the chloroplast', *Current Opinion in Plant Biology* **5**, 43–48.

Nakano, Y. and Asada, K. (1981) 'Hydrogen peroxide is scavenged by ascorbate-specific peroxidase in spinach chloroplasts', *Plant Cell Physiology* **22**, 867–880.

Nakano, Y. and Asada, K. (1987) 'Purification of ascorbate peroxidase in spinach chloroplasts; its inactivation in ascorbate-depleted medium and reactivation by monodehydroascorbate radical', *Plant Cell Physiology* **28**, 131–140.

Neill, S., Desikan, R. and Hancock, J. (2002) 'Hydrogen peroxide signalling', *Current Opinion in Plant Biology* **5**, 388–395.

Noctor, G. and Foyer, C. (1998) 'Ascorbate and glutathione: keeping active oxygen under control', *Annual Review of Plant Physiology and Plant Molecular Biology* **49**, 249–279.

Obara, K., Sumi, K. and Fukuda, H. (2002) 'The use of multiple transcription starts causes the dual targeting of *Arabidopsis* putative monodehydroascorbate reductase to both mitochondria and chloroplasts', *Plant Cell Physiology* **43**, 697–705.

Orvar, B.L. and Ellis, B.E. (1997) 'Transgenic tobacco plants expressing antisense RNA for cytosolic ascorbate peroxidase show increased susceptibility to ozone injury', *The Plant Journal* **11**, 1297–1305.

Overmyer, K., Brosche, M. and Kangasjarvi, J. (2003) 'Reactive oxygen species and hormonal control of cell death', *Trends in Plant Science* **8**, 335–342.

Panchuk, I.I., Volkov, R.A. and Schoffl, F. (2002) 'Heat stress and heat shock transcription factor-dependent expression and activity of ascorbate peroxidase in *Arabidopsis*', *Plant Physiology* **129**, 838–853.

Pappa, H., Patterson, W.H. and Poulos, T.L. (1996) 'The homologous tryptophan critical for cytochrome c peroxidase function is not essential for ascorbate peroxidase activity', *Journal of Biological Inorganic Chemistry* **1**, 61–66.

Patterson, W.R. and Poulos, T.L. (1994) 'Characterization and crystallization of recombinant pea cytosolic ascorbate peroxidase', *The Journal of Biological Chemistry* **269**, 17020–17024.

Patterson, W.R. and Poulos, T.L. (1995) 'Crystal structure of recombinant pea cytosolic ascorbate peroxidase', *Biochemistry* **34**, 4331–4341.

Patterson, W.R., Poulos, T.L. and Goodin, D.B. (1995) 'Identification of a porphyrin p cation radical in ascorbate peroxidase compound I', *Biochemistry* **34**, 4342–4345.

Pei, Z.-M., Murata, Y., Benning, G. *et al.* (2000) 'Calcium channels activated by hydrogen peroxide mediate abscisic acid signaling in guard cells', *Nature* **406**, 731–734.

Pignocchi, C. and Foyer, C.H. (2003) 'Apoplastic ascorbate metabolism and its role in the regulation of cell signalling', *Current Opinion in Plant Biology* **6**, 379–389.

Pnueli, L., Hongjian, L. and Mittler, R. (2003) 'Growth suppression, abnormal guard cell response, and augmented induction of heat shock proteins in cytosolic ascorbate peroxidase (Apx1) – deficient *Arabidopsis* plants', *The Plant Journal* **34**, 187–203.

Polle, A. (2001) 'Dissecting the superoxide dismutase-ascorbate peroxidase-glutathione pathway in chloroplasts by metabolic modeling. Computer simulations as a step towards flux analysis', *Plant Physiology* **126**, 445–462.

Poulos, T.L. and Kraut, J. (1980) 'The stereochemistry of peroxidase catalysis', *The Journal of Biological Chemistry* **255**, 8199–8205.

Raven, E.L. (2003) 'Understanding functional diversity and substrate specificity in haem peroxidases; what can we learn from ascorbate peroxidase?', *Natural Product Reports* **20**, 367–381.

Rizhsky, L., Davletova, S., Liang, H. and Mittler, R. (2004) 'The zinc-finger protein Zat12 is required for cytosolic ascorbate peroxidase 1 expression during oxidative stress in *Arabidopsis*', *The Journal of Biological Chemistry* **279**, 11736–11743.

Rizhsky, L., Hallak-Herr, E., Van Breusegem, F. *et al.* (2002) 'Double antisense plants with suppressed expression of ascorbate peroxidase and catalase are less sensitive to oxidative stress than single antisense plants with suppressed expression of ascorbate peroxidase or catalase', *The Plant Journal* **32**, 329–342.

Rodriguez-Milla, M.A., Maurer, A., Rodriguez-Huete, A. and Gustafson, J.P. (2003) 'Glutathione peroxidase genes in *Arabidopsis* are ubiquitous and regulated by abiotic stresses through diverse signaling pathways', *The Plant Journal* **36**, 602–615.

Rozen, A., Mittler, R. and Tel-Or, E. (1991) 'A unique ascorbate peroxidase activity in the cyanobacterium Synechococcus PCC 7942 (R2)', *Free Radical Research Communications* **17**, 1–8.

Sharp, K.H., Mewies, M., Moody, P.C. and Raven, E.L. (2003) 'Crystal structure of the ascorbate peroxidase–ascorbate complex', *Nature Structural Biology* **10**, 303–307.

Sharp, K.H., Moody, P.C.E., Brown, K.A. and Raven, E.L. (2004) 'Crystal structure of the ascrobate peroxidase–salicylhydroxamic acid complex', *Biochemistry* **43**, 8644–8651.

Shigeoka, S., Ishikawa, T., Tamoi, M. *et al.* (2002) 'Regulation and function of ascorbate peroxidase isoenzymes', *Journal of Experimental Botany* **53**, 1305–1319.

Shigeoka, S., Nakano, Y. and Kitaoka, S. (1980) 'Purification and some properties of L-ascorbic-acid-specific peroxidase in Euglena gracilis Z', *Archives of Biochemistry and Biophysics* **201**, 121–127.

Sivaraja, M., Goodin, D.B., Smith, M. and Hoffman, B.M. (1989) 'Identification by ENDOR of Trp191 as the free-radical site in cytochrome c peroxidase compound ES', *Science* **245**, 738–740.

Tognolli, M., Penel, C., Greppin, H. and Simon, P. (2002) 'Analysis and expression of the class III peroxidase large gene family in *Arabidopsis thaliana*', *Gene* **288**, 129–138.

Torres, M.A., Dangl, J.L. and Jones, J.D. (2002) '*Arabidopsis* gp91phox homologues AtrbohD and AtrbohF are required for accumulation of reactive oxygen intermediates in the plant defense response', *Proceedings of the National Academy of Sciences USA* **99**, 517–522.

Vandenabeele, S., Vanderauwera, S., Vuylsteke, M. *et al.* (2004) 'Catalase deficiency drastically affects high light-induced gene expression in *Arabidopsis thaliana*', *The Plant Journal* **39**, 45–58.

Wada, K., Tada, T., Nakamura, Y. *et al.* (2003) 'Crystal structure of chloroplastic ascorbate peroxidase from tobacco plants and structural insights into its instability', *The Journal of Biochemistry* **134**, 239–244.

Wang, J., Zhang, H. and Allen, R.D. (1999) 'Overexpression of an *Arabidopsis* peroxisomal ascorbate peroxidase gene in tobacco increases protection against oxidative stress', *Plant Cell Physiology* **40**, 725–732.

Yabuta, Y., Motoki, T., Yoshimura, K., Takeda, T., Ishikawa, T. and Shigeoka, S. (2002) 'Thylakoid membrane-bound ascorbate peroxidase is a limiting factor of antioxidative systems under photo-oxidative stress', *The Plant Journal* **32**, 915–925.

Yoshimura, K., Yabuta, Y., Ishikawa, T. and Shigeoka, S. (2000) 'Expression of spinach ascorbate peroxidase isoenzymes in response to oxidative stresses', *Plant Physiology* **123**, 223–234.

Zamocky, M., Janecek, S. and Koller, F. (2000) 'Common phylogeny of catalase-peroxidases and ascorbate peroxidases', *Gene* **256**, 169–282.

5 Catalases in plants: molecular and functional properties and role in stress defence

Jürgen Feierabend

5.1 Introduction

The enzyme catalase converts hydrogen peroxide (H_2O_2) to oxygen and water:

$$2H_2O_2 \rightarrow 2H_2O + O_2$$

Catalases are ubiquitous enzymes among aerobic organisms. H_2O_2 may be generated by several oxidases transferring two electrons to O_2, or by the disproportionation of O_2^- after a univalent reduction of O_2. Superoxide and H_2O_2 are produced in the photosynthetic and respiratory electron transport chains of chloroplasts or mitochondria, and by flavin-containing oxidases. Photosynthesizing plant cells show particularly high rates of H_2O_2 production. The rates of H_2O_2 production were estimated to account for 80–160 $\mu M\ s^{-1}$ in chloroplasts (Asada, 1992) or for up to 10 $\mu mol\ m^{-2}\ s^{-1}$ in photosynthesing leaves (Foyer & Noctor, 2003). H_2O_2 is a toxic reactive oxygen species (ROS) oxidizing cysteine and methionine residues of proteins. Oxidation of cysteines leads to the formation of disulfide bonds. Methionine is oxidized to methionine sulfoxide (Levine et al., 1996). Photosynthesis is rapidly inactivated by low concentrations of H_2O_2 because CO_2 fixation is inhibited (Asada, 1992, 1994). A concentration of 10 μM H_2O_2 caused a 50% inhibition of CO_2 assimilation in isolated chloroplasts (Kaiser, 1976). Sensitive targets of inhibition by H_2O_2 are thiol-containing and thioredoxin-regulated Calvin cycle enzymes, such as NADP-dependent glyceraldehyde-3-phosphate dehydrogenase, fructose-1,6-bisphosphatase, sedoheptulose-1,7-bisphosphatase or phosphoribulokinase (Jacquot et al., 1997). H_2O_2 is a less reactive oxidant when compared to other ROS. Nevertheless, H_2O_2 may be particularly harmful, because it is relatively stable and may therefore spread within or among cells by diffusion. H_2O_2 can give rise to more ROS, which greatly increases its cytotoxicity. In the presence of transition metals, such as Fe^{2+} or Cu^{1+}, the extremely reactive hydroxyl radical can be formed in the Fenton reaction:

$$M^{n+} + H_2O_2 \rightarrow M^{(n+1)+} + {}^{\cdot}OH + OH^-$$

The resulting Fe^{3+} or Cu^{2+} can be re-reduced by O_2^- in the Haber–Weiss reaction so that Fe^{2+} for the Fenton reaction is recycled. The simultaneous presence of H_2O_2, O_2^- and Fe^{3+} is therefore a dangerous source of $^{\cdot}OH$-formation.

$$O_2^- + H_2O_2 \xrightarrow{Fe^{3+}} O_2 + {}^{\cdot}OH + OH^-$$

Due to its direct and indirect cytotoxicity, survival of plants depends on the efficient scavenging of H_2O_2. Except for catalases, additional enzymes catalyzing the removal of H_2O_2 occur in plants. Peroxidases require a reductant for the detoxification of H_2O_2.

$$H_2O_2 + R\text{-}H_2 \rightarrow 2H_2O + R$$

Ascorbate peroxidases occur in the chloroplasts, the cytosol and in peroxisomes, and depend on the re-reduction of ascorbate by glutathione, reduced ferredoxin or NADPH (Chapter 4; Shigeoka et al., 2002). Glutathione peroxidases, which are common in animals, appear to play no major role in the detoxification of H_2O_2 (Eshdat et al., 1997). Peroxiredoxins (see Chapter 2) catalyze the reduction of both alkyl hydroperoxides and H_2O_2 with thioredoxin or glutaredoxin and occur in various compartments of the plant cell (Rouhier & Jacquot, 2002). Cyclic oxidation of surface-exposed methionine residues of proteins and their subsequent re-reduction with thioredoxin by peptide methionine sulfoxide reductase may also contribute to the removal of H_2O_2 (Levine et al., 1996). Peptide methionine sulfoxide reductases occur both in the chloroplasts and in the cytosol of plant cells (Sadanandom et al., 2000).

Relative to the alternative H_2O_2-scavenging systems, catalases are distinguished by very high turnover numbers but rather low affinities toward H_2O_2 (Nicholls et al., 2001). Consequently, catalases provide very efficient tools for the gross removal and control of high H_2O_2 levels, but they are less suited for a fine tuning of sensitive redox balances with low H_2O_2 concentrations that may be important for regulatory mechanisms. A major advantage of catalases is that they do not depend on any additional reductant for the scavenging of H_2O_2.

Comprehensive reviews covering the biochemistry and structure of catalases (Nicholls & Schonbaum, 1963; Bravo et al., 1997; Zámocký & Koller, 1999; Nicholls et al., 2001; Chelikani et al., 2004) and the occurrence and properties of plant catalases (Scandalios, 1994; Willekens et al., 1995; Scandalios et al., 1997; Heinze & Gerhardt, 2002) have appeared. This chapter will concentrate on molecular, functional and regulatory aspects of plant catalases and their roles in stress defence.

5.2   Biochemistry and molecular structure of catalases

5.2.1   Types of catalases

Three major groups of phylogenetically unrelated catalases have been described:

1. The 'typical' monofunctional catalases
2. The bifunctional catalase-peroxidases
3. The non-heme Mn-containing catalases

Low catalase activity was furthermore found to be associated with several other heme-containing proteins but may be mainly due to the fact that the heme alone is able to catalyze a very slow catalase reaction (Nicholls *et al.*, 2001; Zámocký *et al.*, 2001). Non-heme catalases have been identified in *Lactobacillus plantarum* and in two thermophilic bacteria. Their active site contains a manganese-rich reaction center, whereas the other two classes of catalases contain heme groups (Zámocký & Koller, 1999; Nicholls *et al.*, 2001). Catalase-peroxidases (KatG genes) are phylogenetically related to eukaryotic ascorbate peroxidases and to cytochrome c peroxidase and occur in archaebacteria, eubacteria including cyanobacteria, and some fungi (Nicholls *et al.*, 2001; Zámocký *et al.*, 2001). They are, in general, dimers with molecular weights of 120–340 kDa.

The most widespread and most extensively studied group of catalases is the classical monofunctional catalases. They are generally tetrameric enzymes containing four ferric protoporphyrin IX prosthetic groups. Based on their phylogenetic relationships, three main clades of monofunctional catalases have been discriminated (Nicholls *et al.*, 2001). Clade I includes the small subunit (55–60 kDa) enzymes from plants and a branch of bacterial catalases, such as CatF of *Pseudomonas syringae* (Carpena *et al.*, 2003). Clade II contains large subunit (78–84 kDa) catalases from fungi and a second bacterial branch (e.g. *Escherichia coli* hydroperoxidase II HPII). Clade III contains small subunit enzymes from archaebacteria, a third eubacterial subset (e.g. *Proteus mirabilis* catalase PMC), fungi (e.g. *Saccharomyces cerevisiae* catalase A) and animals (e.g. bovine liver catalase BLC).

5.2.2 Molecular structure

Sequences for more than 300 catalases are now available. At present, sequences or sequence fragments for about 130 plant catalases from about 80 different plant species are available in the database, mostly from cDNA or genomic DNA sequences. Crystal structures for a single clade I catalase (CatF of *P. syringae*), 2 clade II catalases (PVC from *Penicillium vitale* and HPII from *E. coli*) and 5 clade III catalases (BLC, HEC from human erythrocytes, PMC, MLC from *Micrococcus lysodeikticus* and CATA from *S. cerevisiae*) have been determined (see Bravo *et al.*, 1997; Nicholls *et al.*, 2001; Zámocký & Koller, 2001; Carpena *et al.*, 2003). Crystal structures for plant catalases have not been determined. Comparisons of the available crystal structures revealed, however, a high degree of conservation of tertiary structures, both among the small subunit catalases of clade I and III and among the small subunit catalases and the core regions of the large subunit catalases of clade II. It is to be expected that structures of plant catalases will also closely resemble those that have been determined up to now since sequence positions of fundamental functional or structural significance are strictly conserved in plant catalases also (Figure 5.1).

A single subunit of a tetrameric small subunit catalase shows a globular structure carrying an extended amino-terminal arm (Figure 5.2). The central core structure of the subunit is formed by an 8-stranded β-barrel domain of about 250 residues, which is preceded by the amino-terminal arm of about 70 amino acids (for details, see Murthy *et al.*, 1981; Fita & Rossmann, 1985a; Gouet *et al.*, 1995;

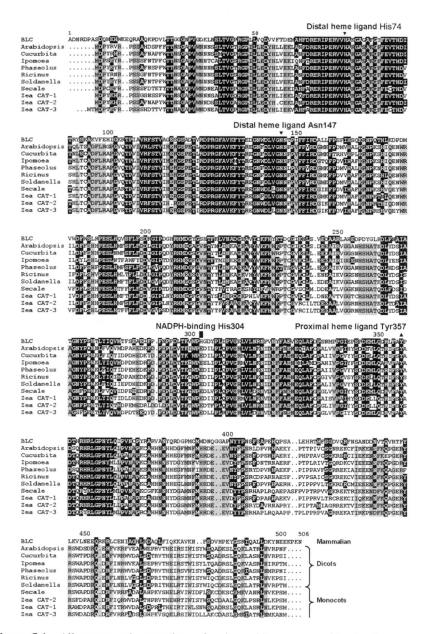

Figure 5.1 Alignment and comparison of amino acid sequences of bovine liver catalase (BLC) and catalases from several dicotyledonous and monocotyledonous plants. Except for maize, for all other plant species sequences of isozymes designated as CAT-1 are shown. Amino acids of plant catalases that are identical to the corresponding positions in BLC are marked in black. Amino acids that differ from BLC but are conserved among all plant catalases shown are marked in grey. The conserved distal heme-binding amino acids His74 and Asn147 (▼; numbering of BLC) and the proximal ligand of the heme-iron Tyr357 (▲) are marked. His304 (■) which is involved in the binding of NADPH in NADPH-binding catalases, like BLC, is replaced by Glu in the plant catalases.

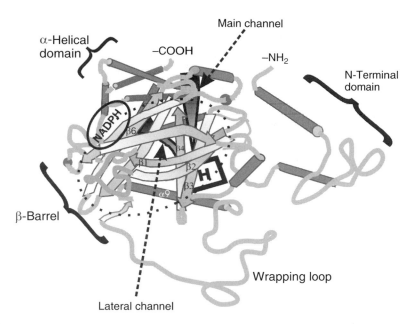

Figure 5.2 Simplified model of the tertiary structure of a single subunit of a small subunit catalase showing its principal domains, the locations of the binding sites for the cofactors heme (H) and NADPH (in NADPH-binding catalases only), and of the main (perpendicular) and minor (lateral) channels leading to the central heme cavity within the central globular β-barrel structure. Adapted from Fita and Rossmann (1985a) and Nicholls *et al.* (2001).

Bravo *et al.*, 1997; Zámocký and Koller, 1999; Nicholls *et al.*, 2001). The first subset of 4 β-strands is separated from the second subset of 4 β-strands by 3 α-helices. The globular β-barrel domain is followed by about 110 residues with little secondary structure, designated as the wrapping loop, which links the β-barrel to a carboxy-terminal α-helical domain of about 60 residues. The α-helical domain docks to the surface of the β-barrel structure, interacting with the 3 α-helices (α3–α5), which separate the 2 subsets of β-strands in the β-barrel domain. The large subunit catalases have an extended carboxy-terminal domain. In the quarternary structure of the holoenzyme, the subunits are interwoven by the interaction of the amino-terminal arm of one subunit with the wrapping loop of an adjacent subunit.

The heme, a protoporphyrin IX (heme b), is deeply buried within the core structure of the enzyme, at least 20 Å apart from the protein surface. It is not covalently bound within a pocket of the protein. The distal side of the heme-binding cavity is formed by the first half of the central β-barrel (β1–β4). A histidine (His74 of BLC on the β1-strand) and an asparagine residue (Asn147 of BLC on the β4-strand) are essential for the interaction with the heme on its distal side. These essential amino acids and the surrounding sequence regions are also strictly conserved among plant catalases (Figure 5.1). The proximal side of the heme-binding pocket is bound by the helix α9 of the wrapping loop domain and residues from the amino-terminal arm of a neighboring subunit. The tyrosine (Tyr 357 of BLC on α9) acting as proximal ligand of the heme-iron (Bravo *et al.*, 1997) is conserved in plant catalases (Figure 5.1). Two channels lead from the molecular surface to the distal side of the heme pocket, a major or perpendicular channel of about 30 Å and a minor or lateral channel of about 13–19 Å (Figure 5.2). The perpendicular channel is regarded as the main substrate channel providing access for the H_2O_2 to the active center. The role of the minor channel is less clear – whether it serves as an additional route for H_2O_2 to the active center or as exit for reaction products.

Small subunit catalases of clade III were found to bind one NADPH per subunit as second cofactor. Binding of NADPH was observed in bacterial (*Micrococcus luteus* and *P. mirabilis* catalases), yeast (SCC-A and SCC-T) and mammalian (BLC) catalases (Fita & Rossmann, 1985b; Zámocký & Koller, 1999; Nicholls *et al.*, 2001). The NADPH binds to a peripheral fold of the enzyme at the entrance of the minor channel (Figure 5.2). Surface-exposed portions of the second subset of four strands of the β-barrel domain and portions (α10) of the α-helical domain participate in the binding of NADPH. A histidine residue (His304 of BLC) interacting with the pyrophosphate group and the nicotinamide portion of the NADPH appears to be most essential and is strictly conserved among strongly NADPH-binding catalases. Yeast catalases have a glutamine instead of the histidine and are only weakly binding NADPH. Catalase from tobacco was reported to bind NADPH (Durner & Klessig, 1996). However, increasing evidence indicates that plant catalases are unable to bind NADPH. Catalases from potato tubers (Beaumont *et al.*, 1990) and from leaves of *Secale cereale* or *Homogyne alpina* (unpublished results) did not bind NADPH. Furthermore, crucial amino acids involved in NADPH binding, such as the NADPH-binding histidine, are not conserved in plant catalase sequences. The position corresponding to His 304 of BLC is replaced by glutamic acid, or, in a few instances, by aspartic acid in plant catalases (see Figure 5.1). Glutamic acid is regarded as unsuitable for NADPH binding (Zámocký & Koller, 1999; Carpena *et al.*, 2003). Plant catalases belong to the small subunit enzymes of clade I. The only catalase of clade I for which a crystal structure was determined, CatF of *P. syringae*, was shown to be unable to bind NADPH, because key residues required for nucleotide binding were missing (Carpena *et al.*, 2003). These observations further suggest that clade I catalases, including those of plants, are presumably unable to bind NADPH. The biochemical function of NADPH as cofactor of catalases is not yet sufficiently clarified. It is involved in redox reactions with the heme and protects the enzyme from inactivation.

5.2.3 Mechanism of the catalytic reaction and kinetic properties

The basic mechanism of the catalase reaction has been elaborated with mammalian and bacterial catalases (Nicholls & Schonbaum, 1963; Deissenroth & Dounce, 1970; Dounce, 1983; Nicholls *et al.*, 2001). Reaction intermediates were identified by specific spectral changes. The catalytic reaction was shown to proceed in two steps. In a first step, the native or resting enzyme (ferri-catalase) is oxidized by

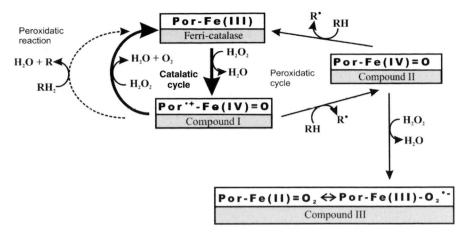

Figure 5.3   Reaction scheme illustrating reactions and intermediate forms of catalase. The intermediate compound I can, in addition to H_2O_2, also oxidize two- or one-electron donating peroxidatic substrates. With one-electron donating peroxidatic substrates, the inactive (for reaction with H_2O_2) intermediate compound II is formed, which can further react with H_2O_2 to form the inactive compound III. Compound III can spontaneously decompose to native ferri-catalase and $O_2^{\cdot-}$ or undergo irreversible inactivation.

a first molecule of H_2O_2, resulting in the formation of an intermediate form, designated as compound I (Figure 5.3). In the two electron oxidation, one electron is abstracted from the heme iron with the formation of an oxoferryl group $Fe(IV)=O$ and the second electron is abstracted from the porphyrin moiety of the heme forming a π-cationic radical. In a second step, a second H_2O_2 serves as donor of two electrons to reduce compound I. Ferri-catalase is regenerated, water and O_2 are formed.

As strong oxidant, compound I may also react with other electron donors if they are accessible and able to compete with H_2O_2. At low H_2O_2 concentrations, catalase can, therefore, exert a peroxidatic activity. Compound I is directly reduced to ferri-catalase by two electron donating peroxidatic substrates, such as short-chain alcohols. One-electron donors (e.g. phenolic compounds, $O_2^{\cdot-}$ or endogenous donors) lead to the formation of compound II (Figure 5.3), which is an inactive form of catalase since it cannot react with H_2O_2. However, the inhibition can be reversed by a second one-electron transfer from a peroxidatic substrate, resulting in the regeneration of ferri-catalase. At high H_2O_2 concentrations, compound II is converted to compound III (Lardinois *et al.*, 1996) which is also designated as oxycatalase [oxyferrous catalase: $Por-Fe(II)=O_2$]. Compound III can also be formed by the reaction of the native ferri-catalase with $O_2^{\cdot-}$ (Shimizu *et al.*, 1984). Compound III is an inactive form. In the absence of H_2O_2 or $O_2^{\cdot-}$, it can revert spontaneously to the active native enzyme, accompanied by the liberation of $O_2^{\cdot-}$. Compound III is rather unstable (Shimizu *et al.*, 1988). The prolonged presence of compound III results in the irreversible inhibition of catalase. This leads to the inactivation of catalase in the presence of $O_2^{\cdot-}$ (Shimizu *et al.*, 1984) or of high concentrations of H_2O_2. Mammalian catalases can be protected by the presence of NADPH against inactivation by H_2O_2 or $O_2^{\cdot-}$. NADPH was found to decrease the rate of compound II formation and to increase the rate of its removal. The exact mode of action of the two-electron donor NADPH is, however, still a matter of debate (Kirkman et al., 1999; Zámocký & Koller, 1999; Nicholls et al., 2001). While a transfer of electrons from NADPH to heme must be postulated, the heme is not directly accessible to NADPH, because the latter is bound in a groove on the surface of the enzyme (Fita & Rossmann, 1985b) and is also too large for the substrate channel. Mechanisms of electron tunneling between the two cofactors have been discussed (Zámocký & Koller, 1999; Nicholls et al., 2001).

The affinity of catalases for H_2O_2 is very low. In spite of high homologies between catalase sequences and a highly conserved structural organization, kinetic properties showed a striking range of variation. In a comparison of purified catalases from 16 different organisms (mammalian, fungal and bacterial enzymes) with representatives of all three clades of monofunctional catalases apparent K_m-values between 38 and 600 mM were observed (Switala & Loewen, 2002). For plant enzymes, apparent K_m-values between about 10 and 140 mM have been reported (Scandalios, 1994; Heinze & Gerhardt, 2002). However, the apparent K_m for catalases does not correspond to the standard Michaelis–Menten parameter, because the catalatic reaction comprises two partial reactions that depend on H_2O_2, both the formation and the reduction of compound I (see Figure 5.3). The catalatic reaction does not follow Michaelis–Menten kinetics but the activity increases linearly with the H_2O_2 concentration without saturation. Therefore, catalase activity is controlled by the existing H_2O_2 concentration. Apparent saturation may be caused by an inhibition at high H_2O_2 concentrations (Beaumont et al., 1990; Lardinois et al., 1996; Nicholls et al., 2001). Catalase is known for its remarkably high turnover number and, consequently, high specific activities were observed. For mammalian catalases, turnover rates at maximal H_2O_2 concentrations were extrapolated to range between 2×10^6 and 2×10^7 s^{-1} (Nicholls et al., 2001). A turnover number of 11 200 s^{-1} was reported for potato tuber catalase (Beaumont et al., 1990). Specific activities of 16 bacterial, fungal or mammalian catalases varied between 345 and 4563 $\mu mol\ s^{-1}\ mg^{-1}$ protein (Switala & Loewen, 2002). Specific activities of plant catalases varied between 10 and 500 $\mu mol\ s^{-1}\ mg^{-1}$ protein (Scandalios, 1994; Heinze & Gerhardt, 2002). Sensitivities to inhibitors varied greatly among even closely related catalases. The effective concentrations of classical catalase inhibitors, such as cyanide, azide or 3-amino-triazole varied over an up to 1000-fold concentration range (Scandalios, 1994; Switala & Loewen, 2002).

5.3   Occurrence and properties of plant catalases

5.3.1   Sources of H_2O_2 production in plant cells

In photosynthesizing plant tissues, the highest rates of H_2O_2-production are observed in the chloroplasts and in leaf peroxisomes. In chloroplasts, a direct electron transfer from photosystem I to O_2 in the Mehler-peroxidase reaction produces $O_2^{\cdot-}$ which is converted to H_2O_2 by superoxide dismutases. Under conditions of impaired CO_2-fixation, the Mehler-peroxidase reaction was estimated to account for up to 10% of the photosynthetic electron flow (Asada, 1994; Foyer, 1997). In photosynthesizing C_3-plants, H_2O_2 is furthermore produced in the photorespiratory pathway by the oxidation of glycolate in the peroxisomes. According to Foyer and Noctor (2003), H_2O_2-production by photorespiration greatly exceeds that resulting from the Mehler-peroxidase reaction.

In non-photosynthetic tissues, the microbodies are either differentiated as 'non-specialized' peroxisomes, e.g. in roots, or as glyoxysomes in fat-storing tissues (Huang et al., 1983). Glyoxysomes, which house the reactions of the β-oxidation of fatty acids and of the glyoxylate cycle, are a major source of H_2O_2 formation. In the course of fatty acid degradation, O_2 is reduced to H_2O_2 by the FAD-containing acetyl-CoA oxidase. Non-specialized peroxisomes, e.g. of roots, and leaf peroxisomes are also able to degrade fatty acids, and β-oxidation appears to be a basic function of all higher plant peroxisomes (Gerhardt, 1992). Urate oxidase is an additional H_2O_2-generating reaction of all peroxisomes (Huang et al., 1983). Furthermore, peroxisomes were shown to produce $O_2^{\cdot-}$ which is subsequently converted to H_2O_2 by superoxide dismutases (del Rio et al., 2002).

The mitochondrial respiratory chain is a source of $O_2^{\cdot-}$ formation, particularly when the normal electron flow through the cytochrome pathway is impaired. H_2O_2 is formed from $O_2^{\cdot-}$ by a Mn-containing mitochondrial superoxide dismutase. The formation of $O_2^{\cdot-}$ is thought to result mainly from ubisemiquinone formation at complex III (ubiquinol–cytochrome c oxidoreductase). Ubisemiquinone autoxidizes in the presence of O_2 and thus gives rise to $O_2^{\cdot-}$ (Raha & Robinson, 2000). In plants, $O_2^{\cdot-}$ formation was enhanced when the alternative oxidase was inhibited. The alternative oxidase is therefore regarded as a protective mechanism avoiding $O_2^{\cdot-}$ production in mitochondria (Wagner & Krab, 1995; Purvis, 1997; Mittler, 2002).

H_2O_2 is produced in the apoplastic cell wall space (see Chapter 9). It plays an essential role in the lignification reaction. In response to a pathogen attack, extracellular H_2O_2 is accumulated during the oxidative burst. Depending on the plant species, the apoplastic H_2O_2 is produced from $O_2^{\cdot-}$ generated by a membrane-localized NADPH oxidase or by an extracellular peroxidase (Lamb & Dixon, 1997; Bolwell et al., 1998). Roots of several seedlings were shown to excrete H_2O_2 into the surrounding medium (Frahry & Schopfer, 1998).

5.3.2   Occurrence and subcellular localization of plant catalases

In higher plants, catalases are predominantly or exclusively localized in peroxisomes. All types of plant peroxisomes, leaf peroxisomes, glyoxysomes and non-specialized peroxisomes contain catalase activity. Association of catalase with peroxisomes was demonstrated by subcellular fractionation with organelle separations on sucrose or Percoll gradients and by electron microscopical detection after cytochemical staining for catalase activity with 3,3′-diaminobenzidine or after immunogold labeling (e.g. Frederick & Newcomb, 1969; Huang et al., 1983;

Holtman *et al.*, 1993; Willekens *et al.*, 1995; Corpas *et al.*, 1999). Frequently, peroxisomes contain amorphous or crystalline inclusions, designated as cores (see Huang *et al.*, 1983). Immunogold labeling and biochemical characterization indicated that catalase represented the predominant protein in cores of sunflower cotyledons (Tenberge & Eising, 1995; Kleff *et al.*, 1997). Several groups of algae do not contain catalase activity. Catalase was not detected in *Euglena*, the Prasinophyceae and in diatoms (Bacillariophyceae) (Stabenau *et al.*, 1989; Igamberdiev & Lea, 2002). Instead of the H_2O_2-generating peroxisomal glycolate oxidase, these algae contain glycolate dehydrogenase, which is localized in mitochondria. Heterokonts contain both a peroxisomal glycolate oxidase and catalase but no other leaf peroxisomal enzymes (Gross *et al.*, 1985). Leaf type peroxisomes with glycolate oxidase and catalase were only found in Charophyceae, such as *Chara*, and in species of the subgroup of Zygnematales, such as *Spirogyra* (see Stabenau *et al.*, 2003).

Non-peroxisomal localizations of catalases are occasionally observed. The maize catalase isozyme, CAT-3, was reported to be associated with mitochondria (Scandalios *et al.*, 1980). Subcellular fractionation studies indicated that the catalase of the green alga *Chlamydomonas* was localized in the mitochondria (Kato *et al.*, 1997). Similarly, a mitochondrial catalase was described in the yeast *S. cerevisiae* that appeared to differ from the cytosolic yeast catalase T and the peroxisomal catalase A (Petrova *et al.*, 2002). Although no catalase activity was detected in the chloroplasts after cytochemical staining of leaf sections with 3,3'-diaminobenzidine (Frederick & Newcomb, 1969) and no appreciable catalase activity was associated with properly purified chloroplasts after subcellular fractionation (e.g. Biekmann & Feierabend, 1982; Huang *et al.*, 1983), Sheptovitsky and Brudvig (1996) described and purified a catalase that was associated with a photosystem-II-containing grana fraction from spinach thylakoids. Catalase activity was found in the apoplastic space of maize roots and secreted into the surrounding medium (Salguero & Böttger, 1995). The pathogenic fungus *Claviceps purpurea* excretes a catalase, CPCAT1, to the extracellular space. The secreted catalase could play a role in the self-defence of the fungus by degradation of the H_2O_2 which is produced in the apoplastic space of the host plant during the oxidative burst. Deletion of the *cpcat1* gene showed, however, no reduction of their virulence (Garre *et al.*, 1998). An apoplastic catalase was also detected by immunofluorescence and immunogold electron microscopy in the wall thickenings of differentiating xylem cells of sunflower cotyledons (Eising *et al.*, 1998).

5.3.3   Properties of plant catalases

Catalases have been purified from various plant tissues, such as leaves, cotyledons, roots and endosperm (for a more comprehensive comparison, see Heinze & Gerhardt, 2002). In general, they exhibit the properties of small subunit monofunctional catalases, as pointed out in Sections 5.2.2 and 5.2.3. The native enzymes are tetrameric heme-containing proteins with molecular weights close to 240 kDa. Rare exceptions have been described. Hirasawa *et al.* (1987) purified an unusual green catalase from spinach leaves with a molecular weight of 125 kDa and subunit molecular weights of 55 kDa, suggesting that the native enzyme was a dimer. Except for one normal heme b, this spinach catalase appeared to contain one modified heme which was, apparently, responsible for the green color. Another unusual catalase was isolated from photosystem-II-enriched thylakoid fragments of spinach chloroplasts. The native thylakoid-associated catalase appeared to be a dimer of 130 kDa (Sheptovitsky & Brudvig, 1996). Similarly, the mitochondrial catalase of *Chlamydomonas reinhardtii* was isolated as a dimer of 120 kDa (Kato *et al.*, 1997).

While the peroxidatic activity, measured at low H_2O_2 concentrations, is usually much lower than the catalatic activity (below 1%), specific catalase isozymes were separated from leaf extracts of tobacco, barley or maize that exhibited greatly enhanced peroxidatic activity. The peroxidatic activity of the maize isozyme, CAT-3, accounted for 10–20% of its catalatic activity (Havir & McHale, 1989). Plant catalases are inhibited by typical catalase inhibitors, such as cyanide, azide or 2-aminotriazole, although the effective concentrations may vary considerably between enzymes from different species or between different isozymes (Scandalios, 1994). Furthermore, plant catalases are reversibly inhibited by micromolar concentrations of nitrite (Streb *et al.*, 1993) and by millimolar concentrations of sulfite (Streb *et al.*, 1993; Veljovic-Jovanovic *et al.*, 1999). Catalases, not only those of plants, are inhibited by high concentrations (>100 μM) of salicylic acid, while low concentrations can also protect the enzyme (Rüffer *et al.*, 1995; Durner & Klessig, 1996). Many catalases, including fungal, mammalian and plant catalases, were shown to interact with singlet oxygen. Singlet oxygen modified the heme and gave rise to shifts of the electrophoretic mobility of the native enzymes, thus imitating the behavior of isozymes (Lledías *et al.*, 1998).

Catalases were shown to interact with other proteins or factors. In peroxisomes of spinach leaves, catalase is contained in a multi-enzyme complex allowing rapid metabolic channeling (Heupel & Heldt, 1994). Yang and Poovaiah (2002) demonstrated that catalases from *Arabidopsis* and tobacco were able to bind calmodulin and that the activity of the tobacco catalase was increased in its presence. With the aid of a yeast two-hybrid screening, Fukamatsu *et al.* (2003) detected an interaction of the nucleoside diphosphate kinase, NDK1, with three catalases of *Arabidopsis*. In non-denaturing two-dimensional electrophoretic separations of crude extracts, catalase and NDK1 activities were comigrating. This interaction is regarded as part of a cellular ROS signaling system.

5.3.4 Multiple forms, gene families and gene evolution

A peculiar feature of plants is that they generally contain multiple forms of catalases and small families of catalase genes. When subunit polypeptides coded by distinct genes are simultaneously expressed in the same cell and differ either in charge or molecular weight, heterotetramers of catalase may be formed that can be discriminated by either electrophoretic separation or isoelectric focusing. When heterotetramers are formed from two types of distinct subunits, five isoenzymes

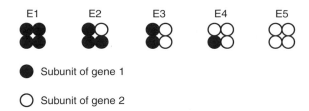

Figure 5.4   Scheme to illustrate the generation of two homotetrameric and three heterotetrameric catalase isozymes from two types of subunit polypeptides encoded by two distinct genes. When both genes equally contribute to the tetrameric catalase composition, a frequency of the five resulting isozymes E1 : E2 : E3 : E4 : E5 of 1 : 4 : 6 : 4 : 1 is expected.

may arise, as illustrated in Figure 5.4. The occurrence of multiple forms was first demonstrated and analyzed by the pioneering work of Scandalios for the maize catalase isozyme system (for references, see Scandalios, 1994; Scandalios *et al.*, 1997). Maize contains three distinct and unlinked nuclear catalase genes, *cat1*, *cat2* and *cat3*. The homotetrameric isozymes CAT-1, CAT-2 and CAT-3 differ in their molecular and biochemical properties (Scandalios, 1994). They also differ in their cell- or organ-specific and temporal patterns of expression and are differentially affected by environmental factors. CAT-1 is the only isozyme expressed in mature pollen and in tissues (milky endosperm, aleurone, scutellum) of the immature developing seed. During germination, the expression of CAT-1 declines in the scutellum and increasing levels of CAT-2 appear. Both isozymes occur in glyoxysomes. During the phase of germination when both catalase genes are expressed, heterotetrameric enzymes with a mixed subunit composition occur and result in the appearance of up to five isozymes (see Figure 5.4). In coleoptiles and etiolated leaves, CAT-1 and CAT-3 are expressed. During greening, CAT-2 is rapidly accumulating in leaves and its formation depends on light. The three isozymes differ in their tissue-specific distribution in mature green leaves. CAT-1 and CAT-3 occur in mesophyll cells. CAT-2 occurs in the peroxisomes of bundle-sheath cells and is assumed to be engaged in photorespiration. The function of CAT-3 is still enigmatic. It differs markedly in its biochemical properties and regulation from the other two catalases and was found to co-purify with mitochondria. CAT-3 occurred only as a homotetramer. Even when expressed in the same cells, heterotetramers between CAT-3 subunits and subunits of the other two-maize catalases were not observed.

Two or three catalase isozyme genes were also identified in other monocotyledonous plants, such as rice (Iwamoto *et al.*, 1999, 2000), barley (Skadsen *et al.*, 1995) or rye (Schmidt & Feierabend, 2000; Schmidt *et al.*, 2002). CAT-1 is the only isozyme detectable in mature rye leaves. CAT-2 is the only isozyme expressed in roots of rye. CAT-2, but not CAT-1, is also expressed in the scutellum of germinating rye seedlings. Immunoblots indicated that a third, yet unidentified, catalase gene is expressed in the scutellum. Among dicotyledonous plants, *Arabidopsis* contains three catalase genes that give rise to at least six isozyme species (Frugoli *et al.*,

1996; McClung, 1997). CAT-1 appeared in all organs, CAT-2 accumulated rapidly after imbibition of seeds and its abundance was increased by light. CAT-2 and CAT-3 were highly expressed in mature leaves, however, CAT-3 was negatively affected by light. In *Nicotiana plumbaginifolia*, three catalase genes were identified (Willekens *et al.*, 1994). *Cat1* is closely homologous to *cat2* (SU2) of cotton and *cat3* is closely homologous to *cat1* (SU1) of cotton (see phylogenetic tree in Frugoli *et al.*, 1998). The expression of both tobacco *cat1* and cotton *cat2* appeared to be associated with a peroxisomal function since the tobacco gene was mainly expressed in leaves and the cotton CAT-2 appeared in the cotyledons only at later stages of germination and depended on light. By contrast, the expression patterns of *N. plumbaginifolia cat3* and cotton *cat1* were consistent with a glyoxysomal function, inasmuch as the mRNA of the tobacco *cat3* accumulated preferentially in seeds and the expression of the cotton *cat1* predominated during the early phase of postgerminative growth of cotton seedlings (Ni & Trelease, 1991a; Willekens *et al.*, 1994). The third *N. plumbaginifolia* gene, *cat2*, was uniformly expressed in different organs with a preference for vascular tissues. Three catalase genes were also identified by their cDNAs in pumpkin (Esaka *et al.*, 1997). *Cat1* mRNA was highly expressed in seeds and early seedling stages and its protein was immunodetected in glyoxysomes. The *cat2* mRNA was highly expressed in green tissues of light-grown pumpkin plants, as expected for a role in photorespiration. *Cat3* expression was constitutive in mature tissues and not dependent on light. It was abundant in green and etiolated cotyledons and roots, but not in young leaves. The mRNAs of two tightly linked catalase genes of castor bean were both expressed in the endosperm, cotyledons and roots of germinating seedlings, but not in mature green leaves (Suzuki *et al.*, 1994). In sunflower cotyledons, cDNA sequences of four catalase genes were determined and enumerated *cat1–cat4*. Subunit polypeptides of CAT-1 and CAT-2 with an apparent molecular mass of 55 kDa formed soluble catalases of the peroxisomal matrix, whereas catalases containing CAT-3 and CAT-4 subunits with an apparent molecular mass of 59 kDa were localized in the crystalline cores of the peroxisomes. By heterotetramer formation between CAT-1 and CAT-2 or CAT-3 and CAT-4, respectively, five matrix catalase isozymes and up to five core catalase isozymes were generated (see Heinze & Gerhardt, 2002). At least ten different catalase isoenzymes were observed in developing mustard cotyledons. Three occurred in dark-grown cotyledons and at least seven appeared only after exposure to light (Drumm & Schopfer, 1974). However, the responsible genes were not yet determined. Corpas *et al.* (1999) discriminated five isoforms of the peroxisomal catalase of pea leaves.

Willekens *et al.* (1995) suggested a classification of plant catalases based on their expression patterns. Class I catalases are those that are expressed in photosynthetic tissues, engaged in photorespiration and positively regulated by light. Class III catalases are expressed in seeds and young seedlings and thought to be responsible for the removal of H_2O_2 during fatty acid degradation in glyoxysomes. Class II includes catalases with presently unclear functions that are expressed in vascular tissue, like *cat2* of *N. plumbaginifolia*, or exhibit unusual properties, like *cat3* of maize. Although the catalase genes of many plants can be assigned to classes I and III, the observed patterns of expression are often more variable and the proposed classes cannot be regarded as three separate clades of phylogenetically related cata- lases. For instance, both CAT-1 and CAT-2 of maize occur in glyoxysomes. Although CAT-1 of rye, based on its sequence homology, is most closely related to CAT-3 of maize (see Frugoli *et al.*, 1998), it represents the only major isozyme of mature leaves, is localized in peroxisomes and related to photorespiration. Evolutionary relationships among plant catalases were analyzed by phylogenetic trees, based on sequence homologies and on comparisons of the exon–intron struc- tures of catalase genes (Frugoli *et al.*, 1998; Iwamoto *et al.*, 1999). It was concluded that all plant catalases were derived from a single putative ancestral catalase gene which had seven introns. Two consecutive gene duplication events gave rise to three (or even four in some dicots) catalase genes followed by a differential loss of introns. The first duplication of the premordial catalase gene of *Arabidopsis*, pre- sumably, generated *cat1* and *cat3*, and *cat2* resulted from a duplication of *cat1*. *Arabidopsis cat2* still contains seven introns whereas *cat1* and *cat3* contain only six, but each lost a different intron. In monocot catalase genes, such as *Zea mays cat3*, up to five of the putative original introns were lost. Following this intron loss, rice *cat1* catalase acquired an additional intron at a novel position.

5.4   Biogenesis and control of expression

5.4.1   *Biosynthesis and import into peroxisomes*

Catalases are encoded by nuclear genes and synthesized on cytosolic 80S ribo- somes. The translation *in vitro* of catalases A and T of *S. cerevisiae* (Hamilton *et al.*, 1982) and of catalase CAT-1 of *S. cereale* (Schmidt *et al.*, 2002) depended on the availability of the heme cofactor. Conceivably, the heme is cotranslationally incor- porated, as observed for globin synthesis (Komar *et al.*, 1993). Most higher plant catalases have to be imported into peroxisomes or glyoxysomes. According to pre- sent knowledge, all peroxisomal matrix proteins are, after their synthesis on free cytosolic ribosomes, posttranslationally transported into the organelle (Gietl, 1996; Olsen, 1998). The import requires the presence of targeting signals on the protein and of receptor proteins mediating its binding to the translocation machinery on the peroxisomal membrane. Two types of targeting signals have been found to direct the transport of peroxisomal proteins. Most frequently, a tripeptide at the carboxyl- terminus serves as protein targeting signal, designated as PTS1. The first identified carboxy-terminal PTS1 was the carboxy-terminal Ser–Lys–Leu–COOH (SKL) tripeptide. This SKL tripeptide or conservative variants of it were sufficient to direct the sorting of peroxisomal proteins of insects, mammals, yeast or try- panosomes into the organelle (Purdue & Lazarow, 1994; Olsen, 1998). The PTS1 sequence is not cleaved after import. In some peroxisomal proteins, a conserved amino-terminal nonapeptide was shown to serve as targeting signal termed PTS2.

The consensus sequence was originally identified as (R/K)(L/I/Q/V) X_5(H/Q)(L/A). It is located at or near the amino-terminus. In plants and mammals, the amino-terminal presequence of the PTS2 is usually cleaved off after import into the organelle. Import studies with modified PTS2 sequences led Flynn et al. (1998) to the conclusion that the action of the PTS2 motif, defined as -R/K-X_6-H/Q-A/L/F-, does not only depend on its specific sequence but also on its structural properties that can also be affected by individual residues within the X_6-region of the non-apeptide. The import cascade for peroxisomal matrix proteins is initiated by the binding of specific soluble receptor proteins to either the PTS1 or PTS2 motifs. After loading to the receptors, these interact with docking proteins of the transloca-tion machinery at the surface of the peroxisomal boundary membrane. Proteins involved in the biogenesis of peroxisomes, including protein import, were desig-nated as peroxins (pex genes). The pex5 gene encodes the PTS1 receptor and the pex7 gene encodes the PTS2 receptor (Olsen, 1998; Holroyd & Erdmann, 2001). Homologues of several peroxins were found in Arabidopsis (Mullen et al., 2001).

Catalases do not carry classical PTS sequences and varying motifs have been suggested as sorting signals (see Trelease et al., 1996). Mullen et al. (1997) investigated the sorting of cottonseed catalase in tobacco BY-2 cells after transient expression induced by particle bombardment. The biolistically introduced catalase gene was distinguished from the endogenous catalase by amino-terminal epitope-tagging with hemagglutinin. Mullen et al. (1997) proposed that the carboxy-terminal PSI-tripeptide served as a degenerate type PTS1 for cottonseed catalase. A car-boxy-terminal PSI-tripeptide can be seen in two of the ten plant catalase sequences shown in Figure 5.1. A PS/TI/M motif is conserved in many plant catalases. Kamigaki et al. (2003) investigated the sorting of pumpkin catalase that was fused with its amino-terminus to the carboxy-terminus of the green fluorescent protein as marker. They confirmed that in a biolistic transient expression system, the carboxy-terminal tripeptide was necessary for sorting into peroxisomes. However, after stable expression in transformed tobacco BY-2 or Arabidopsis, the region of the ten amino acids of the carboxy-terminus of pumpkin catalase was not required for import. By contrast, a Leu at position 11 from the carboxy-terminus was important. Kamigaki et al. (2003) regard the transient expression system as inadequate to ana-lyze the targeting signal. They proposed that the carboxy-terminal region, QKL, from positions 13–11 served as internal PTS1 signal. At these positions, a PTS1-like motif QKL/I/V is conserved in many plant catalases (see Figure 5.1). An involvement of the PTS1-pathway was further supported by the finding that in the yeast two-hybrid system, pumpkin catalase CAT-1 can bind to the Arabidopsis PEX5p PTS1-receptor protein (Kamigaki et al., 2003).

A unique property of the peroxisomal import system is its ability to transport large folded proteins and intact oligomers into the organelle (McNew & Goodman, 1994; Walton et al., 1995; McNew & Goodman, 1996; Olsen, 1998). Evidence for an oligomeric transport of plant peroxisomal matrix proteins was presented by Lee et al. (1997). Isocitrate lyase sorting depends on the presence of a PTS1 motif. However, epitope-tagged isocitrate lyase, lacking the PTS1 tripeptide, was also imported into peroxisomes through 'piggy-backing' when PTS1-carrying subunits were simultaneously expressed. An import of oligomeric proteins was also directed by an amino-terminal PTS2 motif (Kato *et al.*, 1999). The mechanism of the oligomer import is not yet understood (Olsen, 1998). In human skin fibroblasts, assembly of the tetrameric catalase in the cytosol was shown to precede its sorting to the peroxisomes (Middlekoop *et al.*, 1993). On the basis of present knowledge, it is likely that catalase is imported into peroxisomes as folded and assembled protein. Evidence is emerging that indicates a role for chaperones in peroxisomal protein import. Corpas and Trelease (1997) identified two polypeptides in peroxisomal membranes of cucumber seedlings which appeared to be immunorelated to Hsp70 proteins and to plant homologues of DnaJ, a member of the Hsp40 family known to cooperate with Hsp70. Wimmer *et al.* (1997) demonstrated the occurrence of an Hsp70 protein in the matrix of watermelon glyoxysomes. The experiments of Crookes and Olsen (1998) suggested a role of chaperones for peroxisomal protein import. Heat-shock, leading to an upregulation of Hsp70 and Hsp90 chaperones, enhanced the protein import into pumpkin glyoxysomes. Antibodies against Hsp70 or Hsp90 inhibited the import of isocitrate lyase. The specific roles of chaperones during peroxisomal protein import await further elucidation. It also needs to be clarified whether and which chaperones are involved in the assembly of the catalase holoprotein. Chaperonin 60 and α-crystallin were shown to protect bovine liver catalase against thermal inactivation and aggregation *in vitro* (Hook & Harding, 1997). For other heme-proteins, such as cytochrome c, several proteins including a 'heme chaperone' were shown to be involved in the delivery of the heme cofactor to the apoprotein (Ahuja & Thöny-Meyer, 2003).

5.4.2   Control of expression of catalase

As already pointed out in Section 5.3.4, the different catalase isozymes show distinct organ- or cell-specific and temporal patterns of expression and they differ in their response to environmental factors. A comprehensive overview about the regulation of the maize isozymes was provided by Scandalios *et al.* (1997). The maize isozyme *cat1* is expressed in developing seed tissues and in the scutellum of germinating seeds. The maize *cat1* gene has been cloned. The promoter region of the *cat1* gene was fused to the coding region of β-glucuronidase (GUS) as reporter gene and the construct was introduced into *Nicotiana tabacum* by transformation. In transgenic tobacco, high levels of GUS expression were observed in developing and mature seeds and low levels occurred in pollen. This corresponds to the pattern of *cat1* expression in maize and confirms that the promoter is responsible for the specific spatial and temporal expression of *cat1* (Guan & Scandalios, 1993). Analysis of the maize *cat1* promoter revealed the presence of a putative regulatory ARE (antioxidant-responsive element) motif (Polidoros & Scandalios, 1999) and of two abscisic acid responsive elements ABRE1 and ABRE2 (Guan & Scandalios, 2000). In mammalian systems, ARE motifs are known to be involved in the regulation of gene expression in response to oxidative stress. Exposure of maize seedlings to H_2O_2 enhanced the expression of both *cat1* and *cat2*, but only at high concentrations between 10 and 150 mM (Polidoros & Scandalios, 1999). When maize scutella were exposed to singlet oxygen by incubation with the fungal photoactivated toxin cercosporin, the expression of catalases *cat2* and *cat3* was enhanced, however, not that of *cat1* (Williamson & Scandalios, 1992a, 1993).

In developing maize embryos, the accumulation of *cat1* transcript and protein was enhanced in the presence of exogenously applied ABA, while *cat2* expression was suppressed and *cat3* remained unaffected (Williamson & Scandalios, 1992b). Transient expression assays using particle bombardment indicated that of the two ABRE motifs of the *cat1* promoter only ABRE2 was responsible for the induction of *cat1* expression by ABA. Since ABA induced the production of H_2O_2 in cultured maize cell suspensions, the authors suggested that H_2O_2 serves as intermediate signal in the induction of the *cat1* gene by ABA (Guan *et al.*, 2000). In developing embryos and seedlings of maize, the expression of all catalase isozymes was greatly increased in response to auxin treatment. High auxin concentrations induced dramatic increases of total catalase activity. However, CAT-2, which was normally not expressed at this stage of developing embryos, contributed the largest part of the auxin-induced catalase activity (Guan & Scandalios, 2002). A dramatic and preferential accumulation of *cat2* transcript and CAT-2 protein was also induced by 1-mM salicylic acid in scutella of developing maize embryos (Guan & Scandalios, 1995). Genetic investigations indicated the presence of a regulatory locus *Car2* in maize which is closely linked to the *cat1* gene and acts as suppressor of CAT-1 synthesis in the scutellum (Chandlee & Scandalios, 1984).

Control of the development of catalase isozymes that are related to photorespiratory metabolism was studied either in young developing leaves of seedlings or in greening cotyledons. In young leaves, the development of catalase(s) is usually induced or enhanced by light (e.g. Feierabend & Beevers, 1972; Holtman *et al.*, 1998). In developing rye leaves, increases of catalase activity were induced by light via the phytochrome system, however, blue light was more efficient (Feierabend, 1975). The development of catalase activity was also somewhat promoted by cytokinins (de Boer & Feierabend, 1974). In *Arabidopsis* leaves, the abundance of *cat1* and *cat2* transcripts was increased by light, while *cat3* was negatively affected by light (McClung, 1997). The light induction of *cat2* mRNA accumulation appeared to be mediated by phytochrome. In light-grown leaves, the *cat2* mRNA underwent circadian oscillations (Zhong *et al.*, 1994). When *Arabidopsis* leaves were exposed to oxidative stress by exposure to cercosporin which induces singlet oxygen formation or to paraquat which leads to the production of superoxide and H_2O_2, the mRNA levels of both *cat2* and *cat3* were enhanced. However, only the activity of CAT-3 was increased by these treatments indicating that *cat3* was most sensitive to regulation by oxidative stress (Orendi *et al.*, 2001). In barley, an mRNA homologous to the maize *cat2* gene accumulated to high levels in light but not in darkness (Acevedo *et al.*, 1996). In young leaves of maize, *cat1* did not respond to light and the *cat3* transcript exhibited diurnal fluctuations. The mRNA of *cat2* which is expressed in the bundle sheaths and assumed to be involved in photorespiration accumulated to higher levels in light-exposed leaves than in continuous darkness. In leaves of carotenoid-deficient mutants, *cat2* mRNA was barely detectable (Acevedo *et al.*, 1991; Poliodoros & Scandalios, 1997). The expression of the *cat2* gene was, however, also regulated by an unusual translational control (Skadsen & Scandalios, 1987). Whereas *cat2* mRNA was present in etiolated leaves, it was only translated and CAT-2 protein was only accumulated when the leaves were exposed to light. The *cat2* mRNA of dark-grown leaves was not translatable. When total poly[A]$^+$RNA or isolated polysomal RNA from dark-grown or light-exposed leaves were translated in a reticulocyte lysate system, CAT-2 protein was not synthesized from transcripts extracted from etiolated leaves. Within 8 h of light exposure, the *cat2* mRNA was activated and then translatable *in vitro* after extraction from the leaves. The results suggested a translational inactivation of the existing *cat2* mRNA in dark-grown leaves and an activation in the light.

The transcripts of several catalases exhibit diurnal fluctuations under the influence of a circadian rhythm. In leaves of maize seedlings grown under a 12-h photoperiod, the *cat3* transcript showed striking fluctuations. While mRNA levels were very low or undetectable in the late dark and early light period, high levels accumulated in the late light and early dark period. The diurnal fluctuation of the *cat3* mRNA persisted also in continuous light or darkness indicating a control by a circadian rhythm (Redinbough *et al.*, 1990). Entrainment of the rhythm appeared to be complex. A very low fluence phytochrome response was discussed and UV-light could function as an additional environmental cue to entrain the *cat3* circadian rhythm (Boldt & Scandalios, 1995, 1997). Diurnal variations were not observed for the *cat1* or *cat2* transcripts (Redinbough *et al.*, 1990).

Diurnal fluctuations were also described for *cat1* in leaves of *N. plumbaginifolia* with highest levels occurring in the light period. Interestingly, the mRNA expression of all tobacco catalases was suppressed under an increased CO_2-atmosphere (Willekens *et al.*, 1994). In *Nicotiana silvestris*, all three catalase transcripts exhibited diurnal fluctuations. While the abundance of *cat1* mRNA decreased during the light period and increased during the dark period, *cat2* and *cat3* transcripts showed inverse diurnal changes. They accumulated to high levels during the light period and decreased to very low levels during the dark period. The content of immunodetectable catalase protein remained, however, constant over the 24-h period (Dutilleul *et al.*, 2003). In rice leaves, the transcripts of both *catA* (=*cat3*) and *catC* (=*cat2*) exhibited diurnal oscillations. The *catA* transcript, which is mainly expressed in the leaf sheaths, had its peak late in the light period. When the 5′-flanking region of the *catA* gene was fused to a β-glucuronidase (GUS) reporter gene and expressed in transgenic plants, GUS expression did not show the pattern of diurnal oscillations of the *catA* transcript in rice. It was concluded that the 5′-flanking region was not sufficient for the circadian control of the *catA* gene (Iwamoto *et al.*, 2000).

The development of catalase was also studied in cotyledons of germinating mustard or cotton seeds. In mustard cotyledons, three catalase isoforms were observed in darkness and an additional set of isozymes was induced by phytochrome light

(Drumm & Schopfer, 1974). In cotton seedlings, five catalase isozymes were observed that are formed by subunits of two distinct genes *cat1* and *cat2* (see Figure 5.4). During germination, a shift in the pattern of isozymes from tetramers consisting mainly of CAT-1 (SU1) subunits to those consisting mainly of CAT-2 (SU2) subunits took place, indicating a changeover in the expression of CAT-1 which was dominating in glyoxysomes to CAT-2 which was in turn dominating in leaf-type peroxisomes. Accumulation of the *cat1* and *cat2* mRNAs preceeded the accumulation of the corresponding proteins, suggesting a temporally regulated transcription. Run-on transcription assays with isolated nuclei indicated, however, that mRNAs for both genes were transcribed during the whole germination period. It was concluded that the changes of their transcript levels in the cotyledons were determined by posttranscriptional mechanisms (Ni & Trelease, 1991b). Changes in the pattern of catalase isozymes in fat-storing cotyledons of germinating seedlings accompany the changeover from glyoxysomal to a leaf-type peroxisomal metabolism. This gradual functional transition appears to occur within the same organelle and does not require the generation of a new peroxisome population (for literature, see Olsen, 1998).

Changes of catalase expression accompanied senescence of leaves. In senescing excised rye leaves, both the activity of catalase and the capacity for catalase synthesis declined (Kar *et al.*, 1993). Senescing *Arabidopsis* leaves lost the ability for an induction of increased *cat3* expression under oxidative stress conditions (Orendi *et al.*, 2001). In senescing pumpkin cotyledons, *cat2* and *cat3* mRNAs disappeared and *cat1* mRNA levels increased (Esaka *et al.*, 1997). In senescing leaves, a reverse transition from leaf-type peroxisomes to glyoxysomes has been described (see Olsen, 1998).

5.5   Photoinactivation and regulation of turnover

At the end of the first decade of the last century, it was already known that isolated catalases from animal sources, such as bovine liver catalase, were light-sensitive. Talaricio (1909) and Agulhon (1912) already recognized that the photoinactivation of catalase was mediated by blue light and enhanced by the presence of O_2. In plants, photoinactivation of catalases was first observed in intact achlorophyllous tissues, such as albino leaves, etiolated leaves, coleoptiles or roots (Eyster, 1950; Appleman & Pyfrom, 1955; Björn, 1967). Björn (1969) demonstrated that, like bovine liver catalase, isolated bacterial and plant catalases were similarly inactivated in blue light *in vitro*. While Appleman and Pyfrom (1955) obtained conflicting results with green leaves, Feierabend and Engel (1986) demonstrated that photoinactivation of catalase occurred in green leaves of various plant species; however, it only became apparent when new protein synthesis was blocked by translation inhibitors. Photoinactivation appears to be a widespread general property of catalases. Lledías *et al.* (1998) confirmed the occurrence of photoinactivation for catalases from maize, sunflower, bovine liver and human erythrocytes.

However, within a 3-h light exposure, they did not observe any photoinactivation of catalases from *Neurospora crassa, S. cerevisiae, Anacystis nidulans* or *E. coli* (HPII).

Action spectra performed by Cheng *et al.* (1981) and Grotjohann *et al.* (1997) documented that catalases were inactivated by blue visible light which was absorbed by the prosthetic heme. Light of 405 nm was most effective. The photoinactivation depended on the presence of O_2 (Cheng *et al.*, 1981). The necessity of O_2 suggested an involvement of ROS. Among different quenchers or scavengers of ROS that were applied, photoinactivation was efficiently prevented only by formate or short-chain alcohols, such as methanol or ethanol (Cheng *et al.*, 1981; Feierabend & Engel, 1986). Whereas these substances are known as hydroxyl radical scavengers, it is more likely that they acted as peroxidatic substrates converting compound I back to the ferri-catalase form (see Figure 5.3). Such observations suggested that compound I was formed as intermediate during the inactivation by light. Mitchell and Anderson (1965) and Aronoff (1965) demonstrated that photoinactivation of bovine liver catalase *in vitro* followed first order kinetics and was accompanied by a decrease of heme absorbance at 405 nm. They concluded that the destruction of a single heme caused the inactivation of the enzyme. However, Feierabend and Dehne (1996) showed that the heme moiety of catalases remained – at least largely – intact during photoinactivation both *in vitro* and *in vivo*. Investigations of Grotjohann *et al.* (1997) with sunflower catalases confirmed that the destruction of heme groups could not be primarily responsible for their inactivation. The primary event appeared to be a damage of the apoprotein. According to Nakatani (1961), histidine residues of the catalase were oxidized during photoinactivation. Assays performed by Feierabend and Dehne (1996) and Grotjohann *et al.* (1997) indicated, in contradiction to earlier conclusions by Feierabend and Engel (1986), that the heme groups were not dissociated from the enzyme during photoinactivation. In an analysis of spectral changes during the inactivation of bovine liver catalase in UV light *in vitro*, Aubailly *et al.* (2000) observed that UV irradiation caused the formation of compound III. Compound III appeared to be generated by a photoreduction of the Fe(III) of ferri-catalase by an electron transfer from an internal donor and a subsequent binding of oxygen. A tyrosinate ion which serves as ligand for the Fe(III) ion of catalase was discussed as a potential electron source. Compound III is very unstable and results in irreversible inactivation or it may reversibly yield superoxide which can also cause the inactivation of catalase (see Shimizu *et al.*, 1984 and Figure 5.3). Recently, Heck *et al.* (2003) reported that during irradiation with UVB light, catalase of human skin keratinocytes or isolated bovine liver catalase produced some ROS, presumably a peroxide, particularly when the catalatic activity was inhibited.

The *in-vitro* inactivation of catalase was enhanced when irradiation was performed in the presence of mitochondria (Cheng *et al.*, 1981) or chloroplasts (Feierabend & Engel, 1986; Shang & Feierabend, 1999). When catalase was incubated in a suspension, together with isolated chloroplasts in the presence of O_2, it was inactivated in red light which is not absorbed by the catalase but only by chloroplast pigments. While the heme-sensitized photoinactivation is temperature-independent, the chloroplast-mediated inactivation was retarded at low temperature. Comparisons of different scavenger effects suggested that a generation of superoxide in both mitochondria and irradiated chloroplasts caused the inactivation of catalase in the medium surrounding the organelles. The photoinactivation of catalase was also enhanced in the presence of the flavin mononucleotide (FMN)-containing enzyme glycolate oxidase. Glycolate oxidase is inactivated by blue light absorbed by its FMN cofactor. The presence of catalase protected glycolate oxidase, with the result that the photoinactivation of catalase was enhanced (Schäfer & Feierabend, 2000). The observations suggested that catalase exerted a scavenger function protecting glycolate oxidase with which it is closely associated in the peroxisomes (Heupel et al., 1991). In leaves, catalase was only sacrificed for the protection of glycolate oxidase when they were depleted of reduced glutathione, which seems to be the main protectant of glycolate oxidase (Schäfer & Feierabend, 2000).

It is now widely documented that the photoinactivation of catalase occurs also in intact cells and tissues. Inactivation of catalase was one of the most sensitive indicators of intracellular damage induced by visible light in hepatocytes (Cheng & Packer, 1979). In mammalian or human skin, catalase was found to be most susceptible to inactivation by UV light and a loss of catalase may be involved in skin aging or skin disorders (Fuchs et al., 1989; Rhie et al., 2001). Gupta et al. (2001) concluded that the attenuation of catalase activity may play a role in tumor progression in mouse keratinocytes. In green leaves, the photoinactivation of catalase is usually latent, because the loss of activity is concomitantly replaced by the synthesis of new enzyme. When translation was, however, blocked by inhibitors, such as cycloheximide, apparent declines of catalase activity were observed (Feierabend & Engel, 1986; Hertwig et al., 1992; Mishra et al., 1993; Shang & Feierabend, 1999). In both etiolated and mature green rye leaves, the strongest inactivation of catalase occurred in blue light. The rate of inactivation increased with the photon flux. Low intensities of red light were inefficient. However, at higher intensities above 500 μmol m^{-2} s^{-1} of photosynthetically active radiation, a marked inactivation occurred also in red light. Properties of the red-light-induced inactivation, for instance its temperature-dependence, differed from those of the heme-sensitized inactivation by blue light. The results suggested that under high light photooxidative reactions initiated in the chloroplasts can induce an inactivation of catalase, in addition to the heme-sensitized mechanism. It remains, however, unclear which factors or molecules were transmitted to the peroxisomes (Shang & Feierabend, 1999).

In spite of constant steady state levels of activity, catalase exhibits a rapid turnover in light which was demonstrated in mature rye leaves by radioactive labeling and subsequent chase experiments. The rate of this turnover increased in a dose-dependent manner with the light intensity. At a photon flux of 520 μmol m^{-2} s^{-1} PAR, the apparent half time of catalase in rye leaves was 3–4 h (Hertwig et al., 1992). The turnover requires that the rate of synthesis of new catalase must be continuously and flexibly attuned to fluctuating light conditions and resulting

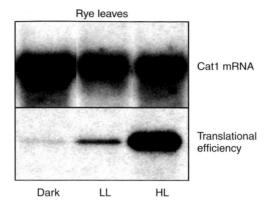

Figure 5.5 Influence of light intensity on the amounts of mRNA and the translational efficiency of catalase (*cat1*) in mature rye leaves. Northern analysis of *cat1* transcripts in rye leaves grown in low light (90 μmol m^{-1} s^{-2} PAR) and subsequently exposed for 7 h to low light (LL), high light (HL; 500 μmol m^{-2} s^{-1}) or darkness (Dark), and *in vitro* translation of CAT-1 in a wheat germ extract (without additional hemin) with poly(A)$^+$RNA extracted from rye leaves after exposure to different light conditions, as indicated above [adapted from Schmidt *et al.* (2002)].

changes of photoinactivation. The control of catalase synthesis in mature rye leaves which express only one single isozyme CAT-1 was exerted by posttranscriptional mechanisms. The content of the *cat1* mRNA did not increase in proportion to the light intensity (Figure 5.5). Highest *cat1* mRNA levels accumulated even in dark-incubated leaves (Schmidt *et al.*, 2002). Investigations of the translation of CAT-1 in cell-free systems with poly(A)$^+$RNA from leaves or with mRNA transcribed *in vitro* from a *cat1*-containing clone revealed two mechanisms of posttranscriptional control (Schmidt *et al.*, 2002). Translation of catalase required the presence of heme. In leaves, the amount of heme that is liberated from inactivated enzyme and not degraded might serve as a control allowing the synthesis of an equivalent amount of new enzyme. As additional and novel mechanism of posttranscriptional control, a light-dependent reversible modulation of the translational efficiency of the *cat1* mRNA was recognized. *Cat1* mRNA from light-exposed leaves was translated much more efficiently than mRNA from dark-exposed leaves (Figure 5.5). After transfer from high light to darkness, the translational efficiency of the *cat1* mRNA declined with a half time of 2 h, although the amount of transcript was increasing. In contrast, when dark-incubated rye leaves were transferred to high light the translational efficiency increased within 3 h to its maximal level after a 1-h lag phase (Figure 5.6). This increase also occurred in the presence of the inhibitor cordycepin which prevented the synthesis of new mRNA.

The light-induced activation of the *cat1* mRNA was specifically suppressed by methylation inhibitors. For *in vitro* translation of *in vitro* synthesized *cat1* mRNA, N7-guanine methylation of the cap structure was crucial. Efficient translation was

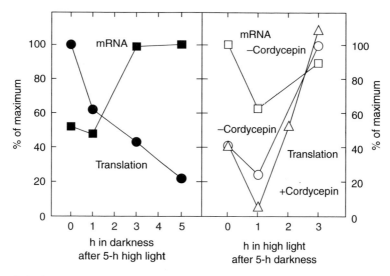

Figure 5.6   Changes of the amounts and of the translational efficiency of the *cat1* mRNA of rye leaves after transfer of leaf sections from high light to darkness (left) or from darkness to high light (right). The increase of the translational efficiency after transfer to high light was also assayed in the presence of the inhibitor cordycepin (400 μM). Translation of *cat1* mRNA was assayed *in vitro* with poly(A)$^+$RNA in a wheat germ extract in the presence of 30-μM hemin.

only observed when the transcript carried a cap and when its guanine was methylated in the N7-position (Schmidt *et al.*, 2002). The increase of the translational efficiency of the rye *cat1* mRNA that was induced by cap methylation did not require the presence of sequences of the untranslated regions of the mRNA which are frequently involved in translational control mechanisms (unpublished results). In darkness, rye leaves accumulate *cat1* mRNA as an inactive store which can be rapidly activated on demand after exposure to light according to the photon flux, and the proportion of active transcript can be continuously attuned to changing light conditions by reversible methylation or demethylation of the existing mRNA. Inasmuch as the activation of the *cat1* mRNA by cap methylation depends on the supply of the methyl group donors glycine and serine, it is indirectly coupled to the functionality of the photorespiratory pathway (Figure 5.7). However, these methyl group donors do not control the rate of mRNA activation. Further investigations (Schmidt & Feierabend, in preparation) indicated that blue light and ROS controlled the rate of activation of the *cat1* mRNA in rye leaves. Red and far-red light were much less or not effective. An activation of the *cat1* mRNA, as seen in high light, was induced in low light when peroxides (H_2O_2, t-butyl hydroperoxide) or compounds inducing an accumulation of peroxides, such as aminotriazole or paraquat, were applied. The redox state of the antioxidants ascorbate and glutathione did not influence the activation of the *cat1* mRNA.

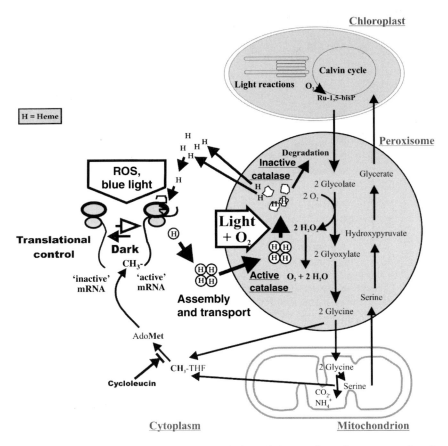

Figure 5.7   Scheme to illustrate the regulation of the repair cycle compensating for the photoinactivation of catalase in rye leaves. Synthesis of new catalase is posttranscriptionally regulated by the availability of heme and by a methylation-dependent translational activation of the *cat1* mRNA. The rate of translational activation of the *cat1* mRNA is controlled by blue light and the level of ROS. AdoMet, S-adenosylmethionine; CH_3–THF, 5-methyl-tetrahydrofolate; cycloleucin, inhibitor of AdoMet synthesis.

A light-induced change of the translational efficiency had also been observed for the *cat2* mRNA of maize (Skadsen & Scandalios, 1987). The *cat2* mRNA accumulated in dark-grown maize leaves but was neither translated *in vivo* nor translatable *in vitro* after extraction from the leaves. Only after exposure to light, the mRNA was activated and became translatable. Whether the translational efficiency of the *cat2* mRNA of green maize leaves was also modulated in response to changing light conditions was not investigated, nor was the mechanism of activation analyzed. Based on the results of Schmidt *et al.* (2002), it is most likely that the translational activities of maize *cat2* or of leaf catalases of other cereals or monocots are also reversibly modulated by methylation, as described for rye. The situation may be different in dicots. The leaf catalases of tobacco suffer also from photoin-activation and a light-dependent turnover must take place. The amounts of mRNA of the major leaf isozyme CAT-1 changed, however, substantially in response to changing light conditions, suggesting that differential mRNA transcription or stability might determine the rate of CAT-1 synthesis (unpublished results).

The light-dependent turnover of catalase makes plants vulnerable to adverse stress conditions which suppress protein synthesis and thus interfere with the repair of catalase in light. Therefore, substantial losses of catalase activity were induced in light but not in darkness when leaves of various plants were exposed to low temperature (Volk & Feierabend, 1989; Feierabend et al., 1992; Mishra et al., 1993), heat shock (Hertwig et al., 1992), salt or chemical stresses (Hertwig et al., 1992; Streb et al., 1993; Streb & Feierabend, 1996). When shade-adapted spruce trees were exposed to full sunlight, repair activities could obviously not cope with the rate of inactivation and catalase activity declined. Under field conditions, catalase activity was higher in branches oriented north than in those oriented south and a diurnal rhythm with lowest catalase activities occurring at noontime was observed (Schittenhelm et al., 1994).

In leaves of cold-acclimated plants, catalases were largely protected against photoinactivation in vivo, although they were mostly as light sensitive as other cata-lases after extraction from the leaves. In one cold-adapted high-mountain plant H. alpina a light-insensitive catalase was detected (Streb et al., 1997a). Light-stable isozymes were also observed in sunflower cotyledons (Grotjohann et al., 1997; Eising et al., 1998). Peroxisomes of sunflower cotyledons contain crystalline cores which mainly consist of specific catalase isozymes (Kleff et al., 1997). While the soluble catalase isozymes of the peroxisomal matrix were photoinactivated like other catalases, the core isozymes were more resistant to photoinactivation (Eising et al., 1998). The catalase isozyme HaCAT-1 from H. alpina and the sunflower core catalase isozyme HNNCATA3 were expressed as active enzymes in insect cells after infection with recombinant Baculovirus containing the respective cDNA sequences (Accession numbers AJ 616025 and AF 243518). The recombinant catalases HaCAT-1 and HNNCATA3 were indeed completely resistant to photoin-activation (Engel et al., in preparation). The deduced amino acid sequences of HaCAT-1 and HNNCATA3 showed a strikingly high identity of 95.3% among each other. While their amino acid sequences were also highly homologous to those of other catalases, they showed a few, altogether only six, unusual amino acid exchanges at positions, which were highly conserved in other catalases. Unusual or unique amino acids in the light-stable catalase sequences were Thr124 (numbering of HaCAT-1) instead of a conservative Val (Val133 of BLC, see Figure 5.1), Ile135 instead of Leu (Leu144 of BLC), Ser206 instead of Gly (Gly125 of BLC), Thr225 instead of His (His235 of BLC) and Met291 instead of Lys (Lys300 of BLC). When the positions of these unusual exchanges were viewed in a model of the three-dimensional struc-ture of catalase, it became evident that they were concentrated in a certain region of the molecule. The most unique amino acid substitutions were localized at the surface of the enzyme in a peripheral groove at the entrance of the minor channel (see Figure 5.2). Most unique was the replacement of a His (His234 of BLC), which is strictly conserved in all eukaryotic catalases (see Figure 5.1), by Thr225 at the entrance of the minor channel. The two light-stable catalases HaCAT-1 and HNNCATA3 have in common that their specific activity is very low. The specific activity of the sunflower core catalase accounted for only 10% of that of the matrix catalase isozymes (Eising & Gerhardt, 1986; Heinze & Gerhardt, 2002). The specific activity of catalase may be related to the width of the lateral channel (see Sevinc *et al.*, 1999). The striking structural and functional similarities of HaCAT-1 and HNNCATA3 strongly suggest that the HaCAT-1 isozyme is also localized in peroxisomal cores which are present in *H. alpina* leaves. It thus appears that the peroxisomal cores of *H. alpina* and sunflower, two species of the Asteraceae family, contain specific catalase isozymes of low activity but high stability that may serve as an emergency reserve to survive extreme light stress conditions. It is, however, quite unlikely that all core catalases are light insensitive.

5.6 Physiological significance and role in stress defence

5.6.1 Responses to deficiencies or overexpression of catalase

The physiological relevance of catalase has been investigated after elimination of catalase activity by the inhibitor 2-aminotriazole, or by the analysis of catalase-deficient or catalase-overexpressing mutants or transgenic plants. Surprisingly, null mutations of the C_4-plant maize for *cat2* (Tsaftaris *et al.*, 1983) or *cat3* (Wadsworth & Scandalios, 1990), and even a double mutant deficient in both *cat2* and *cat3*, were fully viable and did not show any phenotypic symptoms of damage. The expression of *cat1* was not enhanced in these mutants. Null mutants of maize for *cat1* have not been detected (Scandalios *et al.*, 1997). Growth of catalase-deficient C_3-plants was, however, severely affected under photorespiratory conditions. A barley mutant with less than 10% of wild-type catalase activity grew normally in an atmosphere enriched in CO_2, which suppressed photorespiration but poorly in air (Kendall *et al.*, 1983). Catalase-deficiency reduced the survival and agronomic performance of field-grown barley plants (Acevedo *et al.*, 2001). Transgenic tobacco with 10% of wild-type catalase activity showed no visible disorders in low light but developed necrotic lesions under high-light intensity (Willekens *et al.*, 1997).

Catalase-deficiencies affected alternative antioxidative systems of C_3-plants and inhibited photosynthesis. Prolonged H_2O_2 accumulation was not observed after elimination of catalase activity by aminotriazole (Amory *et al.*, 1992; Streb & Feierabend, 1996) or in catalase-deficient transgenic tobacco (Willekens *et al.*, 1997; Rizhsky *et al.*, 2002). However, catalase-deficient mutants of barley (Smith *et al.*, 1984; Palatnik *et al.*, 2002), transgenic tobacco (Willekens *et al.*, 1997) and aminotriazole-treated barley leaves (Smith, 1985) accumulated oxidized glutathione (GSSG). The level of GSSG increased both in the chloroplasts and in the cytosol while the GSH concentration remained relatively constant (Smith *et al.*, 1985). In aminotriazole-treated rye leaves exposed to high light, the levels of both GSH and GSSG increased strikingly; however, the GSH/GSSG ratio was rapidly shifted in favor of the oxidized form. Depletion of catalase in the presence of salt also induced a marked oxidation of the glutathione pool (Streb & Feierabend, 1996). In aminotriazole-treated *Arabidopsis* suspension cells, only the reduced form of glutathione was strongly increased (May & Leaver, 1993). While the ascorbate content declined in catalase-deficient leaves of transgenic tobacco (Willekens *et al.*, 1997), aminotriazole treatment of rye leaves suppressed the accumulation of dehydroascorbate but not that of ascorbate and, thus, resulted in a striking rise of the ratio of reduced to oxidized ascorbate (Streb & Feierabend, 1996). Increased levels of ascorbate peroxidase and glutathione peroxidase appeared to compensate for the low catalase activity in catalase-deficient transgenic tobacco. Nevertheless, the catalase-deficient plants exhibited an increased susceptibility to paraquat, salt and ozone, but not to chilling (Willekens *et al.*, 1997). Remarkably, double antisense plants lacking both catalase and ascorbate peroxidase were less sensitive to oxidative stress imposed by treatment with paraquat than antisense plants lacking only catalase. Photosynthesis was suppressed in these double antisense plants. The reduction of photosynthetic activity together with other metabolic alterations was regarded as strategy to avoid ROS formation (Rizhsky *et al.*, 2002). Most remarkably, the investigations with catalase-deficient transgenic tobacco indicated that catalase functioned as a sink for all cellular H_2O_2, not only for that produced in photorespiration. Bovine liver catalase injected into the intracellular spaces of catalase-deficient leaves could attenuate the development of necrosis under high light. Tobacco leaf discs were able to remove H_2O_2 from an extracellular solution but discs from catalase-deficient leaves were less efficient (Willekens *et al.*, 1997).

A close correlation exists between the levels of catalase activity and the rate of photosynthesis. After inhibition of catalase with aminotriazole, carbon assimilation was inhibited (Amory *et al.*, 1992). Tobacco mutants with enhanced catalase activity showed higher photosynthetic rates under photorespiratory conditions than controls, because these plants were more resistant to O_2-inhibition of photosynthesis (Zelitch, 1990). In catalase-deficient antisense tobacco plants, the CO_2 compensation point Γ increased with decreasing catalase activity and transgenic plants with elevated catalase had a lower Γ than wild-type plants. Increasing catalase activities reduced the photorespiratory loss of CO_2 (Brisson *et al.*, 1998). The removal of H_2O_2 by low or wild-type levels of catalase activity seems to be incomplete. Escaping H_2O_2 may chemically react with glyoxylate, which is thereby converted to CO_2 and formate. Indeed, catalase inhibition by aminotriazole induced a substantial increase of formate (Amory *et al.*, 1992). Other ketoacids like hydroxypyruvate are also rapidly decarboxylated in the presence of H_2O_2. Such H_2O_2-mediated decarboxylations increase the proportion of CO_2 generated per glycolate and reduce the efficiency of carbon assimilation. The more efficient elimination of H_2O_2 through enhanced catalase activities can, therefore, avoid additional H_2O_2-mediated decarboxylations and resulting photorespiratory CO_2 losses (Brisson *et al.*, 1998).

Transgenic tomato plants overexpressing a bacterial catalase were more tolerant than wild-type plants to oxidative damage caused by paraquat, drought or chilling stress (Mohamed *et al.*, 2003). Similarly, overexpression of a wheat catalase in transgenic rice improved the tolerance against low-temperature stress (Matsumura *et al.*, 2002). Transgenic tobacco plants moderately overexpressing the maize *cat2* gene were more tolerant against paraquat; however, they had a reduced capacity to inhibit bacterial growth in the hypersensitive reaction after infection with *P. syringae* strains (Polidoros *et al.*, 2001).

5.6.2   Low- and high-temperature stress

Chilling conditions strongly aggravate photooxidative damage in green photosynthetic tissues, because the balance between photosynthetic energy conversion and consumption is disturbed (Huner *et al.*, 1998). Dark reactions of carbon assimilation are more strongly retarded by low temperatures than energy absorption and electron flow. The resulting overreduction of the photosynthetic electron transport chain enables an enhanced production of ROS. Furthermore, the temperature-independent photoinactivation of catalase is not reduced at low temperature, while protein synthesis, which is needed to replace the photodamaged enzyme, is suppressed. Consequently, substantial losses of catalase activity were observed in leaves of numerous – both cold-sensitive and more cold-tolerant – plants under chilling-conditions even in only moderate light (Omran, 1980; MacRae & Ferguson, 1985; Volk & Feierabend, 1989; Feierabend *et al.*, 1992; Mishra *et al.*, 1993; Gechev *et al.*, 2003). Due to the temperature-dependency of the enzyme activity, the decrease of catalase activity in tobacco leaves was particularly striking, relative to the activity of untreated controls at 25°C, when the enzyme was assayed at 5°C (Gechev *et al.*, 2003). The decline of catalase activity was mostly not accompanied by an apparent rise of the H_2O_2 level (MacRae & Ferguson, 1985; Volk & Feierabend, 1989); however, cold treatments induced increases of antioxidants such as ascorbate or glutathione (Volk & Feierabend, 1989; Gechev *et al.*, 2003). Only in cucumber cotyledons, a chilling-induced increase of the H_2O_2 content was observed (Omran, 1990). The transcript levels of the major leaf catalases *cat1* of *N. plumbaginifolia* (Willekens *et al.*, 1994) or *cat1* of winter rye (Shang *et al.*, 2003) were strongly reduced within a few hours of cold exposure in light but restored during recovery at a moderate temperature. In rye leaves, the cold treatment induced a transitory increase of the catalase mRNA during a subsequent recovery period.

Catalase activities in leaves of cold-acclimated plants, such as cold-hardened winter rye (Streb *et al.*, 1999) or alpine high mountain plants (Streb *et al.*, 1999) were mostly not markedly higher than those of non-acclimated leaves. Correspondingly, the transcript levels of catalases were not increased or even down regulated during cold-acclimation of rye (Shang *et al.*, 2003) or wheat (Baek & Skinner, 2003). However, in leaves of acclimated plants, catalase was largely protected against photoinactivation (Streb *et al.*, 1997a, 1999). After extraction from cold-acclimated leaves, the catalases were mostly as light sensitive as other catalases, but the mechanisms of protection *in vivo* are unknown. Only in leaves of the cold-acclimated alpine plant *H. alpina*, a cold-insensitive catalase was identified (Streb *et al.*, 1997a; see Section 5.5). In cold-hardened rye leaves, both *in vivo* protection against photoinactivation and an enhanced capacity for new catalase synthesis at low temperature contributed to the maintenance of a constant level of enzyme activity at low temperature in light (Shang *et al.*, 2003). While the susceptibility of catalase-deficient transgenic tobacco leaves to chilling stress was not enhanced (Willekens *et al.*, 1997), catalase-overexpressing transgenic rice (Matsumura *et al.*, 2002) or tomato leaves (Mohamed *et al.*, 2003) showed an increased tolerance to chilling-induced photodamage.

Chilling induced light-independent increases of H_2O_2 in embryonic organs of germinating seeds. The chilling-induced H_2O_2-accumulation in shoots and mesocotyls of maize seedlings appeared to be lethal for the plants (Anderson *et al.*, 1995). Chilling evoked differential responses of the maize catalase genes in developing embryos and germinating seedlings (Auh & Scandalios, 1997). Increased activities of the CAT-1 and CAT-2 isozymes were induced in the scutella and embryonic axes of germinating embryos. By a pretreatment at 14°C, maize seedlings can be acclimated to survive at 4°C. By subtraction hybridization and differential screening, *cat3* transcripts were identified as preferentially expressed in acclimated seedlings (Prasad *et al.*, 1994). Activities of the CAT-3 isozyme were elevated in the mesocotyl of acclimated seedlings which was visibly most susceptible to chilling in non-acclimated controls (Anderson *et al.*, 1995). Pretreatment of seedlings with H_2O_2 or menadione – a superoxide-generating compound – induced both an accumulation of *cat3* transcripts and a chilling tolerance of the seedlings. Treatment of acclimating seedlings with the catalase inhibitor aminotriazole abolished the acclimation phenomenon (Prasad, 1997). From these results, it was hypothesized that a transitory H_2O_2 accumulation during acclimation at 14°C induced the marked increase of CAT-3 activity along with other antioxidative enzymes which then suppressed lethal H_2O_2 accumulation under prolonged chilling stress. Thus, catalase seemed to play an essential role for cold-acclimation of maize seedlings (Prasad *et al.*, 1994; Prasad, 1997).

Heat-shock temperatures inactivate the translation of most proteins, except for the synthesis of specific heat-shock proteins. Consequently, translation of catalase was inhibited at 40°C (Hertwig *et al.*, 1992). In the absence of repair, catalase activities of leaves declined at high temperatures in light because of photoinactivation (Feierabend *et al.*, 1992; Anderson, 2002). In the heat-acclimated desert plant *Retama raetam*, exposure to heat-shock temperatures induced only a minor photoinactivation of catalase (Streb *et al.*, 1997b). In maize seedlings, germination under chronic elevated temperatures of 35 or 40°C in darkness reduced the catalase activity in the scutellum. In particular, the amounts of the CAT-2 isozyme and its transcript were diminished (Scandalios *et al.*, 2000).

5.6.3   Salinity stress

Salt stress is accompanied by increased formation of ROS (for overview, see Ashraf & Harris, 2004) as illustrated by the salt-induced changes in the redox state of the antioxidants ascorbate and glutathione in favor of their oxidized forms (Streb & Feierabend, 1996; Savouré *et al.*, 1999). A sublethal NaCl concentration of 150 mM induced increases of catalase activity in leaves of *N. plumbaginifolia* through activation of the *cat2* and *cat3* genes (Savouré *et al.*, 1999). Higher concentrations blocked translation of catalase *in vivo* and induced substantial losses of catalase activity in rye leaves in light, because photoinactivation of the enzyme was exceeding its repair (Hertwig *et al.*, 1992; Streb & Feierabend, 1996; Shim *et al.*, 2003). Shim *et al.* (2003) claim, however, that the decline of catalase activity was caused by a salt-induced accumulation of salicylic acid. High capacities of the antioxidative scavenger systems correlated with increased salt tolerance (Ashraf & Harris, 2004). Salt-tolerant cultivars of cotton (Gosset *et al.*, 1994) and rice (Vaidyanathan *et al.*, 2003) contained higher levels of catalase activity. The higher salt tolerance of the wild *Lycopersicon pennellii* correlated with higher activities of catalase and other anti-oxidative enzymes in their leaves, than in cultivated tomatoes (Shalata & Tal, 1998).

5.6.4   Pathogen defence

The work of Chen *et al.* (1993) identified a catalase as a target for the binding of salicylic acid in tobacco leaves. They proposed that catalase was a receptor as well as transducer for the salicylic acid signal mediating the induction of pathogen defence genes in the systemic acquired resistance (SAR). The inhibition of catalase activity by salicylic acid was assumed to result in an accumulation of H_2O_2 which would then induce the activation of pathogenesis-related genes. Numerous investigations, which cannot all be reviewed in this chapter, have provided conflicting results, both in favor and against this hypothesis. Catalase activity was blocked only by biologically active, but not inactive, analogues of salicylic acid. Rüffer *et al.* (1995) demonstrated, however, that salicylic acid was inhibitory to several iron-containing enzymes by chelation of the Fe. Not all catalase isozymes were equally sensitive to inhibition by salicylic acid (see e.g. Guan & Scandalios, 1995). The levels of salicylic acid found in systemic leaves or infection sites appeared to be lower than required for catalase inactivation. In systemic leaves with high levels of salicylic acid, increases of the H_2O_2 level were not detected during the onset of SAR and H_2O_2 did not activate pathogenesis-related genes in *NahG* transgenic plants in which salicylic acid is eliminated by conversion to catechol (Neuenschwander *et al.*, 1995). Leaves of transgenic tobacco with reduced catalase activity did not constitutively express the pathogenesis-related gene PR-1 at low-light intensity demonstrating that catalase suppression *per se* was not sufficient to signal the induction of pathogenesis-related proteins. However, at moderate or high-light intensity, the expression of PR-1 was induced without pathogen infection and the multiplication of a pathogenic bacterium was repressed. Catalase-deficiency caused necrotic lesions in leaves exposed to high light (Chamnongpol *et al.*, 1996). The authors propose that the primary cause for cell death in *cat1*-deficient tobacco plants was an increase of toxic H_2O_2 in photorespiring cells. Analyses with transgenic tobacco plants expressing *cat1* or *cat2* in antisense orientation confirmed that there was no correlation between modest to high levels of reduction in catalase activity and activation of defence genes. However, antisense plants exhibiting the most severe reduction of catalase activity developed necrosis on their lower leaves, accumulated high levels of pathogenesis-related proteins and showed enhanced resistance to tobacco mosaic virus (Takahashi *et al.*, 1997). It was proposed that the severe oxidative stress caused by the catalase-deficiency induced necrosis and a production of salicylic acid which would subsequently activate pathogenesis-related genes. Several studies appear now to agree that H_2O_2 is acting upstream of salicylic acid in the defence signaling pathway. A decline of catalase with an accompanying increase of H_2O_2 may be important for the hypersensitive response. In the zones of hypersensitive reaction of tobacco-mosaic-virus-infected tobacco plants, the amounts of *cat1* and *cat2* transcripts and catalase activity declined and were paralleled by a strong increase of H_2O_2. In the surrounding zone of localized acquired resistance, *cat1* transcripts were undetectable but *cat2* expression was increased and H_2O_2 did not accumulate. No correlation between catalase activity and salicylic acid was detected (Dorey *et al.*, 1998).

References

Acevedo, A., Paleo, A.D. and Federico, M.L. (2001) 'Catalase deficiency reduces survival and pleiotropically affects agronomic performance in field-grown barley progeny', *Plant Science* **160**, 847–855.

Acevedo, A., Skadsen, R.W. and Scandalios, J.G. (1996) 'Two barley catalase genes respond differentially to light', *Physiologia Plantarum* **96**, 369–374.

Acevedo, A., Williamson, J.D. and Scandalios, J.G. (1991) 'Photoregulation of the *Cat2* and *Cat3* catalase genes in pigmented and pigment-deficient maize: the circadian regulation of *Cat3* is superimposed on its quasi-constitutive expression in maize leaves', *Genetics* **127**, 601–607.

Agulhon, H. (1912) 'Action de la lumiére sur les diastases', *Annales de l'Institut Pasteur* **26**, 38–47.

Ahuja, U. and Thöny-Meyer, L. (2003) 'Dynamic features of a heme delivery system for cytochrome c maturation', *The Journal of Biological Chemistry* **278**, 52061–52070.

Amory, A.M., Ford, L., Pammenter, N.W. and Cresswell, C.F. (1992) 'The use of 3-amino-1,2,4-triazole to investigate the short-term effects of oxygen toxicity on carbon assimilation by *Pisum sativum* seedlings', *Plant, Cell and Environment* **15**, 655–663.

Anderson, J.A. (2002) 'Catalase activity, hydrogen peroxide content and thermotolerance of pepper leaves', *Scientia Horticulturae* **95**, 277–284.

Anderson, M.D., Prasad, T.K. and Stewart, C.R. (1995) 'Changes in isozyme profiles of catalase, peroxidase, and glutathione reductase during acclimation to chilling in mesocotyls of maize seedlings', *Plant Physiology* **109**, 1247–1257.

Appleman, D. and Pyfrom, H.T. (1955) 'Changes in catalase activity and other responses induced in plants by red and blue light', *Plant Physiology* **30**, 543–549.

Aronoff, S. (1965) 'Catalase: kinetics of photooxidation', *Science* **150**, 72–73.

Asada, K. (1992) Production and scavenging of active oxygen in chloroplasts, in *Molecular Biology of Free Radical Scavenging Systems* (ed. J.G. Scandalios), Cold Spring Harbor Laboratory Press, Cold Spring Harbor, NY, pp. 173–192.

Asada, K. (1994) Production and action of active oxygen species in photosynthetic tissues, in *Causes of Photooxidative Stress and Amelioration of Defense Systems in Plants* (ed. C.H. Foyer and P.M. Mullineaux), CRC Press, Boca Raton, pp. 77–104.

Ashraf, M. and Harris, P.J.C. (2004) 'Potential biochemical indicators of salinity tolerance in plants', *Plant Science* **166**, 3–16.

Aubailly, M., Haigle, J., Giordani, A., Morliére, O. and Santus, R. (2000) 'UV photolysis of catalase revisited: a spectral study of photolytic intermediates', *Journal of Photochemistry and Photobiology B: Biology* **56**, 61–67.

Auh, C.-K. and Scandalios, J.G. (1997) 'Spatial and temporal responses of the maize catalases to low temperature', *Physiologia Plantarum* **101**, 149–156.

Baek, K.-H. and Skinner, D.Z. (2003) 'Alteration of antioxidant enzyme gene expression during cold acclimation of near-isogenic wheat lines', *Plant Science* **165**, 1221–1227.

Beaumont, F., Jouve, H.M., Gagnon, J., Gaillard, J. and Pelmont, J. (1990) 'Purification and properties of a catalase from potato tubers (*Solanum tuberosum*)', *Plant Science* **72**, 19–26.

Biekmann, S. and Feierabend, J. (1982) 'Subcellular distribution, multiple forms and development of glutamate–pyruvate (glyoxylate) aminotransferase in plant tissues', *Biochimica et Biophysica Acta* **721**, 268–279.

Björn, L.O. (1967) 'Some effects of light on excised wheat roots with special reference to peroxide metabolism', *Physiologia Plantarum* **20**, 149–170.

Björn, L.O. (1969) 'Photoinactivation of catalases from mammal liver, plant leaves and bacteria. Comparison of inactivation cross sections and quantum yields at 406 nm', *Photochemistry and Photobiology* **10**, 125–129.

Boldt, R. and Scandalios, J.G. (1995) 'Circadian regulation of the *Cat3* catalase gene in maize (Zea mays L.): entrainment of the circadian rhythm of *Cat3* by different light treatments', *The Plant Journal* **7**, 989–999.

Boldt, R. and Scandalios, J.G. (1997) 'Influence of UV-light on the expression of the *Cat2* and *Cat3* catalase genes in maize', *Free Radical Biology and Medicine* **23**, 505–514.

Bolwell, G.P., Davies, D.R., Gerrish, C., Auh, C.-K. and Murphy, T.M. (1998) 'Comparative biochemistry of the oxidative burst produced by rose and french bean cells reveal two distinct mechanisms', *Plant Physiology* **116**, 1379–1385.

Bravo, J., Fita, I., Gouet, P., Jouve, H.M., Melik-Adamyan, W. and Murshudov, G. (1997) Structure of catalases, in *Oxidative Stress and the Molecular Biology of Antioxidant Defences* (ed. J. Scandalios), Cold Spring Harbor Laboratory Press, Plainview, NY, pp. 407–445.

Brisson, L.F., Zelitch, I. and Havir, E.A. (1998) 'Manipulation of catalase levels produces altered photosynthesis in transgenic tobacco plants', *Plant Physiology* **116**, 259–269.

Carpena, X., Soriano, M., Klotz, M.G. *et al.* (2003) 'Structure of the clade 1 catalase, CatF of *Pseudomonas syringae*, at 1.8 Å resolution', *Proteins: Structure, Function, and Genetics* **50**, 423–436.

Chamnongpol, S., Willekens, H., Langebartels, C., Van Montagu, M., Inzé, D. and Van Camp, W. (1996) 'Transgenic tobacco with a reduced catalase activity develops necrotic lesions and induces pathogenesis-related expression under high light', *The Plant Journal* **10**, 491–503.

Chandlee, J.M. and Scandalios, J.G. (1984) 'Analysis of variants affecting the catalase developmental program in maize scutellum', *Theoretical and Applied Genetics* **69**, 71–77.

Chelikani, P., Fita, I. and Loewen, P.C. (2004) 'Diversity of structures and properties among catalases', *Cellular and Molecular Life Sciences* **61**, 192–208.

Chen, Z., Silva, H. and Klessig, D.F. (1993) 'Active oxygen species in the induction of plant systemic acquired resistance by salicylic acid', *Science* **262**, 1883–1886.

Cheng, L., Kellogg III, E.W. and Packer, L. (1981) 'Photoinactivation of catalase', *Photochemistry and Photobiology* **34**, 125–129.

Cheng, L.Y.L. and Packer, L. (1979) 'Photodamage to hepatocytes by visible light', *FEBS Letters* **97**, 124–128.

Corpas, F.J. and Trelease, R.N. (1997) 'The plant 73 kDa peroxisomal membrane protein (PMP73) is immunorelated to molecular chaperones', *European Journal of Cell Biology* **73**, 49–57.

Corpas, F.J., Palma, J.M., Sandalio, L.M. *et al.* (1999) 'Purification of catalase from pea leaf peroxisomes: identification of five different isoforms', *Free Radical Research* **31**, S235–S241.

Crookes, W.J. and Olsen, L.J. (1998) 'The effects of chaperones and the influence of protein assembly on peroxisomal protein import', *The Journal of Biological Chemistry* **273**, 17236–17242.

de Boer, J. and Feierabend, J. (1974) 'Comparison of the effects of cytokinins on enzyme development in different cell compartments of the shoot organs of rye seedlings', *Zeitschrift für Pflanzenphysiologie* **71**, 261–270.

Deisseroth, A. and Dounce, A. (1970) 'Catalase: physical and chemical properties, mechanism of catalysis and physiological role', *Physiological Review* **50**, 319–375.

del Rio, L.A., Corpas, F.J., Sandalio, L.M., Palma, J.M., Gómez, M. and Barroso, J.B. (2002) 'Reactive oxygen species, antioxidant systems and nitric oxide in peroxisomes', *Journal of Experimental Botany* **53**, 1255–1272.

Dorey, S., Baillieul, F., Saindrenan, P., Fritig, B. and Kauffmann, S. (1998) 'Tobacco class I and II catalases are differentially expressed during elicitor-induced hypersensitive cell death and localized acquired resistance', *Molecular Plant–Microbe Interactions* **11**, 1102–1109.

Dounce, A.L. (1983) 'A proposed mechanism for the catalatic action of catalase', *Journal of Theoretical Biology* **105**, 553–567.

Drumm, H. and Schopfer, P. (1974) 'Effect of phytochrome on development of catalase activity and isoenzyme pattern in mustard (*Sinapis alba* L.) seedlings. A re-investigation', *Planta* **120**, 13–30.

Durner, J. and Klessig, D.F. (1996) 'Salicylic acid is a modulator of tobacco and mammalian catalases', *The Journal of Biological Chemistry* **271**, 28492–28501.

Dutilleul, C., Garmier, M., Noctor, G. et al. (2003) 'Leaf mitochondria modulate whole cell redox homeostasis, set antioxidant capacity, and determine stress resistance through altered signaling and diurnal regulation', *The Plant Cell* **15**, 1212–1226.

Eising, R. and Gerhardt, B. (1986) 'Activity and hematin content of catalase from greening sunflower cotyledons', *Phytochemistry* **25**, 27–31.

Eising, R., Heinze, M., Kleff, S. and Tenberge, K.B. (1998) Subcellular distribution and photooxidation of catalase in sunflower, in *Antioxidants in Higher Plants: Biosynthesis, Characteristics, Actions and Specific Functions in Stress Defense* (ed. G. Noga and M. Schmitz), Shaker-Verlag, Aachen, pp. 53–63.

Esaka, M., Yamada, N., Kitabayashi, M. et al. (1997) 'cDNA cloning and differential gene expression of three catalases in pumpkin', *Plant Molecular Biology* **33**, 141–155.

Eshdat, Y., Holland, D., Faltin, Z. and Ben-Hayyim, G. (1997) 'Plant glutathione peroxidases', *Physiologia Plantarum* **100**, 234–240.

Eyster, H.C. (1950) 'Catalase activity in chloroplast pigment deficient types of corn', *Plant Physiology* **25**, 630–638.

Feierabend, J. (1975) 'Developmental studies on microbodies in wheat leaves III. On the photocontrol of microbody development', *Planta* **123**, 63–77.

Feierabend, J. and Beevers, H. (1972) 'Developmental studies on microbodies in wheat leaves I. Conditions influencing enzyme development', *Plant Physiology* **49**, 28–32.

Feierabend, J. and Dehne, S. (1996) 'Fate of the porphyrin cofactors during the light-dependent turnover of catalase and of the photosystem II reaction-center protein D1 in mature rye leaves', *Planta* **198**, 413–422.

Feierabend, J. and Engel, S. (1986) 'Photoinactivation of catalase in vitro and in leaves', *Archives of Biochemistry and Biophysics* **251**, 567–576.

Feierabend, J., Schaan, C. and Hertwig, B. (1992) 'Photoinactivation of catalase occurs under both high- and low-temperature stress conditions and accompanies photoinhibition of photosystem II', *Plant Physiology* **100**, 1554–1561.

Fita, I. and Rossmann, M.G. (1985a) 'The active center of catalase', *Journal of Molecular Biology* **185**, 21–37.

Fita, I. and Rossmann, M.G. (1985b) 'The NADPH binding site on beef liver catalase', *Proceedings of the National Academy of Sciences USA* **82**, 1604–1608.

Flynn, C.R., Mullen, R.T. and Trelease, R.N. (1998) 'Mutational analyses of a type 2 peroxisomal targeting signal that is capable of directing oligomeric protein import into tobacco BY-2 glyoxysomes', *The Plant Journal* **16**, 709–720.

Foyer, C.H. (1997) Oxygen metabolism and electron transport in photosynthesis, in *Oxidative Stress and the Molecular Biology of Antioxidant Defenses* (ed. J.G. Scandalios), Cold Spring Harbor Laboratory Press, Cold Spring Harbor, NY, pp. 587–621.

134 ANTIOXIDANTS AND REACTIVE OXYGEN SPECIES IN PLANTS

Foyer, C.H. and Noctor, G. (2003) 'Redox sensing and signalling associated with reactive oxygen in chloroplasts, peroxisomes and mitochondria', *Physiologia Plantarum* **119**, 355–364.

Frahry, G. and Schopfer, P. (1998) 'Hydrogen peroxide production by roots and its stimulation by exogenous NADH', *Physiologia Plantarum* **103**, 395–404.

Frederick, S.E. and Newcomb, E.H. (1969) 'Cytochemical localization of catalase in leaf microbodies (peroxisomes)', *The Journal of Cell Biology* **43**, 343–353.

Frugoli, J.A., McPeek, M.A., Thomas, T.L. and McClung, C.R. (1998) 'Intron loss and gain during evolution of the catalase gene family in angiosperms', *Genetics* **149**, 355–365.

Frugoli, J.A., Zhong, H.H., Nuccio, M.L. *et al.* (1996) 'Catalase is encoded by a multigene family in *Arabidopsis thaliana* (L.) Heynh', *Plant Physiology* **112**, 327–336.

Fuchs, J., Huflejt, M.E., Rothfuss, L.M., Wilson, D.S., Carcamo, G. and Oacker, L. (1989) 'Acute effects of near ultraviolet and visible light on the cutaneous antioxidant defense system', *Photochemistry and Photobiology* **50**, 739–744.

Fukamatsu, Y., Yake, N. and Hasunuma, K. (2003) '*Arabidopsis* NDK1 is a component of ROS signalling by interacting with three catalases', *Plant Cell Physiology* **44**, 982–989.

Garre, V., Müller, U. and Tudzynski, P. (1998) 'Cloning, characterization, and targeted disruption of *cpcat1*, coding for an in planta secreted catalase of *Claviceps purpurea*', *Molecular Plant–Microbe Interactions* **11**, 772–783.

Gechev, T., Willekens, H., Van Montagu, M. *et al.* (2003) 'Different responses of tobacco antioxidant enzymes to light and chilling stress', *Journal of Plant Physiology* **160**, 509–515.

Gerhardt, B. (1992) 'Fatty acid degradation in plants', *Progress in Lipid Research* **31**, 417–446.

Gietl, C. (1996) 'Protein targeting and import into plant peroxisomes', *Physiologia Plantarum* **97**, 599–608.

Gossett, D.R., Millhollon, E.P. and Lucas, M.C. (1994) 'Antioxidant response to NaCl stress in salt-tolerant and salt-sensitive cultivars of cotton', *Crop Science* **34**, 706–714.

Gouet, P., Jouve, H.-M. and Dideberg, O. (1995) 'Crystal structure of *Proteus mirabilis* PR catalase with and without bound NADPH', *Journal of Molecular Biology* **249**, 933–954.

Gross, W., Winkler, U. and Stabenau, H. (1985) 'Characterization of peroxisomes from the alga *Bumilleriopsis filimormis*', *Plant Physiology* **77**, 296–299.

Grotjohann, N., Janning, A. and Eising, R. (1997) '*In vitro* photoinactivation of catalase isoforms from cotyledons of sunflower (*Helianthus annuus* L.)', *Archives of Biochemistry and Biophysics* **346**, 208–218.

Guan, L. and Scandalios, J.G. (1993) 'Characterization of the catalase antioxidant defense gene *Cat1* of maize, and its developmentally regulated expression in transgenic tobacco', *The Plant Journal* **3**, 527–536.

Guan, L. and Scandalios, J.G. (1995) 'Developmentally related responses of maize catalase genes to salicylic acid', *Proceedings of the National Academy of Sciences USA* **92**, 5930–5934.

Guan, L.M. and Scandalios, J.G. (2002) 'Catalase gene expression in response to auxin-mediated developmental signals', *Physiologia Plantarum* **114**, 288–295.

Guan, L.M., Zhao, J. and Scandalios, J.G. (2000) '*Cis*-elements and *trans*-factors that regulate expression of the maize *Cat1* antioxidant gene in response to ABA and osmotic stress: H_2O_2 is the likely intermediary signaling molecule for the response', *The Plant Journal* **22**, 87–95.

Gupta, A., Butts, B., Kwei, K.A. *et al.* (2001) 'Attenuation of catalase activity in the malignant phenotype plays a functional role in an *in vitro* model for tumor progression', *Cancer Letters* **173**, 115–125.

Hamilton, B., Hofbauer, R. and Ruis, H. (1982) 'Translational control of catalase synthesis by hemin in the yeast *Saccharomyces cerevisiae*', *Proceedings of the National Academy of Sciences USA* **79**, 7609–7613.

Havir, E.A. and McHale, N.A. (1989) 'Enhanced-peroxidatic activity in specific catalase isozymes of tobacco, barley, and maize', *Plant Physiology* **91**, 812–815.

Heck, D.E., Vetrano, A.M., Mariano, T.M. and Laskin, J.D. (2003) 'UVB light stimulates production of reactive oxygen species. Unexpected role of catalase', *The Journal of Biological Chemistry* **278**, 22432–22436.

Heinze, M. and Gerhardt, B. (2002) Plant catalases, in *Plant Peroxisomes* (ed. A. Baker and I.A. Graham), Kluwer Academic Publishers, Dordrecht, NL, pp. 103–140.

Hertwig, B., Streb, P. and Feierabend, J. (1992) 'Light dependence of catalase synthesis and degradation in leaves and the influence of interfering stress conditions', *Plant Physiology* **100**, 1547–1553.

Heupel, R. and Heldt, H.W. (1994) 'Protein organization in the matrix of leaf peroxisomes. A multi-enzyme complex involved in photorespiratory metabolism', *European Journal of Biochemistry* **220**, 165–172.

Hirasawa, M., Gray, K.A., Shaw, R.W. and Knaff, D.B. (1987) 'Spectroscopic properties of spinach catalase', *Biochimica et Biophysica Acta* **911**, 37–44.

Holroyd, C. and Erdmann, R. (2001) 'Protein translocation machineries of peroxisomes', *FEBS Letters* **501**, 6–10.

Holtman, W.L., Graaff, A.M., Lea, P.J. and Kijne, J.W. (1998) 'Light-dependent expression of two catalase subunits in leaves of barley seedlings', *Journal of Experimental Botany* **49**, 1303–1306.

Holtman, W.L., van Duijn, G., Zimmermann, D. *et al.* (1993) 'Monoclonal antibodies for differential recognition of catalase subunits in barley aleuroen cells', *Plant Physiology and Biochemistry* **31**, 311–321.

Hook, D.W.A. and Harding, J.J. (1997) 'Molecular chaperones protect catalase against thermal stress', *European Journal of Biochemistry* **247**, 380–385.

Huang, A.H.C., Trelease, R.N. and Moore Jr., T.S. (1983) *Plant Peroxisomes*, Academic Press, New York, London.

Huner, N.P.A., Öquist, G. and Sarhan, F. (1998) 'Energy balance and acclimation to light and cold', *Trends in Plant Science* **3**, 224–230.

Igamberdiev, A.U. and Lea, P.J. (2002) 'The role of peroxisomes in the integration of metabolism and evolutionary diversity of photosynthetic organisms', *Phytochemistry* **60**, 651–674.

Iwamoto, M., Higo, H. and Higo, K. (2000) 'Differential diurnal expression of rice catalase genes: the 5'-flanking region of *CatA* is not sufficient for circadian control', *Plant Science* **151**, 39–46.

Iwamoto, M., Maekawa, M., Saito, A., Higo, H. and Higo, K. (1999) 'Evolutionary relationship of plant catalase genes inferred from exon-intron structures: isozyme divergence after the separation of monocots and dicots', *Theoretical and Applied Genetics* **97**, 9–19.

Jacquot, J.-P., Lancelin, J.-M. and Meyer, Y. (1997) 'Thioredoxins: structure and function in plant cells', *The New Phytologist* **136**, 543–570.

Kaiser, W.M. (1976) 'The effect of hydrogen peroxide on CO_2 fixation of isolated intact chloroplasts', *Biochimica et Biophysica Acta* **444**, 476–482.

Kamigaki, A., Mano, S., Terauchi, K. *et al.* (2003) 'Identification of peroxisomal targeting signal of pumpkin catalase and the binding analysis with PTS1 receptor', *The Plant Journal* **33**, 161–175.

Kar, M., Streb, P., Hertwig, B. and Feierabend, J. (1993) 'Sensitivity to photodamage increases during senescence in excised leaves', *Journal of Plant Physiology* **141**, 538–544.

Kato, A., Hayashi, M. and Nishimura, M. (1999) 'Oligomeric proteins containing N-terminal targeting signals are imported into peroxisomes in transgenic *Arabidopsis*', *Plant and Cell Physiology* **40**, 586–591.

Kato, J., Yamahara, T., Tanaka, K., Takio, S. and Satoh, T. (1997) 'Characterization of catalase from green algae *Chlamydomonas reinhardtii*', *Journal of Plant Physiology* **151**, 262–268.

Kendall, A.C., Keys, A.J., Turner, J.C., Lea, P.J. and Miflin, B.J. (1983) 'The isolation and characterization of a catalase-deficient mutant of barley (*Hordeum vulgare*)', *Planta* **159**, 505–511.

Kirkman, H.N., Rolfo, M., Ferraris, A.M. and Gaetani, G.F. (1999) 'Mechanisms of protection of catalase by NADPH', *The Journal of Biological Chemistry* **274**, 13908–13914.

Kleff, S., Sandor, S., Mielke, G. and Eising, R. (1997) 'The predominant protein in peroxisomal cores of sunflower cotyledons is a catalase that differs in primary structure from the catalase in the peroxisomal matrix', *European Journal of Biochemistry* **245**, 402–410.

Komar, A.A., Kommer, A., Krasheninnikov, I.A. and Spirin, A.S. (1993) 'Cotranslational heme binding to nascent globin chains', *FEBS Letters* **326**, 261–263.

Lamb, L. and Dixon, R.A. (1997) 'The oxidative burst in plant disease resistance', *Annual Review of Plant Physiology and Plant Molecular Biology* **48**, 251–275.

Lardinois, O.M., Mestdagh, M.M. and Rouxhet, P.G. (1996) 'Reversible inhibition and irreversible inactivation of catalase in presence of hydrogen peroxide', *Biochimica et Biophysica Acta* **1295**, 222–238.

Lee, M.S., Mullen, R.T. and Trelease, R.N. (1997) 'Oilseed isocitrate lyases lacking their essential type 1 peroxisomal targeting signal are piggybacked to glyoxysomes', *The Plant Cell* **9**, 185–197.

Levine, R.L., Mosoni, L., Berlett, B.S. and Stadtman, E.R. (1996) 'Methionine residues as endogenous antioxidants in proteins', *Proceedings of the National Academy of Sciences USA* **93**, 15036–15040.

Lledías, F., Rangel, P. and Hansberg, W. (1998) 'Oxidation of catalase by singlet oxygen', *The Journal of Biological Chemistry* **273**, 10630–10637.

MacRae, E.A. and Ferguson, I.B. (1985) 'Changes in catalase activity and hydrogen peroxide concentration in plants in response to low temperature', *Physiologia Plantarum* **65**, 51–56.

Matsumura, T., Tabayashi, N., Kamagata, Y., Souma, C. and Saruyama, H. (2002) 'Wheat catalase expressed in transgenic rice can improve tolerance against low temperature stress', *Physiologia Plantarum* **116**, 317–327.

May, M.J. and Leaver, C.J. (1993) 'Oxidative stimulation of glutathione synthesis in *Arabidopsis thaliana* suspension cultures', *Plant Physiology* **103**, 621–627.

McClung, C.R. (1997) 'Regulation of catalases in *Arabidopsis*', *Free Radical Biology and Medicine* **23**, 489–496.

McNew, J.A. and Goodman, J.M. (1994) 'An oligomeric protein is imported into peroxisomes *in vivo*', *The Journal of Cell Biology* **127**, 1245–1257.

McNew, J.A. and Goodman, J.M. (1996) 'The targeting and assembly of peroxisomal proteins: some old rules do not apply', *Trends in Biochemical Sciences* **21**, 54–58.

Middelkoop, E., Wiemer, E.A.C., Schoenmaker, D.E.T., Strijland, A. and Tager, J.M. (1993) 'Topology of catalase assembly in human skin fibroblasts', *Biochimica et Biophysica Acta* **1220**, 15–20.

Mishra, N.P., Mishra, R.K. and Singhal, G.S. (1993) 'Changes in the activities of anti-oxidant enzymes during exposure of intact wheat leaves to strong visible light at different temperatures in the presence of protein synthesis inhibitors', *Plant Physiology* **102**, 903–910.

Mitchell, R.L. and Anderson, I.C. (1965) 'Catalase photoinactivation', *Science* **150**, 74.

Mittler, R. (2002) 'Oxidative stress, antioxidants and stress tolerance', *Trends in Plant Science* **7**, 405–410.

Mohamed, E.A., Iwaki, T., Munir, I., Tamoi, M., Shigeoka, S. and Wadano, A. (2003) 'Overexpression of bacterial catalase in tomato leaf chloroplasts enhances photo-oxidative stress tolerance', *Plant, Cell and Environment* **26**, 2037–2046.

Mullen, R.T., Flynn, C.R. and Trelease, R.N. (2001) 'How are peroxisomes formed? The role of the endoplasmic reticulum and peroxins', *Trends in Plant Science* **6**, 256–261.

Mullen, R.T., Lee, M.S. and Trelease, R.N. (1997) 'Identification of the peroxisomal targeting signal for cottonseed catalase', *The Plant Journal* **12**, 313–322.

Murthy, M.R.N., Reid, T.J., Sicignano, A., Tanaka, N. and Rossmann, M.G. (1981) 'Structure of beef liver catalase', *Journal of Molecular Biology* **152**, 465–499.

Nakatani, M. (1961) 'Studies on histidine residues in hemeproteins related to their activities', *The Journal of Biochemistry* **49**, 98–102.

Neuenschwander, U., Vernooij, B., Friedrich, L., Uknes, S., Kessmann, H. and Ryals, J. (1995) 'Is hydrogen peroxide a second messenger of salicylic acid in systemic acquired resistance?', *The Plant Journal* **8**, 227–233.

Ni, W. and Trelease, R.N. (1991a) 'Two genes encode the two subunits of cottonseed catalase', *Archives of Biochemistry and Biophysics* **289**, 237–243.

Ni, W. and Trelease, R.N. (1991b) 'Post-transcriptional regulation of catalase isozyme expression in cotton seeds', *The Plant Cell* **3**, 737–744.

Nicholls, P. and Schonbaum (1963) Catalases, in *The Enzymes* (ed. P.D. Boyer, H. Lardy and K. Myrbäck), Academic Press, New York, London, pp. 147–225.

Nicholls, P., Fita, I. and Loewen, P. (2001) 'Enzymology and structure of catalases', *Advances in Inorganic Chemistry* **51**, 51–106.

Olsen, L.J. (1998) 'The surprising complexity of peroxisome biogenesis', *Plant Molecular Biology* **38**, 163–189.

Omran, R.G. (1980) 'Peroxide levels and the activities of catalase, peroxidase, and indoleacetic acid oxidase during and after chilling cucumber seedlings', *Plant Physiology* **65**, 407–408.

Orendi, G., Zimmermann, P., Baar, C. and Zentgraf, U. (2001) 'Loss of stress-induced expression of catalase3 during leaf senescence in *Arabidopsis thaliana* is restricted to oxidative stress', *Plant Science* **161**, 301–314.

Palatnik, J.F., Valle, E.M., Federico, M.L. *et al.* (2002) 'Status of antioxidant metabolites and enzymes in a catalase-deficient mutant of barley (*Hordeum vulgare* L.)', *Plant Science* **162**, 363–371.

Petrova, V.Y., Rasheva, T.V. and Kujumdzieva, A.V. (2002) 'Catalase enzyme in mitochondria of *Saccharomyces cerevisiae*', *Electronic Journal of Biotechnology* **5**, 29–41.

Polidoros, A.N. and Scandalios, J.G. (1997) 'Response of the maize catalases to light', *Free Radical Biology and Medicine* **23**, 497–504.

Polidoros, A.N. and Scandalios, J.G. (1999) 'Role of hydrogen peroxide and different classes of antioxidants in the regulation of catalase and glutathione S-transferase gene expression in maize (*Zea mays* L.)', *Physiologia Plantarum* **106**, 112–120.

Polidoros, A.N., Mylona, O.V. and Scandalios, J.G. (2001) 'Transgenic tobacco plants expressing the maize *Cat2* gene have altered catalase levels that affect plant–pathogen interactions and resistance to oxidative stress', *Transgenic Research* **10**, 555–569.

Prasad, T.K. (1997) 'Role of catalase in inducing chilling tolerance in pre-emergent maize seedlings', *Plant Physiology* **114**, 1369–1376.

Prasad, T.K., Anderson, M.D., Martin, B.A. and Stewart, C.R. (1994) 'Evidence for chilling-induced oxidative stress in maize seedlings and a regulatory role for hydrogen peroxide', *The Plant Cell* **6**, 65–74.

Purdue, P.E. and Lazarow, P.B. (1994) 'Peroxisomal biogenesis: multiple pathways of protein import', *The Journal of Biological Chemistry* **269**, 30065–30068.

Purvis, A.C. (1997) 'Role of the alternative oxidase in limiting superoxide production by plant mitochondria', *Physiologia Plantarum* **100**, 165–170.

Raha, S. and Robinson, B.H. (2000) 'Mitochondria, oxygen free radicals, disease and ageing', *Trends in Biochemical Sciences* **25**, 502–508.

Redinbaugh, M.G., Sabre, M. and Scandalios, J.G. (1990) 'Expression of the maize *Cat3* catalase gene is under the influence of a circadian rhythm', *Proceedings of the National Academy of Sciences USA* **87**, 6853–6857.

Rhie, G.-E., Seo, J.Y. and Chung, J.H. (2001) 'Modulation of catalase in human skin *in vivo* by acute and chronic UV radiation', *Molecular Cells* **11**, 399–404.

Rizhsky, L., Hallak-Herr, E., Van Breusegem, F. *et al.* (2002) 'Double antisense plants lacking ascorbate peroxidase and catalase are less sensitive to oxidative stress than single antisense plants lacking ascorbate peroxidase or catalase', *The Plant Journal* **32**, 329–342.

Rouhier, N. and Jacquot, J.-P. (2002) 'Plant peroxiredoxins: alternative hydroperoxide scavenging enzymes', *Photosynthesis Research* **74**, 259–268.

Rüffer, M., Steipe, B. and Zenk, M.H. (1995) 'Evidence against specific binding of salicylic acid to plant catalase', *FEBS Letters* **377**, 175–180.

Sadanandom, A., Poghosyan, Z., Fairbairn, D.J. and Murphy, D.J. (2000) 'Differential regulation of plastidial and cytosolic isoforms of peptide methionine sulfoxide reductase in *Arabidopsis*', *Plant Physiology* **123**, 255–264.

Salguero, J. and Böttger, M. (1995) 'Secreted catalase activity from roots of developing maize (*Zea mays* L.) seedlings', *Protoplasma* **184**, 72–78.

Savouré, A., Thorin, D., Davey, M. *et al.* (1999) 'NaCl and CuSO$_4$ treatments trigger distinct oxidative defence mechanisms in *Nicotiana plumbaginifolia* L.', *Plant, Cell and Environment* **22**, 387–396.

Scandalios, J. (1994) Regulation and properties of plant catalases, in *Causes of Photooxidative Stress and Amelioration of Defence Systems in Plants* (ed. Ch.H. Foyer and Ph.M. Mullineaux), CRC Press, Boca Raton, FL, pp. 275–315.

Scandalios, J.G., Acevedo, A. and Ruzsa, S. (2000) 'Catalase gene expression in response to chronic high temperature stress in maize', *Plant Science* **156**, 103–110.

Scandalios, J.G., Guan, L.M. and Polidoros, A. (1997) Catalases in plants: gene structure, properties, regulation, and expression, in *Oxidative Stress and the Molecular Biology of Antioxidant Defences* (ed. J.G. Scandalios), Cold Spring Harbor Laboratory Press, Plainview, NY, pp. 343–406.

Scandalios, J.G., Tong, W.F. and Roupakias, D.G. (1980) '*Cat3*, a third gene locus coding for a tissue-specific catalase in maize: genetics, intracellular location and some biochemical properties', *Molecular and General Genetics* **179**, 33–41.

Schäfer, L. and Feierabend, J. (2000) 'Photoinactivation and protection of glycolate oxidase *in vitro* and in leaves', *Zeitschrift für Naturforschung* **55c**, 361–372.

Schittenhelm, J., Toder, S., Fath, S., Westphal, S. and Wagner, E. (1994) 'Photoinactivation of catalase in needles of Norway spruce', *Physiologia Plantarum* **90**, 600–606.

Schmidt, M. and Feierabend, J. (2000) 'Characterization of cDNA nucleotide sequences encoding two differentially expressed catalase isozyme polypeptides from winter rye

(*Secale cereale* L.) (Accession nos. Z54143, Z99634 and AJ251894) (PGR 00-032)', *Plant Physiology* **122**, 1457.

Schmidt, M., Dehne, S. and Feierabend, J. (2002) 'Post-transcriptional mechanisms control catalase synthesis during its light-induced turnover in rye leaves through the availability of the hemin cofactor and reversible changes of the translation efficiency of mRNA', *The Plant Journal* **31**, 601–613.

Sevinc, M.S., Maté, M.J., Switala, J., Fita, I. and Loewen, P.C. (1999) 'Role of the lateral channel in catalase HPII of *Escherichia coli*', *Protein Science* **8**, 490–498.

Shalata, A. and Tal, M. (1998) 'The effect of salt stress on lipid peroxidation and antioxidants in the leaf of the cultivated tomato and its wild salt-tolerant relative *Lycopersicon pennellii*', *Physiologia Plantarum* **104**, 169–174.

Shang, W. and Feierabend, J. (1999) 'Dependence of catalase photoinactivation in rye leaves on light intensity and quality and characterization of a chloroplast-mediated inactivation in red light', *Photosynthesis Research* **59**, 201–213.

Shang, W., Schmidt, M. and Feierabend, J. (2003) 'Increased capacity for synthesis of the D1 protein and of catalase at low temperature in leaves of cold-hardened winter rye (*Secale cereale* L.)', *Planta* **216**, 865–873.

Sheptovitsky, Y.G. and Brudvig, G.W. (1996) 'Isolation and characterization of spinach photosystem II membrane-associated catalase and polyphenol oxidase', *Biochemistry* **35**, 16255–16263.

Shigeoka, S., Ishikawa, T., Tamoi, M. *et al.* (2002) 'Regulation and function of ascorbate peroxidase isoenzymes', *Journal of Experimental Botany* **53**, 1305–1319.

Shim, I.-S., Momose, Y., Yamamoto, A., Kim, D.-W. and Usui, K. (2003) 'Inhibition of catalase activity by oxidative stress and its relationship to salicylic acid accumulation in plants', *Plant Growth Regulation* **39**, 285–292.

Shimizu, N., Kobayashi, K. and Hayashi, K. (1984) 'The reaction of superoxide radical with catalase', *The Journal of Biological Chemistry* **259**, 4414–4418.

Shimizu, N., Kobayashi, K. and Hayashi, K. (1988) 'Studies on the equilibria and kinetics of the reactions of ferrous catalase with ligands', *Journal of Biochemistry* **104**, 136–140.

Skadsen, R.W. and Scandalios, J.G. (1987) 'Translational control of photo-induced expression of the *Cat2* catalase gene during leaf development in maize', *Proceedings of the National Academy of Sciences USA* **84**, 2785–2789.

Skadsen, R.W., Schulze-Lefert, P. and Herbst, J.M. (1995) 'Molecular cloning, characterization and expression analysis of two catalase isozyme genes in barley', *Plant Molecular Biology* **29**, 1005–1014.

Smith, I.K. (1985) 'Stimulation of glutathione synthesis in photorespiring plants by catalase inhibitors', *Plant Physiology* **79**, 1044–1047.

Smith, I.K., Kendall, A.C., Keys, A.J., Turner, J.C. and Lea, P.J. (1984) 'Increased levels of glutathione in a catalase-deficient mutant of barley (*Hordeum vulgare* L.)', *Plant Science Letters* **37**, 29–33.

Smith, I.K., Kendall, A.C., Keys, A.J., Turner, J.C. and Lea, P.J. (1985) 'The regulation of the biosynthesis of glutathione in leaves of barley (*Hordeum vulgare* L.)', *Plant Science* **41**, 11–17.

Stabenau, H., Säftel, W. and Winkler, U. (2003) 'Microbodies of the alga *Chara*', *Physiologia Plantarum* **118**, 16–20.

Stabenau, H., Winkler, U. and Säftel, W. (1989) 'Compartmentation of peroxisomal enzymes in algae of the group of Prasinophyceae. Occurrence of possible microbodies without catalase', *Plant Physiology* **90**, 754–759.

Streb, P. and Feierabend, J. (1996) 'Oxidative stress responses accompanying photoinactivation of catalase in NaCl-treated rye leaves', *Botanica Acta* **109**, 125–132.

Streb, P. and Feierabend, J. (1999) 'Significance of antioxidants and electron sinks for the cold-hardening-induced resistance of winter rye leaves to photo-oxidative stress', *Plant, Cell and Environment* **22**, 1225–1237.

Streb, P., Feierabend, J. and Bligny, R. (1997a) 'Resistance to photoinhibition of photosystem II and catalase and antioxidative protection in high mountain plants', *Plant, Cell and Environment* **20**, 1030–1040.

Streb, P., Michael-Knauf, A. and Feierabend, J. (1993) 'Preferential photoinactivation of catalase and photoinhibition of photosystem II are common early symptoms under various osmotic and chemical stress conditions', *Physiologia Plantarum* **88**, 590–598.

Streb, P., Shang, W. and Feierabend, J. (1999) 'Resistance of cold-hardened winter rye leaves (*Secale cereale* L.) to photo-oxidative stress', *Plant, Cell and Environment* **22**, 1211–1223.

Streb, P., Tel-Or, E. and Feierabend, J. (1997b) 'Light stress effects and antioxidative protection in two desert plants', *Functional Ecology* **11**, 416–424.

Suzuki, M., Ario, T., Hattori, T., Nakamura, K. and Asahi, T. (1994) 'Isolation and characterization of two tightly linked catalase genes from castor bean that are differentially regulated', *Plant Molecular Biology* **25**, 507–516.

Switala, J. and Loewen, P.C. (2002) 'Diversity of properties among catalases', *Archives of Biochemistry and Biophysics* **401**, 145–154.

Takahashi, H., Chen, Z., Du, H., Liu, Y. and Klessig, D.F. (1997) 'Development of necrosis and activation of disease resistance in transgenic tobacco plants with severely reduced catalase levels', *The Plant Journal* **11**, 993–1005.

Talacrio, G. (1909) 'Comportamento della catalisi del fegato alle luci monochromatishe', *Arch. Pharmacol. Sper.* **8**, 81–109.

Tenberge, K.B. and Eising, R. (1995) 'Immunogold labelling indicates high catalase concentration in amorphous and crystalline inclusions of sunflower (*Helianthus annuus* L.) peroxisomes', *Histochemical Journal* **27**, 184–195.

Trelease, R.N., Xie, W., Lee, M.S. and Mullen, R.T. (1996) 'Rat liver catalase is sorted to peroxisomes by its C-terminal tripeptide Ala-Asn-Leu, not by the internal Ser-Lys-Leu motif', *European Journal of Cell Biology* **71**, 248–258.

Tsaftaris, A.S., Bosabalidis, A.M. and Scandalios, J.G. (1983) 'Cell-type-specific gene expression and acatalasemic peroxisomes in a null *Cat2* catalase mutant of maize', *Proceedings of the National Academy of Sciences USA* **80**, 4455–4459.

Vaidyanathan, H., Sivakumar, P., Chakrabarty, R. and Thomas, G. (2003) 'Scavenging of reactive oxygen species in NaCl-stressed rice (*Oryza sativa* L.) – differential response in salt-tolerant and sensitive varieties', *Plant Science* **165**, 1411–1418.

Veljovic-Jovanovic, S., Milovanovic, L., Oniki, T. and Takahama, U. (1999) 'Inhibition of catalase by sulfite and oxidation of sulfite by H_2O_2 cooperating with ascorbic acid', *Free Radical Research* **31**, S51–S57.

Volk, S. and Feierabend, J. (1989) 'Photoinactivation of catalase at low temperature and its relevance to photosynthetic and peroxide metabolism in leaves', *Plant, Cell and Environment* **12**, 701–712.

Wadsworth, G.J. and Scandalios, J.G. (1990) 'Molecular characterization of a catalase null allele at the *Cat3* locus in maize', *Genetics* **125**, 867–872.

Wagner, A.M. and Krab, K. (1995) 'The alternative respiration pathway in plants: role and regulation', *Physiologia Plantarum* **95**, 318–325.

Walton, P.A., Hill, P.E. and Sabramani, S. (1995) 'Import of stably folded proteins into peroxisomes', *Molecular Biology of the Cell* **6**, 675–683.

Willekens, H., Chamnongpol, S., Davey, M. *et al.* (1997) 'Catalase is a sink for H_2O_2 and is indispensable for stress defence in C_3 plants', *The EMBO Journal* **16**, 4806–4816.

Willekens, H., Inzé, D., Van Montagu, M. and van Camp, W. (1995) 'Catalases in plants', *Molecular Breeding* **1**, 207–228.

Willekens, H., Langebartels, C., Tiré, C., van Montagu, M., Inzé, D. and van Camp, W. (1994) 'Differential expression of catalase genes in *Nicotiana plumbaginifolia* (L.)', *Proceedings of the National Academy of Sciences USA* **91**, 10450–10454.

Williamson, J.D. and Scandalios, J.G. (1992a) 'Differential response of maize catalases and superoxide dismutases to the photoactivated fungal toxin cercosporin', *The Plant Journal* **2**, 351–358.

Williamson, J.D. and Scandalios, J.G. (1992b) 'Differential response of maize catalases to abscisic acid: Vp1 transcriptional activator is not required for abscisic acid-regulated *Cat1* expression', *Proceedings of the National Academy of Sciences USA* **89**, 8842–8846.

Williamson, J.D. and Scandalios, J.G. (1993) 'Response of the maize catalases and superoxide dismutases to cercosporin-containing fungal extracts: the pattern of catalase response in scutella is stage specific', *Physiologia Plantarum* **88**, 159–166.

Wimmer, B., Lottspeich, F., van der Klei, I., Veenhuis, M. and Gietl, C. (1997) 'The glyoxysomal and plastid molecular chaperones (70-kDa heat shock protein) of watermelon cotyledons are encoded by a single gene', *Proceedings of the National Academy of Sciences USA* **94**, 13624–13629.

Yang, T. and Poovaiah, B.W. (2002) 'Hydrogen peroxide homeostasis: activation of plant catalase by calcium/calmodulin', *Proceedings of the National Academy of Sciences USA* **99**, 4097–4102.

Zámocký, M. and Koller, F. (1999) 'Understanding the structure and function of catalases: clues from molecular evolution and *in vitro* mutagenesis', *Progress in Biophysics and Molecular Biology* **72**, 19–66.

Zámocký, M., Regelsberger, G., Jakopitsch, C. and Obinger, C. (2001) 'The molecular peculiarities of catalase-peroxidases', *FEBS Letters* **492**, 177–182.

Zelitch, I. (1990) 'Physiological investigations of a tobacco mutant with O_2-resistant photosynthesis and enhanced catalase activity', *Plant Physiology* **93**, 1521–1524.

Zhong, H.H., Young, J.C., Pease, E.A., Hangarter, R.P. and McClung, C.R. (1994) 'Interactions between light and the circadian clock in the regulation of *CAT2* expression in *Arabidopsis*', *Plant Physiology* **104**, 889–898.

6 Phenolics as antioxidants

Stephen C. Grace

6.1 Introduction

Plants produce a diverse array of organic compounds, the vast majority of which are not directly involved in growth and development. Such compounds, often referred to as 'secondary metabolites,' generally have unknown functions, yet are thought to benefit plants by mediating a wide range of interactions between plants and their environment. Many secondary metabolites act as defensive agents against pathogens and herbivores and provide reproductive advantages as attractants of pollinators and seed dispersers. There is also growing evidence that secondary metabolites have a host of physiological activities related to protection against various forms of environmental stress.

The taxonomic diversity of plants is reflected in the enormous chemical diversity of secondary metabolites, with over 50 000 known chemical structures and many more likely to be identified in the future. These compounds are generally classified into three major groups, based on their biosynthetic origin: terpenoids, alkaloids and phenolics. A full discussion of the physiological and ecological roles of secondary metabolites is beyond the scope of this review. Rather, the purpose of this chapter is to highlight the properties of one particular group of compounds, phenolics (often referred to as polyphenols), and to review the growing body of evidence that phenolics function as antioxidants under certain physiological conditions and, thereby, protect plants against oxidative stress in a variety of environmental contexts.

Plant phenolics are broadly characterized as aromatic metabolites that possess one or more 'acidic' phenolic hydroxyl groups. They range in structure from relatively simple phenols, such as the signaling molecule salicylic acid, to complex polymers such as suberin and lignin. The major classes of phenolics are the hydroxy-cinnamic acids (HCAs), flavonoids, anthocyanins and tannins. These are found in all higher plants, often at high levels. However, certain flavonoid subclasses may be restricted taxonomically, such as the isoflavones, which are limited almost exclusively to legumes. Other minor phenolic classes with limited taxonomic distribution include the stilbenes, coumarins, furanocoumarins and styrylpyrones.

Phenolics are an ecologically significant class of secondary metabolites, accounting for about 40% of the organic carbon circulating in the biosphere (Croteau *et al.*, 2000). Much of this photoassimilated carbon is in the form of lignin, suberin and related structural polymers, which have clearly defined roles in mechanical support, vascular transport and structural reinforcement of long-lived plant tissues. However, the function of 'non-structural' phenolics such as HCAs, flavonoids and tannins has been a subject of vigorous debate for decades. Numerous theories have been proposed that attempt to explain the function of phenolics in purely ecological terms (Feeny, 1976; Coley et al., 1985; Hamilton et al., 2001). However, in addition to influencing plant–animal interactions, phenolics also appear to serve a variety of essential physiological functions associated with acclimation to stressful environments. These include screening high levels of visible and UV light, defense against pathogens, and general protection against oxidative stress. There is also evidence that phenolics may function as signaling molecules that mediate interactions between plants and microbes in the establishment of nutritionally beneficial relationships (Shirley, 1996), and as allelopathic agents that influence competition among plant species (Bais et al., 2003). Over longer time scales, phenolics affect rates of litter decomposition and mineral retention in the soil, and thus affect ecosystem level interactions. From a human perspective, dietary phenolics may affect human health, possibly by acting as antioxidants, anticarcinogens and cardioprotective agents.

6.2   Biosynthetic aspects of phenolic metabolism

The vast majority of plant phenolic compounds are synthesized by the phenylpropanoid pathway, which directs aromatic compounds from the shikimic acid pathway into a host of phenolic-based metabolites (Hahlbrock & Scheel, 1989). The first step in the phenylpropanoid pathway is the deamination of phenylalanine to form trans-cinnamic acid, which contains the C6–C3 phenylpropane skeleton found in all phenolic compounds with the exception of certain benzoic acid derivatives such as the gallic acid esters. This reaction is catalyzed by phenylalanine ammonia lyase (PAL), the key branchpoint enzyme between primary and secondary metabolism that regulates flux into the phenylpropanoid pathway (Howles et al., 1996). Monocots contain the enzyme tyrosine ammonia lyase (TAL), which catalyzes the formation of the monohydroxylated derivative p-coumaric acid directly from tyrosine (Croteau et al., 2000). The major products of the phenylpropanoid pathway are the hydroxycinnamic acids (HCAs), which can accumulate as esters or serve as precursors for other phenolic metabolites including flavonoids and lignin. Although often overlooked, esters of HCAs are extremely widespread in nature. For example, the quinic acid ester of caffeic acid, chlorogenic acid (5-O-caffeoylquinic acid), is found at high levels in many plants and is virtually universal in its distribution (Mølgaard & Ravn, 1988; Macheix & Fleuriet, 1998).

Flavonoid synthesis uses intermediates of both the phenylpropanoid pathway and the polyketide pathway to form the C6'–C3–C6 (flavan) skeleton characteristic of all flavonoids. The first step in flavonoid synthesis involves the condensation of p-coumaroyl-CoA with three molecules of acetate via malonyl-CoA to form naringenin chalcone. This reaction is catalyzed by chalcone synthase (CHS), the major

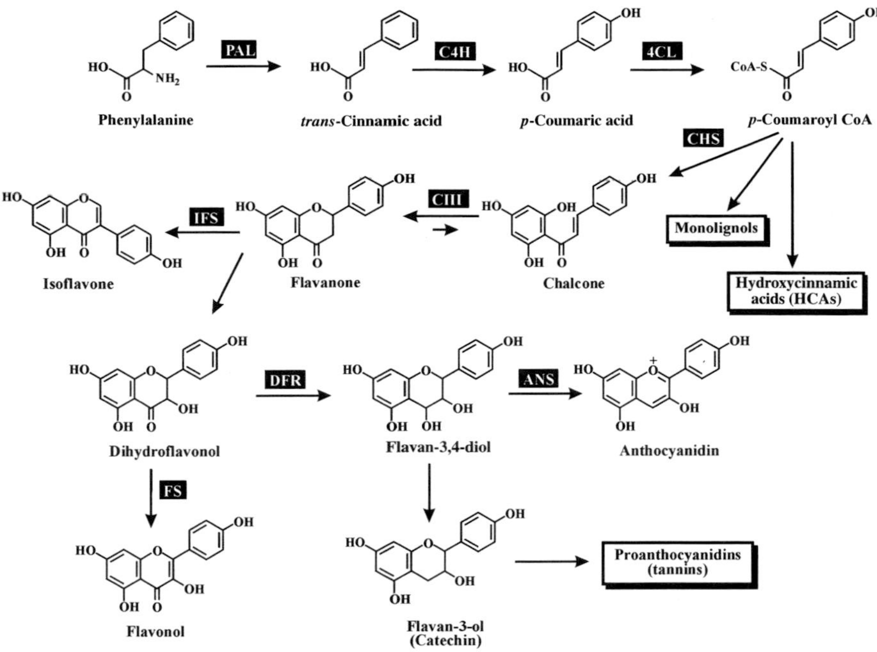

Figure 6.1 Biosynthetic relationships among phenylpropanoids and flavonoids. Key enzymes include phenylalanine ammonia lyase (PAL), cinnamate-4-hydroxylase (C4H), 4-coumarate CoA ligase (4CL), chalcone synthase (CHS), chalcone isomerase (CHI), isoflavone synthase (IFS), dihydroflavonol reductase (DFR), flavonol synthase (FS), anthocyanidin synthase (ANS).

branchpoint enzyme connecting the phenylpropanoid and flavonoid pathways. Flavonoids encompass a large group of products that includes chalcones, flavones, flavonols, catechins, anthocyanins and proanthocyanidins (condensed tannins). Figure 6.1 summarizes biosynthetic relationships and basic structures of HCAs and flavonoids. Details of the biochemistry, genetics and regulation of phenyl-propanoid and flavonoid metabolism have been extensively reviewed elsewhere (Harborne, 1988; Stafford, 1990; Weisshaar & Jenkins, 1998; Winkel-Shirley, 2001).

6.3 Stress-induced phenylpropanoid metabolism

Phenolic compounds are constitutively expressed in all higher plants. However, phenylpropanoid metabolism is often induced when plants are exposed to a wide range of environmental stresses (Dixon & Paiva, 1995). Consequently, phenyl-propanoids have long been associated with multiple stress-related functions in plants, including (i) screening of harmful UV-B radiation, (ii) defense against herbivory, (iii) defense against pathogens, (iv) protection from photoinhibition, and (v) scavenging of reactive oxygen species (ROS).

Plants produce ROS during normal metabolism and at elevated rates during periods of stress (Alscher et al., 1997; Noctor & Foyer, 1998). The most common sources of ROS arise from leakage of electrons to O_2 from the electron transport chains of chloroplasts and mitochondria (Asada, 1996; Møller, 2001). Other sources include photorespiration in peroxisomes, NADPH oxidase in cell membranes and various cell wall peroxidases (Dat et al., 2000). Under certain environmental conditions, or during sensitive developmental stages, plants may experience oxidative stress as a result of increased production of ROS or decreased activity of scavenging enzymes. Many forms of environmental stress have been shown to increase levels of ROS in plant cells and induce the expression of genes involved with protection against oxidative stress. These include high light (Grace & Logan, 1996; Karpinski et al., 1997), drought (Smirnoff, 1993), salinity (Mittova et al., 2003), low temperature (Prasad et al., 1994), heavy metals (Schützendübel & Polle, 2002), UV radiation (Landry et al., 1995), ozone (Pasqualini et al., 2003), pathogen attack (Levine et al., 1994) and wounding (Grantz et al., 1995). The role of oxidative stress as an underlying factor in plant response to environmental stress is now widely accepted (Foyer et al., 1997; Dat et al., 2000; Mittler, 2002). It is therefore noteworthy that many of the same environmental conditions that cause oxidative stress are associated with the induction of phenylpropanoid metabolism in plants. This suggests that phenolics are involved in protection against oxidative stress under adverse environmental conditions.

Plants produce a large number of antioxidants whose major function is to scavenge, or otherwise, detoxify ROS. The major enzymatic scavengers of ROS are superoxide dismutase (SOD), ascorbate peroxidase (APX) and catalase, as well as several enzymes involved in maintaining reduced antioxidant pools such as monodehydroascorbate reductase (MDAR), dehydroascorbate reductase (DHAR) and glutathione reductase (GR) (Noctor & Foyer, 1998). In addition, plants contain several low molecular weight antioxidants such as ascorbate (vitamin C) and glutathione (see Chapters 1 and 3), which are water-soluble, and α-tocopherol (vitamin E) and carotenoids (see Chapter 3), which are lipid-soluble. Until recently, phenolic compounds were not considered part of the antioxidant network of plants. However, studies of the antioxidant properties of phenolic compounds in vitro, combined with the well-characterized activation of phenolic biosynthesis in response to diverse biotic and abiotic stresses, has led to a reevaluation of the physiological function of phenolic compounds in plants (Yamasaki, 1997; Grace & Logan, 2000; Close & McArthur, 2002).

6.3.1   High light

Plants frequently encounter conditions where the rate of light absorption exceeds the capacity for photosynthesis. This often results in a repression of photosynthesis – a phenomenon known as photoinhibition (see Chapter 10). Environmental stresses that lower a plant's photosynthetic capacity, such as water stress, nutrient stress or extremes of temperature, increase the degree to which absorbed light can be excessive (Huner *et al.*, 1998; Niyogi, 1999). Therefore, plants have evolved several mechanisms to prevent over-excitation of the photosynthetic apparatus. These include morphological changes, such as changes in leaf angle or deposition of cuticular waxes, as well as biochemical changes such as an increased capacity for thermal energy dissipation (Björkman & Demmig-Adams, 1994; Müller *et al.*, 2001; Demmig-Adams & Adams, 2002).

Exposure to high light intensities leads to higher intracellular levels of ROS due to increased rates of O_2 photoreduction in chloroplasts and increased flux to H_2O_2 in peroxisomes via photorespiration (Niyogi, 1999; Mittler, 2002). Long-term acclimation to high light is generally associated with elevated activities of a host of antioxidant enzymes, including SOD, APX, GR, as well as low molecular weight antioxidants such as ascorbate and glutathione (Grace & Logan, 1996; Logan *et al.*, 1998a,b). There is also growing evidence that phenolics, acting as antioxidants, may help to prevent photodamage during high light stress (Grace *et al.*, 1998a; Grace & Logan, 2000; Neill & Gould, 2003). Support for this hypothesis comes from studies of light activation of phenylpropanoid biosynthetic genes, as well as field-based studies of phenolic concentrations in plants acclimated to different light environments.

A strong correlation has been demonstrated between light intensity and phenolic levels in a range of plant species (Mole *et al.*, 1988; Dustin & Cooper-Driver, 1992; Waterman & Mole, 1994). To cite two recent examples, Jaakola *et al.* (2004) showed that sun-exposed leaves of bilberry (*Vaccinium myrtillus* L.) had significantly higher concentrations of anthocyanins, flavonols, catechins and HCAs than shade-exposed leaves. Grace *et al.* (1998a) showed that levels of chlorogenic acid – a caffeic acid derivative – were significantly higher in sun-exposed leaves of the evergreen shrub *Mahonia repens* than shaded leaves. These differences were more pronounced in winter when plants were more susceptible to photoinhibition. In addition, there was a strong relationship between chlorogenic acid levels and total antioxidant activity of soluble leaf extracts, as measured in two *in vitro* scavenging assays. These data are consistent with the suggestion that plants increase phenolic production directly in response to oxidative pressure caused by excess light as a physiological response to quench ROS (Close & McArthur, 2002).

In photosynthetic tissues, the synthesis of phenolic compounds and expression of several genes involved in phenylpropanoid metabolism are strongly influenced by light intensity (Hahlbrock & Scheel, 1989; Beggs & Wellman, 1994). Induction of flavonoid biosynthesis by high light has been well characterized, primarily through the analysis of light-dependent anthocyanin accumulation (Mancinelli & Rabino, 1978; Mancinelli, 1985; McClure, 1986; Beggs & Wellman, 1994; Krol *et al.*, 1995; Jenkins *et al.*, 2001; Steyn *et al.*, 2002). The accumulation of flavonoids is due, in part, to the transcriptional regulation of CHS, the enzyme that catalyzes the first committed step in flavonoid biosynthesis (Sakuta, 2000; Jenkins *et al.*, 2001; Rossel *et al.*, 2002). Other phenylpropanoids, including sinapic acid esters in *Arabidopsis*, have also been shown to accumulate in a light-dependent manner (Ruegger *et al.*, 1999; Hemm *et al.*, 2004). Steady state mRNA levels for PAL, CHS and flavanone 3-hydroxylase (F3H) were found to be higher in sun-exposed leaves of bilberry (*Vaccinium myrtillus* L.) than shaded leaves (Jaakola *et al.*, 2004). Recently, Iida *et al.* (2000) identified a gene in *Arabidopsis* involved in acclimation to high light stress. Overexpression of this gene resulted in an increased tolerance to high light, anthocyanin accumulation and adaptive pheno-typic changes, such as thicker leaves with more developed palisade mesophyll. Taken together, these studies suggest that accumulation of HCAs, flavonols and anthocyanins in photosynthetic tissues is an integral part of plant response to high light stress.

6.3.2   UV radiation

Phenolic compounds have long been considered to act as screening agents that protect plants against the damaging effects of UV-B radiation. Several lines of evi-dence support this hypothesis. Flavonoids and HCAs exhibit high UV absorbance and are virtually ubiquitous in higher plants, suggesting an early evolutionary func-tion in UV protection (Rozema *et al.*, 1997; Cooper-Driver, 2001). Flavonols and flavones are present at high concentrations in the epidermal layer of leaves, consis-tent with a role in UV screening (Hutzler *et al.*, 1998; Burchard *et al.*, 2000; Gould *et al.*, 2000). Red cabbage seedlings grown in the presence of the PAL inhibitor 2-amino-indan-2-phosphonic acid (AIP) had reduced flavonoid levels and were more sensitive to the damaging effects of UV-B than plants grown in the absence of inhibitor (Gitz *et al.*, 1998). Flavonoid deficient mutants of *Arabidopsis* were shown to be hypersensitive to UV-B radiation (Li *et al.*, 1993; Lois & Buchanan, 1994), whereas mutants with elevated constitutive levels of flavonoids and HCAs were shown to be tolerant to high levels of UV-B that were otherwise lethal in wild-type plants (Bieza & Lois, 2001). *Arabidopsis* mutants deficient in sinapic acid esters were found to be even more sensitive to the damaging effects of UV-B than flavonoid-deficient mutants, suggesting that HCAs also play an important role in UV protection (Landry *et al.*, 1995). Burchard *et al.* (2000) suggested that epidermal HCAs act as constitutive UV absorbing compounds in the early stages of development in rye leaves (*Secale cereale* L.), whereas epidermal flavonoids play the predominant role in UV screening during later development and acclimation to high levels of UV-B.

Flavonoid synthesis is strongly induced by exposure to UV-B radiation (Lois, 1994; Logemann *et al.*, 2000; Bieza & Lois, 2001). In primary tissues of rye leaves, levels of soluble phenolics increased in conjunction with photosynthetic tolerance to UV-B exposure, suggesting that flavonoids may protect the photosynthetic appara-tus against the damaging effects of UV light (Reuber *et al.*, 1996). In parsley cells, induction of PAL and CHS mRNAs by UV-B was accompanied by expression of genes involved with primary metabolism, including glucose-6-phosphate dehy-drogenase (pentose phosphate pathway), 3-deoxy-arabinoheptulosonate-7-phosphate synthase (shikimate pathway) and acyl-CoA oxidase (fatty acid metabolism), suggesting that primary and secondary metabolic pathways are coordinated in response to UV stress (Logemann *et al.*, 2000).

Exposure to UV-B may cause damage to several key biological targets, including DNA, membranes and the photosynthetic apparatus (Lois & Buchanan, 1994; Strid *et al.*, 1994; Rozema *et al.*, 1997). It has been postulated that flavonoids function, at least in part, by shielding DNA from the damaging effects of UV radiation (Shirley, 1996). However, much of the damage related to UV-B exposure is believed to be caused indirectly through the production of free radicals and other reactive chemical species, although the mechanism of their generation is not known (Takeuchi *et al.*, 1996; Rozema *et al.*, 1997). Rao *et al.* (1996) demonstrated that exposure to both ozone and UV-B increased the activities of several antioxidant enzymes in *Arabidopsis*, including SOD, APX, GR and guaiacol peroxidase. Expression of the PR-1 gene in tobacco was stimulated by ROS, and the induction of PR-1 following exposure to UV-B was prevented by the application of antioxidants, suggesting that stimulation of defense genes by UV-B may involve oxidative stress (Green & Fluhr, 1995). These studies provide evidence that, in addition to their role as UV screening agents, phenolic compounds may also provide protection against UV-B by acting as antioxidants.

Several studies have suggested that anthocyanins are important as UV-screening compounds in plants (Stapleton & Walbot, 1994; Burger & Edwards, 1996). However, a screening role for anthocyanins seems unlikely due to their poor UV-absorbing properties and their often-transient expression in non-epidermal tissues (Gould *et al.*, 2000; Steyn *et al.*, 2002). Gould *et al.* (2000) suggested that HCAs, rather than anthocyanins, protect nascent leaf primordia from the damaging effects of UV-B radiation. It is possible that both HCAs and anthocyanins, acting as antioxidants, are involved in protecting plant tissues against oxidative stress caused by exposure to UV-B.

6.3.3 Low temperatures

Acclimation to low temperatures is a multifaceted process that involves coordinated changes in a host of metabolic activities (Graham & Patterson, 1982; Guy, 1990; Ndong *et al.*, 2001). One of the most well-characterized responses to low temperatures is the induction of phenylpropanoid metabolism leading to increased synthesis of phenolic compounds. A familiar example is the annual autumn reddening in leaves of deciduous plants resulting from seasonal accumulation of anthocyanins during cold acclimation (Chalker-Scott, 1999). This response is not limited to deciduous species, but also occurs in overwintering tissues of evergreen species such as *Pinus banksiana* and *Mahonia repens* (Nozzolillo *et al.*, 1990; Grace *et al.*, 1998a). Low temperatures have been shown to induce the expression of PAL and CHS in a variety of species (Christie *et al.*, 1994; Leyva *et al.*, 1995; Solecka & Kacperska, 1995). Christie *et al.* (1994) demonstrated a correlation between low-temperature-induced anthocyanin synthesis and the accumulation of PAL and

CHS mRNAs in maize. In *Arabidopsis*, the accumulation of PAL and CHS mRNAs in response to low temperature was shown to be light dependent and confined to photosynthetically active cells (Leyva *et al.*, 1995).

An important aspect of cold acclimation is the development of mechanisms that prevent photoinhibition of photosynthesis. Low temperatures severely inhibit the assimilatory reactions of photosynthesis but have little effect on the processes of light capture and electron transport (Huner *et al.*, 1993). This can lead to over-excitation of the photosynthetic apparatus resulting in photoinhibitory damage. Resistance to photoinhibition is related to cold tolerance, and changes in the redox state of PSII have been proposed as a temperature-sensing mechanism for cold acclimation (Huner *et al.*, 1998).

Nature has provided several solutions to the problems associated with low-temperature-induced photoinhibition. Chilling tolerant evergreens in northern temperate climates often show an increased capacity for thermal energy dissipation during winter (Adams & Demmig-Adams, 1995; Öquist & Huner, 2003). This process is associated with elevated levels of zeaxanthin and reorganization of the light harvesting complexes of Photosystems I and II (Adams & Demmig-Adams, 1995; Gilmore & Ball, 2000; Öquist & Huner, 2003). In contrast, chilling tolerant cereals such as rye and wheat maintain a high capacity for photosynthesis and are, therefore, less susceptible to low-temperature-induced photoinhibition than chilling sensitive species (Huner *et al.*, 1993). As with increases in the capacity for thermal energy dissipation, the maintenance of high photosynthetic capacities reduces excitation pressure on Photosystem II.

In chilling sensitive species such as maize, growth at suboptimal temperatures is associated with chronic photoinhibition, increased levels of H_2O_2 and symptoms of oxidative stress including membrane damage (Wise & Naylor, 1987; Hodgson & Raison, 1991; Okuda *et al.*, 1991; Prasad *et al.*, 1994; Pastori *et al.*, 2000). Exposure to low temperatures leads to increased levels of ROS due to increased rates of electron transport to O_2 by the electron transport chains of chloroplasts and mitochondria (Fryer *et al.*, 1998; DeSantis *et al.*, 1999). Consequently, antioxidant defense is critical to survival at low temperature. Several studies have shown that prolonged exposure to low temperatures in chilling tolerant species causes changes in the activities of a host of antioxidant enzymes, including APX, MDAR, GR, catalase and guaiacol peroxidase (Prasad, 1996; Hodges *et al.*, 1997; Logan *et al.*, 1998b; Pastori *et al.*, 2000). Jahnke *et al.* (1991) reported higher GR activities and ascorbate levels in *Zea diploperennis*, a chilling tolerant species, than *Zea mays*, a chilling sensitive species. Overexpression of SOD in maize led to increased protection from low-temperature-induced oxidative stress (Van Breusegem *et al.*, 1999). Higher antioxidant activities were observed in field-grown maize during periods when plants were exposed to chilling (Fryer *et al.*, 1998), and increased glutathione levels and GR activity were found to contribute to chilling tolerance in maize (Kocsy *et al.*, 1996).

Several recent studies have suggested that anthocyanins, acting as visible light screens, may protect photosynthetic tissues against low-temperature-induced photoinhibition (Krol *et al.*, 1995; Gould *et al.*, 2000; Feild *et al.*, 2001; Starr & Oberbauer, 2002; Steyn *et al.*, 2002; Neill & Gould, 2003). Anthocyanins often accumulate in epidermal cells during cold acclimation (Krol *et al.*, 1995), and the expression of PAL and CHS in response to low temperature was shown to be light dependent (Christie *et al.*, 1994; Leyva *et al.*, 1995). Feild *et al.* (2001) showed that anthocyanin-containing sun-exposed leaves of red-osier dogwood are significantly less susceptible to photoinhibition than shaded leaves, which do not accumulate anthocyanins. Similarly, oilseed rape seedlings grown in the presence of the PAL inhibitor AIP showed greater damage to PSII when exposed to low temperatures than plants grown in the absence of the inhibitor (Solecka & Kacperska, 2003). Although a visible light-screening role for anthocyanins has been proposed, anthocyanins may also protect the photosynthetic apparatus against chilling-induced photoinhibition by acting as antioxidants (Yamasaki, 1997; Neill *et al.*, 2002; Neill & Gould, 2003).

Anthocyanins are not the only group of phenolic compounds that have been implicated in protection against low-temperature-induced photoinhibition. In oilseed rape (*Brassica napus* L.), concentrations of both anthocyanins and HCAs were shown to increase in response to low, subfreezing temperatures (Solecka & Kacperska, 2003). A subsequent exposure to freezing temperatures led to a further increase in HCA concentrations but not in anthocyanin levels. Infiltration of leaves with the PAL inhibitor AIP led to chronic photoinhibition of PSII at low temperatures but not at normal temperatures (Solecka & Kacperska, 2003). Subsequent exposure to freezing temperatures caused severe photoinhibition in AIP-treated plants but not in control plants. These data suggest that HCAs play an important role in protecting the photosynthetic apparatus against low-temperature-induced photoinhibition. Since HCAs do not absorb visible light, a light screening role cannot explain their photoprotective effects. Given that HCAs have been shown to act as powerful antioxidants *in vitro* (Rice-Evans *et al.*, 1996; Grace *et al.*, 1998a), they may protect photosynthetic tissues against low-temperature-induced photoinhibition by enhancing the antioxidant capacity of cells.

6.3.4 Pathogens

Plants respond to pathogen attack through a variety of active and passive defense mechanisms. A common and well-characterized response to pathogen infection is an increase in phenylpropanoid metabolism leading to localized synthesis of phenolic compounds (Dalkin *et al.*, 1990; Nicholson & Hammerschmidt, 1992). Although the functional significance of this response is not fully understood, increased phenolic synthesis may result in the deposition of lignin and other high molecular weight phenolic polymers at the infection site, which is believed to prevent further spread of the pathogen. Furthermore, the induction of phenylpropanoid metabolism is often associated with the synthesis of phytoalexins, a chemically diverse group of antimicrobial defense compounds that are thought to be directly toxic to the invading pathogen. In addition to various phenylpropanoid derivatives (flavonoids, isoflavonoids, stilbenes), phytoalexins also include a host of non-phenolic compounds including terpenes, indoles and polyketides (Nicholson & Hammerschmidt, 1992; Dixon, 2001). Recently, transgenic plants have been used to evaluate the effects of altered phytoalexin profiles on plant disease resistance. Constitutive overexpression of isoflavone o-methyltransferase (IOMT) in alfalfa (*Medicago sativa*) led to more rapid and extensive production of the isoflavonoid medicarpin in response to infection by the leaf spot pathogen *Phoma medicaginis*, resulting in amelioration of disease symptoms (He & Dixon, 2000). Introduction of a novel phytoalexin, resveratrol, into alfalfa, by constitutive expression of a grapevine stilbene synthase gene, led to reduced symptoms following infection by *P. medicaginis* (Hipskind & Paiva, 2000).

In addition to stilbene and isoflavonoid phytoalexins, constitutively produced phenylpropanoids such as chlorogenic acid (CGA) have also been implicated in plant disease resistance. CGA is the major phenylpropanoid in tobacco leaves and its accumulation is regulated by PAL (Howles *et al.*, 1996). In studies of transgenic plants with altered PAL activity, susceptibility to pathogen attack was correlated with pre-formed levels of CGA (Maher *et al.*, 1994; Shadle *et al.*, 2003). PAL overexpression significantly increased CGA concentrations and reduced the severity of symptoms in plants infected with the virulent fungal pathogen *Cercospora nicotianae* (Shadle *et al.*, 2003). In contrast, PAL suppression led to reduced levels of CGA and increased susceptibility to infection by *C. nicotianae* (Maher *et al.*, 1994). The introduction of a gene encoding tryptophan decarboxylase led to a dramatic reduction in CGA levels in potato tubers by redirecting shikimic-acid-derived precursors into alternative metabolic sinks. These plants had increased susceptibility to infection by the fungal pathogen *Phytopthora infestans* (Yao *et al.*, 1995). Taken together, these studies suggest that CGA plays a role in disease resistance against fungal pathogens, although the mechanism of enhanced disease resistance in plants with high constitutive levels of CGA is not known. However, since pathogen infection is associated with oxidative stress, the ability of CGA to act as a scavenger of ROS may contribute to increased disease resistance, as has been suggested for other phenolic metabolites (Chong *et al.*, 2002).

In incompatible plant–pathogen interactions, pathogen recognition by the host initiates a hypersensitive response (HR), which is characterized by the rapid generation of superoxide and the accumulation of H_2O_2 at the infection site (Mehdy, 1994; Lamb & Dixon, 1997). This 'oxidative burst' coincides with the induction of localized cell death, which is thought to limit the spread of the invading pathogen. In parsley cells, elicitor-stimulated ROS production was found to be a prerequisite for activation of defense genes, including those involved in phenylpropanoid metabolism (Jabs *et al.*, 1997). Inhibition of elicitor-stimulated ROS production by antioxidants prevented defense gene activation and phytoalexin accumulation. In transgenic tobacco with reduced levels of catalase, cell death and defense genes could be induced by high light, which, as noted earlier, is known to increase H_2O_2 levels (Chamnongpol *et al.*, 1998). Thus, ROS appear to be an integral part of the signaling cascade leading from pathogen infection to local expression of phenylpropanoid defense genes, providing further support for the hypothesis that phenylpropanoids are involved in protecting plant tissues against oxidative stress in response to pathogens.

6.3.5   Ozone

Tropospheric ozone (O_3) is a significant air pollutant that has been shown to reduce crop yields throughout the world. Ozone is a strong oxidant that can enter leaves through stomata and oxidize a variety of cellular targets, either directly or indirectly through the production of secondary reactive species such as superoxide and H_2O_2. This eventually leads to visible symptoms of injury including chlorosis and necrosis (Pell et al., 1997). The localized lesions observed in O_3-treated leaves are similar to those produced during the hypersensitive response in incompatible plant–pathogen interactions (Schraudner et al., 1998; Wohlgemuth et al., 2002; Pasqualini et al., 2003; Chapter 11). Studies with different plant species have shown that O_3 induces expression of several defense-related genes (Sharma & Davis, 1994). Although the specific mechanisms by which O_3 causes changes in gene expression are not known, it is generally believed that O_3 initiates an oxidative burst leading to the formation of ROS, which activates a signaling cascade associated with induction of plant defense genes (see Chapter 11).

Molecular and biochemical studies have shown that O_3 stimulates phenylpropanoid metabolism in plants. The O_3-induced oxidative burst coincides with a marked increase in PAL expression and synthesis of phenolic compounds (Schraudner et al., 1998; Wohlgemuth et al., 2002; Pasqualini et al., 2003). PAL activity was also shown to increase in response to treatment with O_3 in soybean (Booker & Miller, 1998), Scots pine (Rosemann et al., 1991) and parsley (Eckey-Kaltenbach et al., 1994). Rapid increases in transcript levels for PAL and 4-coumarate CoA ligase (4CL) have been observed in soybean (Booker & Miller, 1998), Arabidopsis (Sharma & Davis, 1994), tobacco (Bahl et al., 1995) and parsley (Eckey-Kaltenbach et al., 1994). In bean leaves (Phaseolus vulgaris L.), exposure to realistic concentrations of O_3 led to a decrease in HCA levels but an increase in isoflavonoid levels (Kanoun et al., 2001). In the O_3-sensitive tobacco cultivar, Bel W3, levels of chlorogenic acid and rutin declined following O_3-exposure, possibly due to scavenging reactions against O_3-induced ROS (Pasqualini et al., 2003). Thus, it is clear that long-term exposure to moderate concentrations of O_3 can induce significant changes in the phenolic profile in leaves, which could potentially modify plant–pathogen interactions and plant response to abiotic stress.

6.4   Antioxidant properties of phenolic compounds

Over the last 15 years, a wealth of information has emerged from studies of the chemical and pharmacological properties of phenolic compounds present in fruits, vegetables, wine and tea. Although not generally considered as nutrients, a growing body of evidence suggests that dietary phenolics may contribute to chemoprevention of a variety of human diseases, including coronary heart disease and certain cancers (Hertog *et al.*, 1993; Yang *et al.*, 2001). The antioxidant activity of these compounds is believed to account, in part, for their beneficial health effects (Ames, 1995; Rietveld & Wiseman, 2003). In exploring the potential health benefits of dietary phenolics, these studies have also highlighted a potentially significant physiological function for phenolic compounds in plants.

Antioxidants are broadly defined as molecules that, when present at low concentrations compared to those of an oxidizable substrate, significantly delay or prevent oxidation of that substrate (Halliwell & Gutteridge, 1999). Phenolic compounds are excellent antioxidants by virtue of the electron donating activity of the 'acidic' phenolic hydroxyl group. Indeed, many natural and synthetic antioxidants, including α-tocopherol (Vitamin E) and butylated hydroxytoluene (BHT), are phenolic compounds. Two properties of phenolic compounds account for their radical scavenging properties. First, the one-electron reduction potentials of phenolic (phenoxyl) radicals are typically lower than those of oxygen radicals such as superoxide (O_2^-), peroxyl ($ROO^\cdot$), alkoxyl ($RO^\cdot$) and hydroxyl ($HO^\cdot$) radicals, meaning that these species will readily oxidize phenolics to their respective phenoxyl radicals (Bors *et al.*, 1990; Buettner, 1993; Jovanovic *et al.*, 1994). Second, phenoxyl radicals are generally less reactive than oxygen radicals (Bors *et al.*, 1994). Consequently, phenolic compounds can directly scavenge harmful reactive oxygen intermediates and inactivate them without promoting further oxidative reactions.

Many studies have demonstrated the radical scavenging properties of plant phenolic compounds, and some of the most relevant are briefly described here. Rice-Evans *et al.* (1996) measured the radical scavenging activity of flavonoids and phenolic acids based on their ability to scavenge a preformed radical cation chromophore of 2,2′-azinobis-(3-ethylbenzothiazoline-6-sulfonic acid) ($ABTS^{+\cdot}$) at pH 7.4. In a survey of the antioxidant properties of more than 30 flavonoids and phenolic acids, these workers found that nearly all of the plant phenolics studied exhibited higher scavenging activity than Trolox, the water soluble Vitamin E analog used as an antioxidant standard in this assay (Rice-Evans *et al.*, 1996). By way of comparison, ascorbate (Vitamin C) has an activity equivalent to Trolox, consistent with observations that flavonoids and HCAs are generally more effective than ascorbate in scavenging aqueous phase radicals (Grace *et al.*, 1998a; Fukumoto & Mazza, 2000). Among flavonoids, the highest scavenging activities were found for the flavonol quercetin, the anthocyanidins cyanidin and delphinidin, and the green tea flavan-3-ols epicatechin gallate and epigallocatechin gallate (Rice-Evans *et al.*, 1996, 1997). Figure 6.2 shows the structures of several phenolic compounds that exhibit high antioxidant activity.

Phenolic compounds with the *ortho*-dihydroxy (catechol) structure are considerably more active as hydrogen-donating antioxidants than monohydroxy phenolics (Bors *et al.*, 1990; Rice-Evans *et al.*, 1996; Chen & Ho, 1997). This is due to the fact that an additional hydroxyl group in the *ortho* position lowers the one-electron

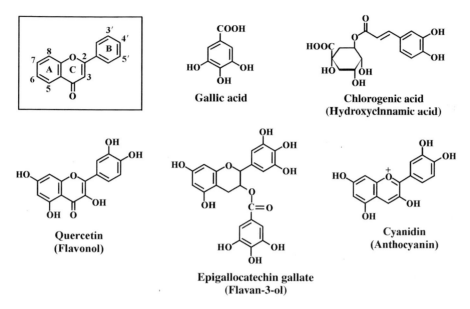

Figure 6.2   Numbering system for the flavonoid ring and examples of plant phenolic compounds with demonstrated antioxidant activity.

reduction potential of the phenolic group by approximately 300–400 mV and increases the stability of the corresponding phenoxyl (semiquinone) radical (Jovanovic *et al.*, 1994; Pietta, 2000). Consequently, phenolics containing the catechol structure are more easily oxidized, and, therefore, are much better radical scavengers than monohydroxy phenolics. Among flavonoids, antioxidant activity is further enhanced by the presence of a 2,3 double bond on the C ring, a free hydroxyl group at the 3 position on the C ring and the presence of hydroxyl groups in the 3 and 5 positions on the A ring (Figure 6.2) (Bors *et al.*, 1990; Rice-Evans *et al.*, 1996; Pietta, 2000).

Although flavonoids have been the focus of most dietary antioxidant studies, the antioxidant properties of nonflavonoid phenolic compounds have also been investigated in a number of radical scavenging assays. Gallic acid (Figure 6.2) and its methyl esters show significant antioxidant activity against the $ABTS^{+\bullet}$ radical cation, whereas simple monohydroxybenzoic acids, such as salicylic acid, do not show appreciable radical scavenging activity (Rice-Evans *et al.*, 1996). In a comparison of the antioxidant activity of several HCAs, Grace *et al.* (1998a) showed that chlorogenic acid had the highest scavenging activities against the $ABTS^{+\bullet}$ and $O_2^{-\bullet}$ radicals compared to caffeic, cinnamic and *p*-coumaric acids. These studies show that, as in the case of flavonoids, the catechol structure is essential for the radical scavenging activity of HCAs (Laranjinha *et al.*, 1994; Nardini *et al.*, 1995; Chen & Ho, 1997; Grace *et al.*, 1998a). Caffeic acid was also found to possess greater scavenging activity against the 2,2-diphenyl-1-picrylhydrazyl (DPPH) radical than both α-tocopherol and cysteine (Moon & Terao, 1998). High molecular weight phenolics such as condensed and hydrolyzable tannins have also been shown to act as strong antioxidants, which may point to a function of these compounds in long-lived tissues such heartwood and seed coats (Hagerman et al., 1998).

Apart from studies of their direct radical scavenging activities, phenolic compounds have also been shown to protect biological targets such as DNA and lipids against oxidative damage caused by ROS. Many studies have defined the antioxidant potential of plant phenolics in terms of their ability to retard oxidation of low density lipoprotein (LDL) by acting as chain-breaking antioxidants. Flavonoids and HCAs are highly effective in preventing lipid peroxidation and LDL oxidation, both of which have been implicated in the etiology of coronary heart disease (Laranjinha et al., 1994; Castelluccio et al., 1995; Nardini et al., 1995; Salah et al., 1995; Moon & Terao, 1998). Flavonoids are potent scavengers of alkyl peroxyl radicals and generally show greater antioxidant activity on a mole-for-mole basis than the 'classical' lipid-phase antioxidant α-tocopherol in membrane models of lipid peroxidation (Torel et al., 1986; Chimi et al., 1991; Terao et al., 1994; Sawa et al., 1999). It has been suggested that flavonoid aglycones located near the surface of membranes may protect membrane lipids by scavenging oxygen radicals generated in the aqueous phase (Terao et al., 1994). The metal chelating properties of phenolic compounds containing the catechol structure may also contribute to their ability to prevent lipid peroxidation (Afanas'ev et al., 1989; Nardini et al., 1995).

Over the last decade, considerable attention has been focused on the antioxidant properties of anthocyanins, the water-soluble flavonoid pigments that impart the characteristic red, blue and purple coloration seen in various plant tissues. These compounds have long been recognized to play an important ecological role in attracting pollinators and seed dispersers to aid in plant reproduction. However, the function of anthocyanins in vegetative tissues remains unclear. As noted earlier, anthocyanin synthesis shows a strong light dependence in a wide range of plant species (Mancinelli & Rabino, 1978; Mancinelli, 1985; McClure, 1986; Beggs & Wellman, 1994; Steyn et al., 2002). Anthocyanin accumulation is often a visible sign of stress, particularly in response to high light, low temperatures, wounding, pathogen infection and phosphate deficiency. There is growing evidence that anthocyanins may serve a dual role in leaves as antioxidants and as screening agents that attenuate high levels of visible light and, therefore, shield the photosynthetic apparatus against the damaging effects of high light (Chalker-Scott, 1999; Feild et al., 2001; Hoch et al., 2001; Steyn et al., 2002; Neill & Gould, 2003). The antioxidant and light screening functions of anthocyanins are not mutually exclusive since high light stress is associated with increased levels of ROS in plant cells (see Chapter 10).

Anthocyanins are among the most potent antioxidants in vitro, with capacities to quench oxygen radicals three–four times greater than the water-soluble Vitamin E analog Trolox (Rice-Evans et al., 1996; Wang et al., 1997). Cyanidin glucosides, which are widespread in plants and impart the characteristic red color of leaves and other vegetative plant tissues, can directly scavenge the $O_2^{\cdot-}$ radical (Yamasaki et al., 1996), and have been shown to reduce $O_2^{-\bullet}$ concentrations in isolated chloroplasts due to a combination of light screening and radical scavenging properties (Neill & Gould, 2003). Anthocyanins extracted from fruits and seed coats have also been shown to reduce lipid peroxidation in a concentration-dependent manner (Tsuda *et al.*, 1996; Gabrielska *et al.*, 1999). Neill *et al.* (2002) demonstrated that red leaves of *Elatostema rugosum* had significantly higher antioxidant activities than green leaves as measured by their DPPH radical scavenging activity and cyclic voltammetry assays of leaf extracts. In addition to anthocyanins, red leaves also contained higher concentrations of caffeic acid derivatives than green leaves, which would also contribute to higher antioxidant activities.

Flavonoids and HCAs have also been shown to scavenge peroxynitrite (PN), a highly toxic oxidant formed by the rapid reaction between $O_2^{-\bullet}$ and nitric oxide (NO$^{\bullet}$). Macrophages, neutrophils and endothelial cells have all been shown to generate PN, and evidence is accumulating that PN may also be produced in plants (Yamasaki & Sakihama, 2000). At physiological pH, PN is protonated to form peroxynitrous acid, which decomposes into a variety of secondary reactive species (Squadrito & Pryor, 1998). The oxidizing properties of peroxynitrous acid are similar to those of the $^{\bullet}$OH radical, as shown by the ability of PN to cause a spectrum of oxidative damage to DNA and lipids akin to the $^{\bullet}$OH radical (Radi *et al.*, 1991; Burney *et al.*, 1999). PN is also a potent nitrating species, particularly towards aromatic molecules such as tyrosine and guanine. The formation of 3-nitrotyrosine has been used as an indicator of PN formation *in vivo*, although recent studies have shown that 3-nitrotyrosine may be a biomarker for several reactive nitrogen species in addition to PN (Beckman & Koppenol, 1996; Brennan *et al.*, 2002).

Pannala *et al.* (1997) showed that flavonoids can scavenge PN and prevent the formation of 3-nitrotyrosine. The major flavonoids found in green tea (epicatechin, epigallocatechin, epicatechin gallate and epigallocatechin gallate) were found to be more effective than Trolox in scavenging PN. Haenan *et al.* (1997) observed that a catechol group on the B-ring and a hydroxyl group at position 3 were required for PN scavenging activity. However, Tsuda *et al.* (2000) reported that pelargonidin, an anthocyanin containing a single B-ring hydroxyl group, can also react with PN and prevent tyrosine nitration. HCA derivatives with the *o*-dihydroxy substitution pattern (caffeic acid, chlorogenic acid) or the *o*-methylated analog ferulic acid were more effective in scavenging PN than Trolox or several monohydroxy isomers of coumaric acid (Pannala *et al.*, 1998). Spectral analysis indicates that ferulic acid and isomers of coumaric acid undergo nitration by PN whereas the catechols caffeic acid and chlorogenic acid undergo oxidation but not nitration (Grace *et al.*, 1998b; Pannala *et al.*, 1998). The PN scavenging activity of HCAs was confirmed by Grace *et al.* (1998b), who showed that chlorogenic acid inhibits the formation of strand breaks in DNA exposed to PN. This activity was enhanced in the presence of horseradish peroxidase, which can act as an enzymatic scavenger of PN if a suitable electron donor such as chlorogenic acid is present to maintain the active form of the enzyme.

6.5   Biological targets of phenolic antioxidants

The ability of purified phenolics to scavenge ROS does not necessarily translate to enhanced antioxidant protection *in vivo*. Since most forms of ROS are short-lived, antioxidants must be localized close to sites of ROS production to prevent oxidative damage to sensitive biological targets. However, evidence for an antioxidant function of phenolic compounds *in vivo* was recently provided by Gould *et al.* (2002), who measured ROS levels in *Pseudowintera colorata* leaves following mechanical injury using dichlorofluorescein (DCFH) and scopoletin as histochemical probes for H_2O_2. Wounding produced a localized burst of H_2O_2 from palisade mesophyll cells, the intensity and duration of which were lower in red leaf laminae than green laminae. Red leaves were enriched in anthocyanins, flavonols, dihydroflavonols and HCAs, all of which exhibit antioxidant properties *in vitro*. These data are consistent with the hypothesis that cells with high concentrations of phenolics have an elevated antioxidant status *in vivo*, which may help to lower steady state concentrations of ROS during exposure to environmental stress.

Plant phenolics show considerable variability with respect to their tissue and subcellular localization. Flavonoids and HCAs are synthesized in the cytoplasm from shikimic-acid-derived precursors and may be transported into the vacuole as glycosides or glutathione conjugates (Alfenito *et al.*, 1998; Walczak & Dean, 2000; Frangne *et al.*, 2002) or deposited in the cell wall as covalently linked co-polymers (Harris & Hartley, 1976; Strack *et al.*, 1988; Hutzler *et al.*, 1998). Studies of flavonoid deficient mutants of *Arabidopsis* suggest that flavonoids are synthesized in the same cells in which they accumulate (Peer *et al.*, 2001). Hutzler *et al.* (1998) analyzed the tissue and subcellular distribution of phenolics in leaves of several plant species using confocal laser scanning microscopy. In broad bean (*Vicia faba*) and rye (*Secale cereale*), flavonol glycosides were localized primarily in vacuoles of epidermal cells, but were also present in mesophyll cells (Hutzler *et al.*, 1998; Burchard *et al.*, 2000). Primary leaves of rye were also found to contain high concentrations of soluble HCAs, which were localized exclusively in epidermal cells (Schultz & Weissenböck, 1986). Flavonoids have also been reported in chloroplasts of barley (Saunders & McClure, 1976). In needles of Norway spruce (*Picea abies*), flavonols were detected in the nucleus and cytoplasm of epidermal cells (Hutzler *et al.*, 1998). Flavonols have also been detected in nuclei of *Flaveria chloraefolia* and *Arabidopsis* cells, where it has been suggested they may function as antioxidants or transcriptional regulatory factors (Grandmaison & Ibrahim, 1996; Peer *et al.*, 2001).

Gould *et al.* (2000) found that different flavonoid classes accumulated in specific locations in leaves of *Quintinia serrata*. Flavones were found exclusively in the epicuticular waxes, flavonols were found in the vacuoles of epidermal cells and anthocyanins were associated predominantly with leaf mesophyll cells. This pattern of spatial localization suggests that different flavonoid classes serve distinct roles in plants. For example, flavones are optimally located to intercept UV-B radiation, but presumably do not play a role in scavenging ROS generated in mesophyll cells. Similarly, the location of anthocyanins is unsuitable for screening UV-B, but is ideal for scavenging ROS produced by chloroplasts.

Chloroplasts are considered to be the major source of ROS in leaf cells (Asada, 1996; Foyer, 1997), whereas soluble phenolics are thought to be localized primarily in the vacuole. Vacuolar phenolics are unlikely to encounter $O_2^{-\cdot}$ radicals originating from chloroplasts, since $O_2^{-\cdot}$ cannot readily diffuse across the phospholipid membranes surrounding the chloroplast and vacuole (Takahashi & Asada, 1983). However, under physiological conditions, $O_2^{-\cdot}$ is rapidly protonated to the neutral hydroperoxyl radical ($HO_2^{\cdot}$), or dismutated by SOD to H_2O_2, both of which can freely permeate the tonoplast and diffuse into the vacuole. Since H_2O_2 is less chemically reactive than oxygen radicals, it has the potential to diffuse over greater distances within the cell than most other forms of ROS.

Although plant phenolics are potent antioxidants, their spatial separation from sites of oxidant production in chloroplasts and mitochondria seems inconsistent with an antioxidant function *in vivo*. However, high levels of ROS formed during severe stress may result in leakage of H_2O_2 out of these organelles into the cytoplasm, vacuole and apoplast. In contrast to their radical scavenging properties, phenolics do not directly scavenge H_2O_2. However, Takahama (1989) demonstrated that flavonols can be oxidized by H_2O_2 in a process requiring a vacuolar form of peroxidase in leaves. Other studies have confirmed that flavonols, anthocyanins and HCAs are readily oxidized by H_2O_2 in the presence of guaiacol peroxidase, which is localized in various cellular compartments but is especially abundant in vacuoles and in the apoplast (Takahama & Oniki, 1992; Yamasaki, 1997; Yamasaki et al., 1997; Yamasaki & Grace, 1998; Zancani & Nagy, 2000). Upon oxidation, phenolic compounds are regenerated to their reduced, active form by ascorbate and possibly by thiols as well (Takahama & Oniki, 1992; Yamasaki & Grace, 1998; Zancani & Nagy, 2000). Sakihama et al. (2000) recently provided evidence indicating that MDAR may also catalyze the reduction of phenoxyl radicals to their parent phenols. Thus, phenolic compounds, acting as electron donors to guaiacol peroxidases in vacuoles and in the apoplast, may serve an auxiliary antioxidant role to complement the ascorbate peroxidase-based H_2O_2 scavenging system. It has been proposed that the phenolic peroxidase-based scavenging system would assume increasing importance under conditions of severe stress when H_2O_2 produced in chloroplasts and mitochondria overwhelms the scavenging capacity of these organelles and diffuses into other cellular compartments (Yamasaki et al., 1997; Yamasaki & Grace, 1998; Grace & Logan, 2000).

6.6 Prooxidant properties of phenolic compounds

Plant phenolics are generally considered to act as antioxidants due to their ability to scavenge ROS and chelate metals. However, phenolics can also exhibit prooxidant activity under certain conditions, which may result in oxidative damage to several biological targets (Hodnick et al., 1986; Laughton et al., 1989; Cao et al., 1997;

Yamanaka *et al.*, 1997). This dual action is related to the fact that hydrogen atom donation by *o*-diphenols, the primary basis for antioxidant activity, results in the formation of a phenoxyl (semiquinone) radical that can undergo secondary reactions of a prooxidant nature. Phenoxyl radicals can react with O_2 to produce $O_2^-\cdot$ (Reaction 6.1) which can be reduced further to H_2O_2 by the $O_2^-\cdot$ scavenging activity of the parent phenolic (Reaction 6.2).

$$Ph–O^-\cdot + O_2 \rightarrow Ph{=}O + O_2^-\cdot \qquad (6.1)$$

$$Ph–OH + O_2^-\cdot + H^+ \rightarrow Ph–O^-\cdot + H_2O_2 \qquad (6.2)$$

In addition, certain phenolics are prone to autoxidation (Reaction 6.3), which can generate $O_2^-\cdot$ and lead to more reactive species such as H_2O_2 and the $\cdot$OH radical.

$$Ph–OH + O_2 \rightarrow Ph–O^-\cdot + O_2^-\cdot + H^+ \qquad (6.3)$$

The combination of these reactions accounts for the redox cycling activity of some flavonoids and nonflavonoid phenolics, with the concomitant production of ROS (Hodnick *et al.*, 1988; Canada *et al.*, 1990).

Flavonoids and phenolic acids often demonstrate prooxidant activity in the presence of transition metals such as iron or copper (Cao *et al.*, 1997; Yamanaka *et al.*, 1997; Sakihama *et al.*, 2002). Although phenolics have been shown to chelate redox active metals (Afanas'ev *et al.*, 1989; Nardini *et al.*, 1995), they can also reduce them, thereby increasing their ability to form free radicals from peroxides (Decker, 1997). DNA is an especially important biological target of metal-catalyzed oxidative damage. It has been shown that the interaction of plant phenolics with DNA-associated copper can result in a spectrum of DNA lesions, including oxidative base modifications, strand breaks and formation of DNA adducts (Li & Trush, 1994). In a recent study, Sakihama *et al.* (2002) investigated the prooxidative effects of several *o*-dihydroxycinnamic acids by measuring their ability to generate $\cdot$OH radicals and induce strand breaks in DNA in the presence of Cu(II). These workers found significant differences in the prooxidant activity of phenolic 'antioxidants' and proposed that the ability of their respective semiquinones to generate $O_2^-\cdot$ (Reaction 6.1) was the major factor in the oxidative potential of catechol–copper complexes.

6.7   Anti-herbivore properties of phenolics

Flavonoids and tannins have long been associated with plant defense against herbivory, a property that has been attributed to their ability to bind proteins and increase astringency, thus lowering the digestibility and palatability of leaves. The ability of phenolics to form complexes with proteins via both covalent and non-covalent interactions may impair enzymatic functions, reduce protein digestibility and reduce the bioavailability of amino acids (Robbins *et al.*, 1987). Oxidation of phenolics usually occurs as a result of tissue damage caused by insects due to activation of the enzyme polyphenol oxidase (PPO). Oxidation of catechols by PPO enhances the anti-nutritive properties of phenolics by producing highly reactive o-quinones, which undergo polymerization and cross-linking to produce the commonly observed browning of injured plant tissues. Quinones are also able to covalently modify free amino and sulfhydryl groups in dietary proteins within the mouth and gut of the insect (Felton *et al.*, 1989).

In addition to forming reactive quinones, it has been proposed that phenolics can induce oxidative stress by generating $O_2^{\cdot}$ and other forms in ROS in the gut of phytophagous insects (Appel, 1993; Summers & Felton, 1994). This process may be enhanced by alkaline pH, oxidative enzymes such as peroxidase and PPO and redox active metals (Appel, 1993; Sakihama *et al.*, 2002). To test the prediction that high levels of phenolics increase resistance to insect herbivory by inducing oxidative stress, Johnson and Felton (2001) investigated biochemical markers of oxidative stress and antioxidant capacity of midgut fluid and hemolymph in the tobacco budworm, *Heliothis virescens*, in feeding experiments using transgenic tobacco with elevated levels of chlorogenic acid. Contrary to predictions that high phenolic levels would exacerbate oxidative stress, these workers found that elevated levels of chlorogenic acid improved the antioxidant activity of larval fluids. Thus, high levels of phenolics may actually benefit the antioxidant status of phytophagous insects, much as dietary phenolics do in mammals.

Although the concept that phenolics evolved as plant defense compounds in response to herbivory pressures is well established in the ecological literature, several studies have failed to demonstrate a significant negative effect of plant phenolics on insect herbivory. For example, in a survey of 16 woody plant species and 6 herbivorous insect species, Ayers *et al.* (1997) could find little evidence that tannins provide an evolutionarily stable plant defense, and concluded that 'selective pressures from folivorous insects cannot explain the diversion of so much carbon, in so many plant species, into the synthesis of condensed tannins'. Recently, Close and McArthur (2002) challenged the concept that phenolics evolved as plant defense compounds against herbivory, and proposed instead that the main role of plant phenolics may be to protect leaves from photodamage, not herbivores, by acting as antioxidants. According to this hypothesis, plants invest carbon in phenolic compounds in response to oxidative pressures produced from exposure to excess light as a physiological response to quench ROS.

6.8 Conclusions

Phenylpropanoids and flavonoids play important structural roles in plants, as well as in defense against a variety of biotic and abiotic stresses. Research over the last decade has clearly demonstrated that a wide range of environmental factors that predispose plants to oxidative stress, i.e. high light, UV radiation, low temperatures, ozone and pathogens, induce the synthesis of phenolic metabolites with antioxidant properties. Taken together, information obtained from biochemical studies of gene expression and physiological studies of plant stress responses provides ample evidence that certain phenolics serve a primary role as antioxidants in plants. It is likely that phenolics act in concert with other protective molecules in plant cells, including enzymatic and non-enzymatic scavengers of ROS, perhaps compensating for deficiencies of such molecules during periods of stress. A future challenge will be to identify phenolic metabolites that are indispensable in providing protection against oxidative stress, and to elucidate mechanisms by which antioxidant activity is coordinated in cellular environments. Another important area of research will be to identify how oxidative signals regulate the expression of phenylpropanoid defense genes. Such research may facilitate efforts to engineer secondary metabolism to improve the quality and quantity of phenolic antioxidants in agriculturally important plants.

References

Adams, W.W., III and Demmig-Adams, B. (1995) 'The xanthophyll cycle and sustained thermal energy dissipation activity in *Vinca minor* and *Euonymus kiautschovicus* in winter', *Plant, Cell and Environment* **18**, 117–127.

Afanas'ev, I.B., Dorozhko, A.I., Brodskii, A.V., Kostyuk, A. and Potapovitch, A.I. (1989) 'Chelating and free radical scavenging mechanisms of inhibitory action of rutin and quercetin in lipid peroxidation', *Biochemical Pharmacology* **38**, 1763–1769.

Alfenito, M.R., Souer, E., Goodman, C.D. *et al.* (1998) 'Functional complementation of anthocyanin sequestration in the vacuole by widely divergent glutathione S-transferases', *The Plant Cell* **10**, 1135–1149.

Alscher, R.G., Donahue, J.L. and Cramer, C.L. (1997) 'Reactive oxygen species and antioxidants: relationships in green cells', *Physiologia Plantarum* **100**, 224–233.

Ames, B.N. (1995) 'The causes and prevention of cancer', *Proceedings of the National Academy of Sciences USA* **92**, 5258–5265.

Appel, H.M. (1993) 'Phenolics in ecological interactions: the importance of oxidation', *Journal of Chemical Ecology* **19**, 1521–1552.

Asada, K. (1996) Radical production and scavenging in the chloroplasts, in *Photosynthesis and the Environment* (ed. N.R. Baker), Kluwer Academic Publishers, Dordrecht, pp. 123–150.

Ayres, M.P., Clausen, T.P., MacLean, S.F., Redman, A.M. and Reichardt, P.B. (1997) 'Diversity of structure and antiherbivore activity in condensed tannins', *Ecology* **78**, 1696–1712.

Bahl, A., Loitsch, S.M. and Kahl, G. (1995) 'Transcriptional activation of plant defence genes by short-term air pollutant stress', *Environmental Pollution* **89**, 221–227.

Bais, H.P., Vepachedu, R., Gilroy, S., Callaway, R.M. and Vivanco, J.M. (2003) 'Allelopathy and exotic plant invasion: from molecules and genes to species interactions' *Science* **301**, 1377–1380.

Beckman, J.S. and Koppenol, W.H. (1996) 'Nitric oxide, superoxide, and peroxynitrite: the good, the bad, and ugly', *American Journal of Physiology and Cell Physiology* **271**, C1424–C1437.

Beggs, C.J. and Wellman, E. (1994) Photocontrol of flavonoid biosynthesis, in *Photomorphogenesis in Plants*, 2nd edn. (ed. R.E. Kendrick and G.H.M. Kronenberg), Kluwer Academic, Dordrecht, pp. 733–751.

Bieza, K. and Lois, R. (2001) 'An *Arabidopsis* mutant tolerant to lethal ultraviolet-B levels shows constitutively elevated accumulation of flavonoids and other phenolics', *Plant Physiology* **126**, 1105–1115.

Björkman, O. and Demmig-Adams, B. (1994) Regulation of photosynthetic light energy capture, conversion, and dissipation in leaves of higher plants, in *Ecophysiology of Photosynthesis* (ed. E.-D. Schulze and M.M. Caldwell), Springer, Berlin, pp. 17–47.

Booker, F.L. and Miller, J.E. (1998) 'Phenylpropanoid metabolism and phenolic composition of soybean [*Glycine max* (l. Merr.)] leaves following exposure to ozone', *Journal of Experimental Botany* **49**, 1191–1202.

Bors, W., Heller, W., Michel, C. and Saran, M. (1990) 'Flavonoids as antioxidants: determination of radical scavenging efficiencies', *Methods in Enzymology* **186**, 343–355.

Bors, W., Michel, C. and Saran, M. (1994) 'Flavonoid antioxidants: rate constants for reactions with oxygen radicals', *Methods in Enzymology* **234**, 420–429.

Brennan, M.L., Wu, W., Fu, X. *et al.* (2002) 'A tale of two controversies: defining both the role of peroxidases in nitrotyrosine formation in vivo using eosinophil peroxidase and myeloperoxidase-deficient mice, and the nature of peroxidase-generated reactive nitrogen species', *The Journal of Biological Chemistry* **277**, 17415–17427.

Buettner, G.R. (1993) 'The pecking order of free radicals and antioxidants: lipid peroxidation, alpha-tocopherol, and ascorbate', *Archives of Biochemistry and Biophysics* **300**, 535–543.

Burchard, P., Bilger, W. and Weissenböck, G. (2000) 'Contribution of hydroxycinnmates and flavonoids to epidermal shielding of UV-A and UV-B radiation in developing rye primary leaves as assessed by ultraviolet-induced chlorophyll fluorescence measurements', *Plant, Cell and Environment* **23**, 1373–1380.

Burger, J. and Edwards, G.E. (1996) 'Photosynthetic efficiency, and photodamage by UV and visible radiation, in red versus green leaf Coleus varieties', *Plant & Cell Physiology* **37**, 395–399.

Burney, S., Caulfield, J.L., Niles, J.C., Wishnok, J.S. and Tannenbaum, S.R. (1999) 'The chemistry of DNA damage from nitric oxide and peroxynitrite', *Mutation Research* **424**, 37–49.

Canada, A.T., Giannella, E., Nguyen, T.D. and Mason, R.P. (1990) 'The production of reactive oxygen species by dietary flavonols', *Free Radical Biology & Medicine* **9**, 441–449.

Cao, G., Sofic, E. and Prior, R.L. (1997) 'Antioxidant and prooxidant behavior of flavonoids: structure–activity relationships', *Free Radical Biology & Medicine* **22**, 749–760.

Castelluccio, C., Paganga, G., Melikian, N. *et al.* (1995) 'Antioxidant potential of intermediates in phenylpropanoid metabolism in higher plants', *FEBS Letters* **368**, 188–192.

Chalker-Scott, L. (1999) 'Environmental significance of anthocyanins in plant stress responses', *Photochemistry and Photobiology* **70**, 1–9.

Chamnongpol, S., Willekens, H., Moeder, W. *et al.* (1998) 'Defense activation and enhanced pathogen tolerance by H_2O_2 in transgenic tobacco', *Proceedings of the National Academy of Sciences USA* **95**, 5818–5823.

Chen, J.H. and Ho, C.-T. (1997) 'Antioxidant activities of caffeic acid and its related hydroxycinnamic acid compounds', *Journal of Agricultural and Food Chemistry* **45**, 2374–2378.

Chimi, H., Cillard, J., Cillard, P. and Rahmani, M. (1991) 'Peroxyl radical scavenging activity of some natural phenolic antioxidants', *Journal of the American Oil Chemists' Society* **68**, 307–312.

Chong, J., Baltz, R., Schmitt, C., Beffa, R., Fritig, B. and Saindrenan, P. (2002) 'Downregulation of a pathogen-responsive tobacco UDP-Glc : phenylpropanoid glucosyltransferase reduced scopoletin glucoside accumulation, enhances oxidative stress, and weakens virus resistance', *The Plant Cell* **14**, 1093–1107.

Christie, P.J., Alfenito, M.R. and Walbot, V. (1994) 'Impact of low-temperature stress on general phenylpropanoid and anthocyanin pathways: enhancement of transcript abundance and anthocyanin pigmentation in maize seedlings', *Planta* **194**, 541–549.

Close, D.C. and McArthur, C. (2002) 'Rethinking the role of many plant phenolics – protection from photodamage not herbivores?', *Oikos* **99**, 166–172.

Coley, P.D., Bryant, J.P. and Chapin, F.S. (1985) 'Resource availability and plant antiherbivore defense', *Science* **230**, 895–899.

Cooper-Driver, G. (2001) Biological roles for phenolic compounds in the evolution of early land plants, in *Plants Invade the Land: Evolutionary and Environmental Perspectives* (ed. P.G. Gensel and D. Edwards), Columbia University Press, New York, pp. 159–172.

Croteau, R., Kutchan, T.M. and Lewis, N.G. (2000) Natural products (secondary metabolites), in *Biochemistry and Molecular Biology of Plants* (ed. B. Buchanan, W. Gruissen and R. Jones), American Society of Plant Physiologists, Beltsville, USA, pp. 1250–1318.

Dalkin, K., Edwards, R., Edington, B. and Dixon, R.A. (1990) 'Stress responses in alfalfa (*Medicago sativa* L.) I. Induction of phenylpropanoid biosynthesis and hydrolytic enzymes in elicitor-treated cell suspension cultures', *Plant Physiology* **92**, 440–446.

Dat, J., Vandenabeele, S., Vranová, E., Van Montagu, M., Inzé, D. and Van Breusegem, F. (2000) 'Dual action of the active oxygen species during plant stress responses', *Cellular and Molecular Life Sciences* **57**, 779–795.

Decker, E.A. (1997) 'Phenolics: prooxidants or antioxidants?', *Nutrition Reviews* **55**, 396–398.

Demmig-Adams, B. and Adams, W.W., III (1992) 'Photoprotection and other responses of plants to high light stress', *Annual Review of Plant Physiology and Plant Molecular Biology* **43**, 599–626.

De Santis, A., Landi, P. and Genchi, G. (1999) 'Changes in mitochondrial properties in maize seedlings associated with selection for germination at low temperatures. Fatty acid composition, cytochrome c oxidase, and adenine nucleotide translocase activities', *Plant Physiology* **119**, 743–754.

Dixon, R.A. (2001) 'Natural products and plant disease resistance', *Nature* **411**, 843–847.

Dixon, R.A. and Paiva, N.L. (1995) 'Stress-induced phenylpropanoid metabolism', *The Plant Cell* **7**, 1085–1097.

Dustin, C. and Cooper-Driver, G. (1992) 'Changes in phenolic production in the hay-scented fern (*Dennstaedtia punctilobula*) in relation to resource availability', *Biochemical Systematics and Ecology* **20**, 99–106.

Eckey-Kaltenbach, H., Ernst, D., Heller, W. and Sandermann, Jr., H. (1994) 'Biochemical plant responses to ozone. IV. Cross-induction of defensive pathways in parsley (*Petroselinium crispum* L.) plants', *Plant Physiology* **104**, 67–74.

Feeny, P.P. (1976) 'Plant apparency and chemical defense', *Recent Advances in Phytochemistry* **10**, 1–40.

Feild, T.S., Lee, D.W. and Holbrook, N.M. (2001) 'Why leaves turn red in autumn. The role of anthocyanins in senescing leaves of red-osier dogwood', *Plant Physiology* **127**, 566–574.

Felton, G.W., Donato, K.K., Del Vecchio, R.J. and Duffey, S.S. (1989) 'Activation of foliar oxidases by insect feeding reduces nutritive quality of dietary protein for foliage for noctuid herbivores', *Journal of Chemical Ecology* **15**, 2667–2693.

Foyer, C.H. (1997) Oxygen metabolism and electron transport in photosynthesis, in *Oxidative Stress and the Molecular Biology of Antioxidant Defenses* (ed. J.G. Scandalios), Cold Spring Harbor Laboratory Press, Cold Spring Harbor, NY, pp. 587–621.

Foyer, C.H., Lopez-Delgado, H., Dat, J.F. and Scott, I.M. (1997) 'Hydrogen peroxide and glutathione-associated mechanisms of acclimatory stress tolerance and signaling', *Physiologia Plantarum* **100**, 241–254.

Frangne, N., Eggmann, T., Koblischke, C., Weissenböck, G., Martinoia, E. and Klein, M. (2002) 'Flavone glucoside uptake into barley mesophyll and *Arabidopsis* cell culture vacuoles. Energization occurs by H^+-antiport and ATP-binding cassette-type mechanisms', *Plant Physiology* **128**, 726–733.

Fryer, M.J., Andrews, J.R., Oxborough, K., Blowers, D.A. and Baker, N.R. (1998) 'Relationships between CO_2 assimilation, photosynthetic electron transport and active O_2 metabolism in leaves of maize in the field during periods of low temperature', *Plant Physiology* **116**, 571–580.

Fukumoto, L.R. and Mazza, G. (2000) 'Assessing antioxidant and prooxidant activities of phenolic compounds', *Journal of Agricultural and Food Chemistry* **48**, 3597–3604.

Gabrielska, J., Oszmianski, J., Komorowska, M. and Langner, M. (1999) 'Anthocyanin extracts with antioxidant radical scavenging effect', *Zeitschrift fur Naturforschung* **54**, 314–324.

Gilmore, A.M. and Ball, M.C. (2000) 'Protection and storage of chlorophyll in overwintering evergreens', *Proceedings of the National Academy of Sciences USA* **97**, 11098–11101.

Gitz, D.C., Liu, L. and McClure, J.W. (1998) 'Phenoloic metabolism, growth, and UV-B tolerance in phenylalanine ammonia-lyase-inhibited red cabbage seedlings', *Phytochemistry* **49**, 377–386.

Gould, K.S., Markham, K.R., Smith, R.H. and Goris, J.J. (2000) 'Functional role of anthocyanins in the leaves of *Quintinia serrata* A.', *Journal of Experimental Botany* **51**, 1107–1115.

Gould, K.S., McKelvie, J. and Markham, K.R. (2002) 'Do anthocyanins function as antioxidants in leaves? Imaging of H_2O_2 in red and green leaves after mechanical injury', *Plant, Cell and Environment* **25**, 1261–1269.

Grace, S.C. and Logan, B.A. (1996) 'Acclimation of foliar antioxidant systems to growth irradiance in three broad-leaved evergreen species', *Plant Physiology* **112**, 1631–1640.

Grace, S.C. and Logan, B.A. (2000) 'Energy dissipation and radical scavenging by the plant phenylpropanoid pathway', *Philosophical Transactions of the Royal Society of London Series B* **355**, 1499–1510.

Grace, S.C., Logan, B.A. and Adams, W.W., III (1998a) 'Seasonal differences in foliar content of chlorogenic acid, a phenylpropanoid antioxidant, in *Mahonia repens*', *Plant, Cell and Environment* **21**, 513–522.

Grace, S.C., Salgo, M.G. and Pryor W.A. (1998b) 'Scavenging of peroxynitrite by a phenolic-peroxidase system prevents oxidative damage to DNA', *FEBS Letters* **426**, 24–28.

Graham, D. and Patterson, B.D. (1982) 'Responses of plants to low, non-freezing temperatures: proteins, metabolism and acclimation', *Annual Review of Plant Physiology* **33**, 347–372.

Grandmaison, J. and Ibrahim, J.K. (1996) 'Evidence for nuclear binding of flavonol sulphate esters of *Flaveria chloraefolia*', *Journal of Plant Physiology* **147**, 653–660.

Grantz, A.A., Brummell, D.A. and Bennett, A.B. (1995) 'Ascorbate free radical reductase mRNA levels are induced by wounding', *Plant Physiology* **108**, 411–418.

Green, R. and Fluhr, R. (1995) 'UV-B-induced PR-1 accumulation is mediated by active oxygen species', *The Plant Cell* **7**, 203–212.

Guy, C.L. (1990) 'Cold acclimation and freezing tolerance: role of protein metabolism', *Annual Review of Plant Physiology and Plant Molecular Biology* **41**, 187–223.

Haenen, G.R.M., Paquay, J.B.G., Korthouwer, R.E.M. and Bast, A. (1997) 'Peroxynitrite scavenging by flavonoids', *Biochemical and Biophysical Research Communications* **236**, 591–593.

Hagerman, A.E., Riedl, K.M. and Jones, G.A. (1998) 'High molecular weight plant polyphenolics (tannins) as biological antioxidants', *Journal of Agricultural and Food Chemistry* **46**, 1887–1892.

Hahlbrock, K. and Scheel, D. (1989) 'Physiology and molecular biology of phenylpropanoid metabolism', *Annual Review of Plant Physiology and Plant Molecular Biology* **40**, 347–369.

Halliwell, B. and Gutteridge, J.M.C. (1999) *Free Radicals in Biology and Medicine*, 3rd edn., Oxford University Press, London, UK.

Hamilton, J.G., Zangerl, A.R., DeLucia, E.H. and Berenbaum, M.R. (2001) 'The carbon-nutrient balance hypothesis: its rise and fall', *Ecological Letters* **4**, 86–95.

Harborne, J.B. (1988) *The Flavonoids. Advances in Research since 1980*, Chapman and Hall, New York, USA.

Harris, P.J. and Hartley, R.D. (1976) 'Detection of bound ferulic acid in cell walls of the Gramineae by ultraviolet fluorescence microscopy', *Nature* **259**, 508–510.

He, X.-Z. and Dixon, R.A. (2000) 'Genetic manipulation of isoflavone7-O-methyltransferase enhances the biosynthesis of 4'-O-methylated isoflavonoid phytoalexins and disease resistance in alfalfa', *The Plant Cell* **12**, 1689–1702.

Hemm, M.R., Rider, S.D., Ogas, J., Murry, D.J. and Chapple, C. (2004) 'Light induces phenylpropanoid metabolism in *Arabidopsis* roots', *The Plant Journal* **38**, 765–778.

Hertog, M.G., Feskens, E.J., Hollman, P.C., Katan, M.B. and Kromhout D. (1993) 'Dietary antioxidant flavonoids and risk of coronary heart disease: the Zutphen elderly study', *Lancet* **342**, 1007–1011.

Hipskind, J.D. and Paiva, N.L. (2000) 'Constitutive accumulation of a resveratrol-glucoside in transgenic alfalfa increases resistance to *Phoma medicaginis*', *Molecular Plant–Microbe Interactions* **13**, 551–562.

Hoch, W.A., Zeldin, E.L. and McCown, B.H. (2001) 'Physiological significance of anthocyanins during autumnal leaf senescence', *Tree Physiology* **21**, 1–8.

Hodges, D.M., Andrews, C.J., Johnson, D.A. and Hamilton, R.I. (1997) 'Antioxidant enzyme responses to chilling stress in differentially sensitive inbred maize lines', *Journal of Experimental Botany* **48**, 1105–1113.

Hodgson, R.A.J. and Raison, J.K. (1991) 'Superoxide production by thylakoids during chilling and its implication in the susceptibility of plants to chilling-induced photoinhibition', *Planta* **183**, 222–228.

Hodnick, W.F., Kalynaraman, B., Pritsos, C.A. and Pardini, R.S. (1988) 'The production of hydroxyl and semiquinone free radicals during the autoxidation of redox active flavonoids', *Basic Life Sciences* **49**, 149–152.

Hodnick, W.F., Kung, F.S., Roettger, W.J., Bohmont, C.W. and Pardini, R.S. (1986) 'Inhibition of mitochondrial respiration and production of toxic oxygen radicals by flavonoids. A structure–activity study', *Biochemical Pharmacology* **35**, 2345–2357.

Howles, P.A., Sewalt, V.J.H., Paiva, N.L. *et al.* (1996) 'Overexpression of L-phenylalanine ammonia-lyase in transgenic tobacco reveals control points for flux into phenylpropanoid biosynthesis', *Plant Physiology* **112**, 1617–1624.

Huner, N.P.A., Öquist, G., Hurry, V.M., Krol, M., Falk, S. and Griffith, M. (1993) 'Photosynthesis, photoinhibition and low temperature acclimation in cold tolerant plants', *Photosynthesis Research* **37**, 19–39.

Huner, N.P.A., Öquist, G. and Sarhan, F. (1998) 'Energy balance and acclimation to light and cold', *Trends in Plant Sciences* **3**, 224–230.

Hutzler, P., Fischbach, R., Heller, W. *et al.* (1998) 'Tissue localization of phenolic compounds in plants by confocal laser microscopy', *Journal of Experimental Botany* **323**, 953–965.

Iida, A., Kazuoka, T., Torikai, S., Kikuchi, H. and Oeda, K. (2000) 'A zinc finger protein RHL41 mediates the light acclimatization response in *Arabidopsis*', *The Plant Journal* **24**, 191–203.

Jaakola, L., Määttä-Riihinen, K., Kärenlampi, S. and Hohtola, A. (2004) 'Activation of flavonoid biosynthesis by solar radiation in bilberry (*Vaccinium myrtillus* L.) leaves', *Planta* **218**, 721–724.

Jabs, T., Tschöpe, M., Colling, C., Hahlbrock, K. and Scheel, D. (1997) 'Elicitor-stimulated ion fluxes and O_2 from the oxidative burst are essential components in triggering defense gene activation and phytoalexin synthesis in parsley', *Proceedings of the National Academy of Sciences USA* **94**, 4800–4805.

Jahnke, L.S., Hull, M.R. and Long, S.P. (1991) 'Chilling stress and oxygen metabolising enzymes in *Zea mays* and *Zea diploperennis*', *Plant, Cell and Environment* **14**, 97–104.

Jenkins, G.I., Long, J.C., Wade, H.K., Shenton, M.R. and Bibikova, T.N. (2001) 'UV and blue light signalling: pathways regulating chalcone synthase gene expression in *Arabidopsis*', *New Phytologist* **151**, 121–131.

Johnson, K.S. and Felton, G.W. (2001) 'Plant phenolics as dietary antioxidants for herbivorous insects: a test with genetically modified tobacco', *Journal of Chemical Ecology* **27**, 2579–2597.

Jovanovic, S.V., Steenken, S., Tosic, M., Marjanovic, B. and Simic, M.G. (1994) 'Flavonoids as antioxidants', *Journal of American Chemical Society* **116**, 4846–4851.

Kanoun, M., Goulas, M.J.P. and Biolley, J.P. (2001) 'Effect of a chronic and moderate ozone pollution on the phenolic pattern of bean leaves (*Phaseolus vulgaris* L. cv Nerina): relations with visible injury and biomass production', *Biochemical Systematics and Ecology* **29**, 443–457.

Karpinski, S., Escobar, C., Karpinska, B., Creissen, G. and Mullineaux, P.M. (1997) 'Photosynthetic electron transport regulates the expression of cytosolic ascorbate peroxidase genes in *Arabidopsis* during excess light stress', *The Plant Cell* **9**, 627–640.

Kocsy, G., Brunner, M., Rüegsegger, A., Stamp, P. and Brunold, C. (1996) 'Glutathione synthesis in maize genotypes with different sensitivities to chilling', *Planta* **198**, 365–370.

Krol, M., Gray, G.R., Hurry, V.M., Öquist, G., Malek, L. and Huner, N.P. (1995) 'Low-temperature stress and photoperiod effect an increased tolerance to photoinhibition in *Pinus banksiana* seedlings', *Canadian Journal of Botany* **73**, 1119–1127.

Lamb, D. and Dixon, R.A. (1997) 'The oxidative burst in plant disease resistance', *Annual Review of Plant Physiology and Plant Molecular Biology* **48**, 251–275.

Landry, L.G., Chapple, C.C.S. and Last, R.L. (1995) '*Arabidopsis* mutants lacking phenolic sunscreens exhibit enhanced ultraviolet-B injury and oxidative damage', *Plant Physiology* **109**, 1159–1166.

Laranjinha, J.A., Almeida, L.M. and Madeira, V.M. (1994) 'Reactivity of dietary phenolic acids with peroxyl radicals: antioxidant activity upon low density lipoprotein peroxidation', *Biochemical Pharmacology* **48**, 487–494.

Laughton, M.J., Halliwell, B., Evans, P.J. and Hoult, J.R.S. (1989) 'Antioxidant and prooxidant actions of the plant phenolics quercetin, gossypol and myricetin: effects on lipid peroxidation, hydroxyl radical generation and bleomycin-dependent damage to DNA', *Biochemical Pharmacology* **38**, 2859–2865.

Levine, A., Tenhaken, R., Dixon, R. and Lamb, C. (1994) 'H_2O_2 from the oxidative burst orchestrates the plant hypersensitive disease resistance response', *Cell* **79**, 583–593.

Leyva, A., Jarillo, A., Salinas, J. and Martinez-Zapater, J.M. (1995) 'Low temperature induces the accumulation of phenylalanine ammonia lyase and chalcone synthase mRNAs of *Arabidopsis thaliana* in a light-dependent manner', *Plant Physiology* **108**, 39–46.

Li, J., Ou-Lee, T.M., Raba, R., Amundson, R.G. and Last, R.L. (1993) '*Arabidopsis* flavonoid mutants are hypersensitive to UV-B irradiation', *The Plant Cell* **5**, 171–179.

Li, Y. and Trush, M.A. (1994) 'Reactive oxygen-dependent DNA damage resulting from the oxidation of phenolic compounds by a copper-redox cycle mechanism', *Cancer Research* **51**(Suppl), 1895S–1898S.

Logan, B.A., Demmig-Adams, B., Adams, W.W., III and Grace, S.C. (1998a) 'Antioxidants and xanthophyll cycle-dependent energy dissipation in *Curcurbita pepo* and *Vinca major* acclimated to four growth PFDs in the field', *Journal of Experimental Botany* **49**, 1869–1879.

Logan, B.A., Grace, S.C., Adams, W.W., III and Demmig-Adams, B. (1998b) 'Seasonal differences in xanthophyll cycle characteristics and antioxidants in *Mahonia repens* growing in different light environments', *Oecologia* **116**, 9–17.

Logemann, E., Tavernaro, A., Schulz, W., Somssich, I.E. and Hahlbrock, K. (2000) 'UV light selectively coinduces supply pathways from primary metabolism and flavonoid secondary product formation in parsley', *Proceedings of the National Academy of Sciences USA* **97**, 1903–1907.

Lois, R. (1994) 'Accumulation of UV-absorbing flavonoids induced by UV-B radiation in *Arabidopsis thaliana*. I. Mechanisms of UV-resistance in *Arabidopsis*', *Planta* **4**, 498–503.

Lois, R. and Buchanan, B.B. (1994) 'Severe sensitivity to ultraviolet radiation in an *Arabidopsis* mutant deficient in flavonoid accumulation. II. Mechanisms of UV-resistance in *Arabidopsis*', *Planta* **4**, 504–509.

Macheix, J.J. and Fleuriet, A. (1998) Phenolic acids in fruits, in *Flavonoids in Health and Disease* (ed. C.A. Rice-Evans and L. Packer), Marcel Dekker, New York, USA, pp. 35–59.

Maher, E.A., Bate, N.J., Ni, W., Elkind, Y., Dixon, R.A. and Lamb, C.J. (1994) 'Increases disease susceptibility of transgenic tobacco plants with suppressed levels of preformed phenylpropanoid products', *Proceedings of the National Academy of Sciences USA* **91**, 7802–7806.

Mancinelli, A.L. (1985) 'Light-dependent anthocyanin synthesis – a model system for the study of plant photomorphogenesis', *Botanical Reviews* **51**, 107–157.

Mancinelli, A.L. and Rabino, I. (1978) 'High irradiance responses of plant photomorphogenesis', *Botanical Reviews* **44**, 129–180.

McClure, J.W. (1986) 'Physiology of flavonoids in plants, in *Plant Flavonoids in Biology and Medicine. Biochemical, Pharmacological and Structure–Activity Relationships* (ed. V. Cody, E. Middleton and J.B. Harborne), Alan R. Liss, New York, USA, pp. 77–85.

Mehdy, M.C. (1994) 'Active oxygen species in plant defense against pathogens', *Plant Physiology* **105**, 467–472.

Mittler, R. (2002) 'Oxidative stress, antioxidants, and stress tolerance', *Trends in Plant Sciences* **7**, 405–410.

Mittova, V., Tal, M., Volokita, M. and Guy, M. (2003) 'Upregulation of the leaf mitochondrial and peroxiosomal antioxidative systems in reponse to salt-induced oxidative stress in the wild salt-tolerant tomato species, *Lycopersicon pennellii*', *Plant, Cell and Environment* **26**, 845–856.

Mole, S., Ross, J.A.M. and Waterman, P.G. (1988) 'Light-induced variation in phenolic levels in foliage of rain-forest plants. I. Chemical changes', *Journal of Chemical Ecology* **14**, 1–21.

Mølgaard, P. and Ravn, H. (1988) 'Evolutionary aspects of caffeoyl ester distribution in dicotyledons', *Phytochemistry* **27**, 2411–2421.

Møller, I.M. (2001) 'Plant mitochondria and oxidative stress: electron transport, NADPH turnover, and metabolism of reactive oxygen species', *Annual Review of Plant Physiology and Plant Molecular Biology* **52**, 561–591.

Moon, J.-H. and Terao, J. (1998) 'Antioxidant activity of caffeic acid and dihydrocaffeic acid in lard and low-density lipoprotein', *Journal of Agricultural and Food Chemistry* **46**, 5062–5065.

Müller, P., Li, X.-P. and Niyogi, K.K. (2001) 'Non-photochemical quenching: a response to excess light energy', *Plant Physiology* **125**, 1558–1566.

Nardini, M., D'Aquino, M., Tomassi, G., Gentili, V., Di Felice, M. and Scaccini, C. (1995) 'Inhibition of low density lipoprotein oxidation by caffeic acid and other hydrocinnamic acid derivatives', *Free Radical Biology and Medicine* **19**, 541–552.

Ndong, C., Danyluk, J., Huner, N.P.A. and Sarhan, F. (2001) 'Survey of gene expression in winter rye during changes in growth temperature, irradiance or excitation pressure', *Plant Molecular Biology* **45**, 691–703.

Neill, S.O. and Gould, K.S. (2003) 'Anthocyanins in leaves: light attenuators or antioxidants?', *Functional Plant Biology* **30**, 865–873.

Neill, S.O., Gould, K.S., Kilmartin, P.A., Mitchell, K.S. and Markham, K.R. (2002) 'Antioxidant activity of red versus green leaves in *Elatostema rugosum*', *Plant, Cell and Environment* **25**, 539–547.

Nicholson, R.L. and Hammerschmidt, R. (1992) 'Phenolic compounds and their role in disease resistance', *Annual Review of Phytopathology* **30**, 369–389.

Niyogi, K.K. (1999) 'Photoprotection revisited: genetic and molecular approaches', *Annual Review of Plant Physiology and Plant Molecular Biology* **50**, 333–359.

Noctor, G. and Foyer, C.H. (1998) 'Ascorbate and glutathione: keeping active oxygen under control', *Annual Review of Plant Physiology and Plant Molecular Biology* **49**, 249–279.

Nozzolillo, C., Isabelle, P. and Das, G. (1990) 'Seasonal changes in the phenolic constituents of jack pine seedlings (*Pinus banksiana*) in relation to the purpling phenomenon', *Canadian Journal of Botany* **68**, 2010–2017.

Okuda, T., Masuda, Y., Yamanaka, A. and Sagiska, S. (1991) 'Abrupt increase in the level of hydrogen peroxide in leaves of winter wheat is caused by cold treatment', *Plant Physiology* **97**, 1265–1267.

Öquist, G. and Huner, N.P.A. (2003) 'Photosynthesis of overwintering evergreen plants', *Annual Review of Plant Biology* **54**, 329–355.

Pannala, A.S., Razaq, R., Halliwell, B., Singh, S. and Rice-Evans, C. (1998) 'Inhibition of peroxynitrite dependent tyrosine nitration by hydroxycinnamates: nitration or electron donation?', *Free Radical Biology and Medicine* **24**, 594–606.

Pannala, A.S., Rice-Evans, C.A., Halliwell, B. and Singh, S. (1997) 'Inhibition of peroxynitrite-mediated tyrosine nitration by catechin polyphenols', *Biochemical and Biophysical Research Communications* **232**, 164–168.

Pasqualini, S., Piccioni, C., Reale, L., Ederli, L., Della Torre, G. and Ferranti, F. (2003) 'Ozone-induced cell death in tobacco cultivar Bel W3 plants. The role of programmed cell death in lesion formation', *Plant Physiology* **133**, 1122–1134.

Pastori, G., Foyer, C.H. and Mullineaux, P. (2000) 'Low temperature-induced changes in the distribution of H_2O_2 and antioxidants between the bundle sheath and mesophyll cells of maize leaves', *Journal of Experimental Botany* **51**, 107–113.

Peer, W.A., Brown, D.E., Tague, B.W., Muday, G.K., Taiz, L. and Murphy, A.S. (2001) 'Flavonoid accumulation patterns of transparent testa mutants of *Arabidopsis*', *Plant Physiology* **126**, 536–548.

Pell, E.J., Schlagnhaufer, C.D. and Arteca, R.N. (1997) 'Ozone-induced oxidative stress: Mechanisms of action and reaction', *Physiologia Plantarum* **100**, 264–273.

Pietta, P.-G. (2000) 'Flavonoids as antioxidants', *Journal of Natural Products* **63**, 1035–1042.

Prasad, T. (1996) 'Mechanisms of chilling-induced oxidative stress injury and tolerance in developing maize seedlings: changes in antioxidant system, oxidation of proteins and lipids, and protease activities', *The Plant Journal* **10**, 1017–1026.

Prasad, T.K., Anderson, M.D., Martin, B.A. and Stewart, C.R. (1994) 'Evidence for chilling-induced oxidative stress in maize seedlings and a regulatory role for hydrogen peroxide', *The Plant Cell* **6**, 65–74.

Radi, R., Beckman, J.S., Bush, K.M. and Freeman, B.A. (1991) 'Peroxynitrite-induced membrane lipid peroxidation: the cytotoxic potential of superoxide and nitric oxide', *Archives of Biochemistry and Biophysics* **288**, 481–487.

Rao, M.V., Paliyah, G. and Ormrod, D.P. (1996) 'Ultraviolet-B- and ozone-induced biochemical changes in antioxidant enzymes of *Arabidopsis thaliana*', *Plant Physiology* **110**, 125–136.

Reuber, S., Bornman, J.F. and Weissenböck, G. (1996) 'Phenylpropanoid compounds in primary leaf tissues of rye (*Secale cereale*). Light response of their metabolism and the possible role in UV-B protection', *Physiologia Plantarum* **97**, 160–168.

Rice-Evans, C.A., Miller, N.J. and Paganga, G. (1996) 'Structure–activity relationships of flavonoids and phenolic acids', *Free Radical Biology and Medicine* **20**, 933–956.

Rice-Evans, C.A., Miller, N.J. and Paganga, G. (1997) 'Antioxidant properties of phenolic compounds', *Trends in Plant Sciences* **2**, 152–159.

Rietveld, A. and Wiseman, S. (2003) 'Antioxidant effects of tea: evidence from human clinical trials', *Journal of Nutrition* **133**, 3285S–3292S.

Robbins, C.T., Hanley, T.A. and Hagerman, A.E. (1987) 'Role of tannins in defending plants against ruminants: reduction in protein availability', *Ecology* **68**, 98–107.

Rosemann, D., Heller, W. and Sandermann, Jr., H. (1991) 'Biochemical plant responses to ozone. II. Induction of stilbene synthesis in Scots pine (*Pinus sylvestris* L.) seedlings', *Plant Physiology* **97**, 1280–1286.

Rossel, J.B., Wilson, I.W. and Pogson, B.J. (2002) 'Global changes in gene expression in response to high light in *Arabidopsis*', *Plant Physiology* **130**, 1109–1120.

Rozema, J., Van de Staaij, J., Bjorn, L.O. and Caldwell, M. (1997) 'UV-B as an environmental factor in plant stress: stress and regulation', *Trends in Ecological Evolution* **12**, 22–28.

Ruegger, M., Meyer, K., Cusumano, J.C. and Chapple, C. (1999) 'Regulation of ferulate-5-hydroxylase expression in *Arabidopsis* in the context of sinapate ester biosynthesis', *Plant Physiology* **119**, 101–110.

Sakihama, Y., Cohen, M.F., Grace, S.C. and Yamasaki, H. (2002) 'Plant phenolic antioxidant and prooxidant activities: phenolics-induced oxidative damage mediated by metals in plants', *Toxicology* **177**, 67–80.

Sakihama, Y., Mano, J., Sano, S., Asada, K. and Yamasaki, H. (2000) 'Reduction of phenoxyl radicals mediated by monodehydroascorbate reductase', *Biochemical and Biophysical Research Communications* **279**, 949–954.

Sakuta, M. (2000) 'Transcriptional control of chalcone synthase by environmental stimuli', *Journal of Plant Research* **113**, 327–333.

Salah, N., Miller, N.J., Paganga, G., Tijburg, L., Bolwell, G.P. and Rice-Evans, C. (1995) 'Polyphenolic flavonols as scavengers of aqueous phase radicals and as chain breaking antioxidants', *Archives of Biochemistry and Biophysics* **322**, 339–346.

Saunders, J.A. and McClure, J.W. (1976) 'The occurrence and photoregulation of flavonoids in barley plastids', *Phytochemistry* **15**, 805–807.

Sawa, T., Nakao, M., Akaike, T., Ono, K. and Maeda, H. (1999) 'Alkylperoxyl radical-scavenging activity of various flavonoids and other phenolic compounds: implications for the anti-tumor-promoter effect of vegetables', *Journal of Agricultural and Food Chemistry* **47**, 397–402.

Schraudner, M., Moeder, W., Wiese, C. *et al.* (1998) 'Ozone-induced oxidative burst in the ozone biomonitor plant, tobacco Bel W3', *The Plant Journal* **16**, 235–245.

Schultz, M. and Weissenböck, G. (1986) 'Isolation and separation of epidermal and mesophyll protoplasts from rye primary leaves – tissue-specific characteristics of secondary phenolic product accumulation', *Zeitschrift fur Naturforschung* **41c**, 22–27.

Schützendübel, A. and Polle, A. (2002) 'Plant responses to abiotic stresses: heavy metal-induced oxidative stress and protection by mycorrhization', *Journal of Experimental Botany* **53**, 1351–1365.

Shadle, G.L., Wesley, S.V., Korth, K.L., Chen, F., Lamb, C. and Dixon, R.A. (2003) 'Phenylpropanoid compounds and disease resistance in transgenic tobacco with altered expression of L-phenylalanine lyase', *Phytochemistry* **64**, 153–161.

Sharma, Y.K. and Davis, K.R. (1994) 'Ozone-induced expression of stress-related genes in *Arabidopsis thaliana*', *Plant Physiology* **105**, 1089–1096.

Shirley, B.W. (1996) 'Flavonoid biosynthesis: "new" functions for an "old" pathway', *Trends in Plant Sciences* **1**, 377–382.

Smirnoff, N. (1993) 'Role of active oxygen in the response of plants to water deficits and dessication', *The New Phytologist* **125**, 27–58.

Solecka, D. and Kacperska, A. (1995) 'Phenylalanine ammonia lyase activity in leaves of winter oilseed rape plants as affected by acclimation of plants to low temperature', *Plant Physiology and Biochemistry* **33**, 585–591.

Solecka, D. and Kacperska, A. (2003) 'Phenylpropanoid deficiency affects the course of plant acclimation to cold', *Physiologia Plantarum* **119**, 253–262.

Squadrito, G.L. and Pryor, W.A. (1998) 'The nature of reactive species in systems that produce peroxynitrite', *Chemical Research in Toxicology* **11**, 718–719.

Stafford, H.A. (1990) *Flavonoid Metabolism*, CRC Press. Inc., New York, USA.

Stapleton, A.E. and Walbot, V. (1994) 'Flavonoids can protect maize DNA from the induction of ultraviolet radiation damage', *Plant Physiology* **105**, 881–889.

Starr, G. and Oberbauer, S.F. (2002) 'The role of anthocyanins for photosynthesis of Alaskan arctic evergreens during snow melt', *Advances in Botanical Research* **37**, 129–145.

Steyn, W.J., Wand, S.J.E., Holcroft, D.M. and Jacobs, G. (2002) 'Anthocyanins in vegetative tissues: a proposed unified function in photoprotection', *The New Phytologist* **155**, 349–361.

Strack, D., Heilemann, J. and Klinkott, J.S. (1988) 'Cell wall bound phenolics from spruce needles', *Zeitschrift fur Naturforschung* **43c**, 37–41.

Strid, Å., Chow, W.S. and Anderson, J.M. (1994) 'UV-B damage and protection at the molecular level in plants', *Photosynthesis Research* **39**, 475–489.

Summers, C.B. and Felton, G.W. (1994) 'Prooxidant effects of phenolics acids on the generalist herbivore *Helicoverpa zea* (Lepidoptera: Noctuidae): potential mode of action for phenolic compounds in plant anti-herbivore chemistry', *Insect Biochemistry and Molecular Biology* **24**, 943–953.

Takahama, U. (1989) 'A role of hydrogen peroxide in the metabolism of phenolics in mesophyll cells of *Vicia faba* L', *The Plant Cell Physiology* **30**, 295–301.

Takahama, U. and Oniki, T. (1992) 'Regulation of peroxidase-dependent oxidation of phenolics in the apoplast of spinach leaves by ascorbate', *The Plant Cell Physiology* **33**, 379–387.

Takahashi, M.S. and Asada, K. (1983) 'Superoxide anion permeability of phospholipid membranes and chloroplast thylakoids', *Archives of Biochemistry and Biophysics* **226**, 558–566.

Takeuchi, Y., Kubo, H., Kasahara, H. and Sakaki, T. (1996) 'Adaptive alterations in the activities of scavengers of active oxygen in cucumber cotyledons irradiated with UV-B', *Journal of Plant Physiology* **147**, 589–592.

Terao, J., Piskuli, M. and Yao, Q. (1994) 'Protective effect of epicatechin, epicatechin gallate, and quercetin on lipid peroxidation in phospholipid bilayers', *Archives of Biochemistry and Biophysics* **308**, 278–284.

Torel, J., Cillard, J. and Cillard, P. (1986) 'Antioxidant activity of flavonoids and reactivity with peroxyl radical', *Phytochemistry* **25**, 383–385.

Tsuda, T., Kato, Y. and Osawa, T. (2000) 'Mechanism for the peroxynitrite scavenging by anthocyanins', *FEBS Letters* **484**, 207–210.

Tsuda, T., Shiga, K., Ohshima, K., Kawakishi, S. and Osawa, T. (1996) 'Inhibition of lipid peroxidation and the active oxygen scavenging effect of anthocyanin pigments isolated from *Phaseolus vulgaris* L', *Biochemical Pharmacology* **52**, 1033–1039.

Van Breusegem, F., Slooten, L., Stassart, J.-M. *et al.* (1999) 'Effects of over-production of tobacco MnSOD in maize chloroplasts on foliar tolerance to cold and oxidative stress', *Journal of Experimental Botany* **50**, 71–78.

Walczak, H.A. and Dean, J.V. (2000) 'Vacuolar transport of the glutathione conjugate of *trans*-cinnamic acid', *Phytochemistry* **53**, 441–446.

Wang, H., Cao, G. and Prior, R.L. (1997) 'Oxygen radical absorbing capacity of anthocyanins', *Journal of Agricultural and Food Chemistry* **45**, 304–309.

Waterman, P.G. and Mole, S. (1994) *Analysis of Phenolic Plant Metabolites*, Blackwell Scientific, Oxford, UK.

Weisshaar, B. and Jenkins, G. (1998) 'Phenylpropanoid biosynthesis and its regulation', *Current Opinion in Plant Biology* **1**, 251–257.

Winkel-Shirley, B. (2001) 'Flavonoid biosynthesis, a colorful model for genetics, biochemistry, cell biology, and biotechnology', *Plant Physiology* **126**, 485–493.

Wise, R.R. and Naylor, A.W. (1987) 'Chilling-enhanced photooxidation. Evidence for the role of singlet oxygen and superoxide in the breakdown of pigments and endogenous antioxidants', *Plant Physiology* **83**, 278–282.

Wohlgemuth, H., Mittelstrass, K., Kschieschan, S. *et al.* (2002) 'Activation of an oxidative burst is a general feature of sensitive plants exposed to the air pollutant ozone', *Plant, Cell and Environment* **25**, 717–726.

Yamanaka, N., Oda, O. and Nagao, S. (1997) 'Prooxidant activity of caffeic acid, dietary non-flavonoid phenolic acid, on Cu^{2+}-induced low density lipoprotein oxidation', *FEBS Letters* **405**, 186–190.

Yamasaki, H. (1997) 'A function of color', *Trends in Plant Sciences* **2**, 7–8.

Yamasaki, H. and Grace, S.C. (1998) 'EPR detection of phytophenoxyl radicals stabilized by zinc ions: evidence for the redox coupling of plant phenolics in the H_2O_2-peroxidase system', *FEBS Letters* **422**, 377–380.

Yamasaki, H. and Sakihama, Y. (2000) 'Simultaneous production of nitric oxide and peroxynitrite by plant nitrate reductase: *in vitro* evidence for the NR-dependent formation of active nitrogen species', *FEBS Letters* **468**, 89–92.

Yamasaki, H., Sakihama, Y. and Ikehara, N. (1997) 'Flavonoid-peroxidase reaction as a detoxification mechanism of plant cells against H_2O_2', *Plant Physiology* **115**, 1405–1412.

Yamasaki, H., Uefuji, H. and Sakihama, Y. (1996) 'Bleaching of the red anthocyanin induced by superoxide radical', *Archives of Biochemistry and Biophysics* **332**, 183–186.

Yang, C.S., Landau, J.M., Huang, M.T. and Newmark, H.L. (2001) 'Inhibition of carcinogenesis by dietary polyphenolic compounds', *Annual Review of Nutrition* **21**, 381–406.

Yao, K., De Luca, V. and Brisson, N. (1995) 'Creation of a metabolic sink for tryptophan alters the phenylpropanoid pathway and the susceptibility of potato to *Phytopthora infestans*', *The Plant Cell* **7**, 1787–1799.

Zancani, M. and Nagy, G. (2000) 'Phenol-dependent H_2O_2 breakdown by soybean root plasma membrane-bound peroxidase is regulated by ascorbate and thiols', *Journal of Plant Physiology* **156**, 295–299.

7 Reactive oxygen species as signalling molecules

Radhika Desikan, John Hancock and Steven Neill

7.1 Introduction

Reactive oxygen species (ROS) is the term used to describe forms of oxygen that are energetically more reactive than molecular oxygen. Typically ROS (sometimes also referred to as AOS, active oxygen species, or ROI, reactive oxygen intermediates) are molecular species that have undergone electron addition(s) and are thus reduced forms of oxygen. Such ROS include the two free radical species, the superoxide anion O_2^- and its protonated form the perhydroxyl radical $HO_2^\cdot$, the uncharged, non-radical species hydrogen peroxide (H_2O_2) and the highly reactive hydroxyl radical $OH^\cdot$ (Figure 7.1). ROS also include singlet oxygen ($^1O_2$) generated by photoexcitation of chlorophyll. ROS are all toxic molecules, their particular destructiveness depending on their reactivity. High concentrations can result in non-controlled oxidation of a variety of cellular structures, including DNA, proteins and membrane lipids, that may result in the disruption of metabolism and destroy cellular structures. Because plant cells are continuously generating ROS as products of normal aerobic metabolism, it is not surprising that plants also possess a range of enzymatic and non-enzymatic antioxidant mechanisms to prevent ROS from reaching destructive levels, thus maintaining cellular redox balance. Moreover, some of these antioxidant processes are inducible, responding to elevated concentrations of ROS, thereby indicating that cellular mechanisms have evolved to recognise and ameliorate elevated ROS content. ROS are far more important to plant biology, however, than simply as toxic by-products and agents of cellular damage. At low concentrations, they are key signalling molecules that act at the interface between abiotic and biotic stresses and function as central co-ordinators of cell biology and responses to numerous developmental and environmental stimuli. Their importance as cellular regulators is reflected by the rapidly growing number of primary research papers and recent reviews found in the literature (Mittler, 2002; Neill *et al.*, 2002; Vranova *et al.*, 2002b; Desikan *et al.*, 2004a,b; Laloi *et al.*, 2004; Mori & Schroeder, 2004).

ROS are generated in plants in many ways in several cellular compartments (Mittler, 2002; Desikan *et al.*, 2004b). These include non-enzymatic mechanisms such as electron transfer to molecular oxygen during photosynthesis and respiration in chloroplasts and mitochondria, respectively, and as by-products of various enzymes such as photorespiratory glycolate oxidase in peroxisomes, amine oxidase and oxalate oxidase in the apoplast and xanthine oxidase and enzymes of fatty acid oxidation in peroxisomes (Figure 7.1). It is clear that several abiotic stresses such

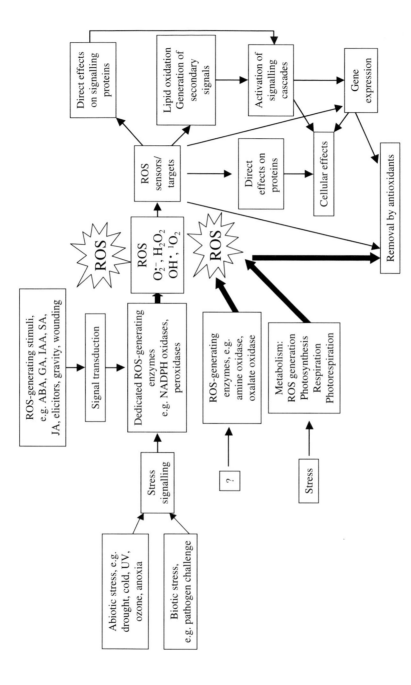

Figure 7.1 The central role of ROS in signalling. Abiotic or biotic stress, hormonal signalling and environmental stimuli can all lead to ROS generation, inducing a multitude of effects in the cell. ROS are fine-tuned by the enzymes producing them and the antioxidant mechanisms removing them. ROS are generated and consumed in several different cellular compartments, and their 'overspill' may result in damage and aberrant responses. It is also possible that ROS-generating systems and the signalling proteins that activate them, and are activated by them, are co-localised, perhaps as macromolecular complexes, to specific microdomains within the cell.

as low and high temperatures, excess irradiation, drought and salinity can disrupt cellular homeostasis and enhance ROS generation in cells. It is also clear that plant cells contain enzymes for which the specific function appears to be ROS generation. These include the NADPH oxidases and cell wall peroxidases (Bolwell *et al.*, 2002; Mittler, 2002; Neill *et al.*, 2002). Generation of superoxide and H_2O_2 by these enzymes is increased by various abiotic stresses such as wounding, temperature extremes, anoxia, air pollutants, UV irradiation and high light. Challenge by potentially phytopathogenic micro-organisms or elicitors generated during plant–pathogen interactions also activate ROS production via these enzymes (Figure 7.1). Such activation must mean that cells possess signalling mechanisms that relay the stress signals into enhanced amounts and activities of ROS-generating enzymes. In turn, there must also be in place the intracellular signal transduction machinery necessary to recognise elevated ROS and switch on ameliorative processes that include antioxidant mechanisms and activation of defence responses such as programmed cell death and expression of genes encoding proteins conferring resistance against pathogens. Enhanced ROS generation in response to both abiotic and biotic challenges probably helps to explain the phenomena of acclimation and cross-protection, in which pre-exposure to the same or different stress can enhance tolerance of subsequent stress. As ROS are both generated and consumed in several cellular compartments, the particular sub-cellular locations in which they are generated and perceived are likely to be important in determining the spectrum of cellular responses activated; it may be that 'abnormal' responses occur when ROS levels in a particular compartment swamp the co-localised antioxidants, with subsequent spill-over into other regions of the cell. Such effects on cell biology would go some way to explain why ROS concentrations are normally kept low. However, ROS concentrations in specific cellular compartments also need to be kept within certain limits because ROS are not only signals that function at the extremes of cell biology during stress, but also signals that are very much part of normal cell functioning. Hormonal signals such as abscisic acid (ABA), methyl jasmonate (Me-JA) and auxin; nod factors and phytotoxins all stimulate generation of ROS that, at least partly, mediate some of the biological effects of these stimuli. Stomatal closure, root hair initiation, root growth and function and leaf extension are examples of biological processes requiring ROS generation and action (Figure 7.1). In some cases, ROS may act as true signal molecules, initiating a subsequent cascade of signalling events that results in altered cell function; in others, ROS might act as effector molecules, e.g. by altering directly the properties of proteins or other structures in response to a primary signal. In either case, controlled synthesis of ROS, biochemical responses and rapid utilisation and/or removal are essential.

External application of various forms of ROS to plants has been shown to induce a number of responses. Of course, this does not by itself prove that endogenous ROS mediate the particular process under study. Externally applied ROS may cause a response in cells not normally exposed to it, and may fail to reach those cells in which endogenous ROS do have a role – i.e. responses may be pharmacological as opposed to physiological. In fact, as with plant hormones, various criteria should be satisfied in order to provide convincing evidence that a particular ROS does indeed have an endogenous signalling role. These include induction or modulation of the particular process following ROS application; inhibition correlated with removal of ROS (inhibition of synthesis/stimulation of metabolism) brought about by chemical or genetic means; changes correlated with changes in ROS sensitivity; and correlation of ROS synthesis/concentration with the particular developmental/physiological process. In recent years, an increasing number of studies have provided data to meet these criteria. Thus, in addition to the effects of exogenous ROS, experimental approaches include pharmacological manipulation of endogenous ROS via the application of ROS scavengers and inhibitors of ROS-generating enzymes, transgenic analyses in which the expression of ROS-generating enzymes (in particular NADPH oxidase) and ROS-scavenging enzymes (such as catalase and ascorbate peroxidase) have been altered, and assays of endogenous ROS that include real time cell imaging as well as destructive quantitative assays.

In this chapter, we focus on the signalling mechanisms by which ROS control cellular processes. We discuss the various types of ROS and their potential micro-locales or micro-domains within cells, perhaps reflecting co-localisation of ROS-generating enzymes, activators of ROS generation and, at least, some of the downstream targets of ROS. We discuss how ROS may be sensed in cells via direct oxidative effects on proteins and describe recent data implicating calcium fluxes and reversible protein phosphorylation as central mediators of ROS signalling, and finish by asking how ROS might act in concert with other signalling molecules such as nitric oxide, that is often generated with similar kinetics and in response to the same stimuli as are ROS.

7.2   ROS chemistry

ROS are derived by the sequential reduction of molecular oxygen, with varying number of electrons resulting in different ROS (Figure 7.2). The single electron reduction of O_2 results in the generation of the superoxide anion, $O_2^{\cdot-}$. This is usually, but not always, the first ROS to be made. It is relatively reactive and as the added single electron is in the unpaired state, superoxide is classified as a free radical. Dismutation of superoxide is inevitable, especially at low pH, with one superoxide anion giving up its added electron to another superoxide anion, and then with protonation resulting in the generation of H_2O_2. H_2O_2 is a weak oxidising agent, and will react with biological molecules such as proteins (see later). The dismutation reaction is also very efficiently catalysed by superoxide dismutase (SOD). Several types of SOD exist in plants, located in different cellular compartments. Their fast turnover rates suggests that rapid removal of superoxide from cellular compartments is essential for normal cell function.

Both superoxide and H_2O_2 are potential signalling molecules in plants, although most work has concentrated on the latter. H_2O_2 is relatively small, uncharged, can

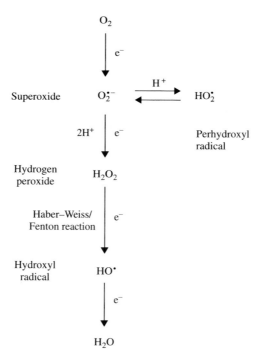

Figure 7.2   The chemistry of ROS. ROS are usually generated by the sequential reduction of molecular oxygen.

traverse membranes and is rapidly destroyed. These are all characteristics favourable for a cell signalling molecule. Superoxide, on the other hand, has a relatively short half-life and is charged. Its actions, therefore, may be restricted to the close proximity to its site of generation. However, superoxide can be protonated to form the perhydroxyl radical, $HO_2^{\bullet}$, so it does have the potential to permeate membranes, and a signalling role is possible (Jabs, 1999).

   In the presence of transition metals such as iron and copper, further reactions can take place, e.g. through the Haber–Weiss mechanism or the Fenton reaction, to yield hydroxyl radicals ($^{\bullet}$OH) (Figure 7.2). These are extremely reactive and will react at a diffusion-limited rate. The presence and concentration of transition metals at different parts of the cell might determine the formation of these hydroxyl radicals, potentially generating 'hot-spots' and localised action. Therefore, the damaging effects of these radicals could also be localised to defined regions of the cell. Hence, the first molecule they encounter will be attacked, be it protein, lipid or nucleic acid. It is difficult to envisage such a molecule having a controlled role in signalling but it may well be generated in response to a primary signal (e.g. Schopfer *et al.*, 2002; Foreman *et al.*, 2003). Superoxide can react with another very influential signalling free radical species, nitric oxide ($NO^{\bullet}$), to yield peroxynitrite ($OONO^-$). $OONO^-$ is highly reactive, and although it may be a toxic agent in mammalian systems, the evidence, to date, indicates that it is not so in plants. Indeed, formation of $OONO^-$ may actually abrogate cellular functions of NO and H_2O_2 (Delledonne et al., 2001, 2002).

Singlet oxygen ($^1O_2$) is another form of reactive oxygen but here there is no addition of an extra electron to molecular oxygen; rather, an electron is elevated to a higher energy orbital, thereby freeing oxygen from its spin-restricted state. This removal of spin restriction causes singlet oxygen to react rapidly with organic molecules, potentially causing damage. In plants, singlet oxygen can be formed by photoexcitation of chlorophyll and its reaction with oxygen, which can indeed result in lipid damage. Normally, efficient photoprotective agents such as carotenoids quench the excited state of chlorophyll, thus preventing the formation of singlet oxygen. However, recent work has indicated that singlet oxygen can also act as a signalling molecule, mediating the expression of a number of genes (Op den Camp et al., 2003).

7.3 ROS signalling

7.3.1 Specificity of ROS

As described earlier, various forms of ROS are likely to be present concurrently. As the equilibrium between these forms is unlikely to be static, it is sometimes difficult to discern exactly which ROS is biologically active in any given situation. Indeed, it may well be that, for example, both superoxide and H_2O_2 are active, with perhaps overlapping yet distinct responses. Thus, whilst H_2O_2 clearly signals in some situations, in others, superoxide or perhydroxyl radicals appear to be the active signal. Superoxide release was found to be essential for phytoalexin accumulation during defence responses in tobacco cells (Perrone et al., 2003); hydroxyl radicals initiated root hair growth in Arabidopsis (Foreman et al., 2003); superoxide, but not H_2O_2, has been shown to initiate cell death in soybean (Jabs, 1999), and the distinction between H_2O_2 and other ROS in the elongation of maize leaves is not clear (Rodriguez et al., 2002).

Clearly, it is desirable to assay ROS specifically, ideally within cells and in specific sub-cellular locations. Various histochemical assays have been used, e.g. using nitroblue tetrazolium for superoxide and DAB for H_2O_2. Intracellular real-time imaging using fluorescent dyes such as H_2–DCFDA is also a powerful tool to probe the sites and kinetics of ROS generation. However, although H_2–DCFDA is typically used to assay H_2O_2, it is not specific. Physico-chemical methods to quantify ROS include chemiluminescence and fluorescence, but again, these methods are not totally specific. Electron paramagnetic resonance (EPR) spectroscopy is used to detect free radicals, particularly superoxide, but is not amenable for real-time sub-cellular localisation (further discussion of methods for detecting ROS can be found in Chapters 8 and 9). Application of specific ROS, ROS generators and/or scavengers is an approach often used to determine which particular ROS is active, although the question of specificity remains. Identification of ROS-generating systems and generation of mutants specifically impaired in ROS generation (or response) is potentially an excellent way to delineate ROS specificity and is discussed later.

7.3.2 Perception and direct effects of ROS

In order to function as signalling molecules, ROS must be perceived in some way, so that the signal can be propagated to mediate cellular responses. Potentially, ROS might be perceived by several (or many) different target proteins in parallel, such that several responses could be initiated, depending on the specific cell type. Ideally, a change must be effected on the perceiving molecule, so that the next component in the signalling cascade can become involved, through inhibition, activation or recruitment. The next component in the signalling cascade could be a protein linked to, or 'tethered' with, the target protein such that H_2O_2 effects on the target protein (which is not necessarily a signalling protein) in turn alters the conformation/function of the partner protein that mediates a signalling cascade. H_2O_2 is a very small molecule and it seems unlikely that a receptor exists to recognise it in a classical receptor/ligand binding manner. However, H_2O_2 is a weak oxidising agent, and it could be that through this activity it transduces its 'message' (Rhee et al., 2000; Cooper et al., 2002; Danon, 2002). It has been suggested that proteins can take two states – a signalling state and a quiescent state – and the term nanotransducers has been coined for this potential group of ROS signalling proteins (Cooper et al., 2002). Moreover, it is also possible that the microlocale of ROS within a cell determines the efficiency with which they act as signals. ROS may be produced at discrete regions within the cell, in specified compartments or via the same protein present on different faces of the cell. Depending on their cellular location, ROS could target different proteins, possibly in macromolecular complexes. Such 'signalosomes' might then be localised to defined regions within the cell where they mediate an effect. Therefore, the ultimate defining factor for the effects of ROS within a cell is the location – i.e. where they are produced, where they act and what they interact with in their immediate neighbourhood.

Upstream 'sensor' proteins which are likely to sense the presence of ROS and transduce this signal to a cellular response have been identified in yeasts. The histidine kinase (HK) two-component system has been proposed to act as a H_2O_2 receptor, with the perception of ROS leading to a phosphorelay event that ends in cell survival response to oxidative stress (Singh, 2000; Buck et al., 2001). Several HKs exist in plants whose functions range from osmosensing to hormone perception (Hwang et al., 2002). Whether any of these plant HKs possess ROS-sensing functions remains to be determined.

Other potential ROS sensors, which are not necessarily proteins at the head of a signalling cascade, might include cytosolic proteins or partners of upstream proteins (such as HKs), e.g. thiol peroxidases, protein kinases or phosphatases. In yeast, a cytosolic thiol peroxidase has been identified as an enzyme which is a direct target for oxidation by H_2O_2 (Delaunay *et al.*, 2002). Transcription factors (TFs) can also act as targets for redox signalling (see later); therefore, targets of ROS exist in every part of the cell, and it is possible that multiple proteins sense and transduce ROS signals in various microlocales, to effect a specific cellular response.

Whatever the location of ROS synthesis and action, ROS are likely to target proteins that contain a thiol, or –SH, group of a cysteine or perhaps a methionine residue. In many proteins, the –SH may already be oxidatively modified to a cystine group, –S–S–, which aids in the conformational stability of the protein polypeptide. However, oxidation of –SH by its de-protonation through H_2O_2 may aid in the formation of cystine bridges in proteins, with a resultant change in conformational shape. Such a change in shape may then modulate the activity of the protein or allow it to interact with a binding partner that was unable to recognise it earlier. Such reactivity relies on two factors: the reactivity of the –SH with H_2O_2, itself determined by the accessibility of the cysteine to H_2O_2 and its relative redox mid-point potential; and the availability of another cysteine to enable the –S–S– bridge to be formed. Reduction of the –S–S– group would allow the return of the protein to its original conformation, but again this relies on the spatial availability of the group to the reducing agent/mechanism. An excellent example of such a mechanism is seen with the control of the interaction between the NPR1 protein (nonexpressor of pathogenesis-related protein 1) with a member of the TGA family of basic domain/Leu zipper (bZIP) transcription factors. Here –S–S– bridges in TGA1 are reduced to –SH via cellular redox changes that occur during increased accumulation of salicylic acid (SA) following pathogen challenge (Despres *et al.*, 2003). Further controlled interaction of NPR1 with TGA1 occurs by redox changes conferred on the NPR1 protein itself. NPR1 is usually oligomeric, with the protein/protein interactions being reliant on the formation of cystine bridges. Accumulation of SA leads to cellular redox changes (possibly via a series of processes involving ROS generation), which result in the reduction of cysteine residues in the protein, causing it to become monomeric, and hence active (Mou *et al.*, 2003).

Other oxidation reactions may also take place that could potentially bestow on the protein the conformational change needed to transduce the signal (Figure 7.3). –SH can be oxidised to the sulphenic acid (–SOH) group. This would alter the spatial requirements of this residue's side group, and so force a change in the position of amino acids in the vicinity. This –SOH group, as with the –S–S– group, can be re-reduced to reverse the change. Further oxidations are also possible. The –SOH group may react with another H_2O_2 to yield the sulphinic acid group (–SO_2H), and then with a further H_2O_2 to create the sulphonic acid group (–SO_3H) (Figure 7.3). Both of these higher oxidation states are thought to be irreversible modifications. In the majority of cases, signalling is a transient event in which the signalling needs to be terminated relatively quickly, and, therefore, it may be unlikely that the formation of –SO_2H and –SO_3H are involved in signalling. However, re-reduction of the sulphinic acid group to the sulphenic acid group has been reported (Biteau *et al.*, 2003). To investigate the oxidation of the cysteine –SH groups in the mammalian

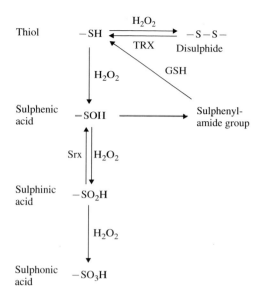

Figure 7.3   Protein modification by ROS. The oxidising properties of ROS mean that thiol groups can be oxidised within proteins, leading to conformational changes and signalling events. Possible modifications can lead to the formation of cystine bridges, or sulphenic acid, sulphinic acid or sulphonic acid groups. Reversal of such oxidations might involve reduced glutathione (GSH), thioredoxin (TRX) or sulphiredoxin (Srx).

phosphatase PP1B, an activity inhibited by H_2O_2, the protein was H_2O_2-treated and analysed by peptic digestion and mass spectrometry. The $-SO_2H$ and $-SO_3H$ were found to exist if high concentrations of H_2O_2 were used, but no $-SOH$ form of the cysteine could be distinguished. A structural analysis approach identified a new form of the protein created after treatment with H_2O_2 (Salmeen *et al.*, 2003; van Montfort *et al.*, 2003). It was found that the $-SOH$ was not formed in a stable manner as predicted, but rather a cyclisation of the amino acids within the protein resulted, giving what was termed the sulphenyl-amide group (Figure 7.3). Reversal of the cyclisation was mediated by the reduced form of glutathione (GSH), a reducing compound known to be at relatively high concentrations in plant cells. It remains to be seen whether such structural changes are commonplace in proteins following redox changes. In plants, the protein phosphatases (PPs) ABI1 and ABI2 have also been found to be inhibited by H_2O_2 *in vitro* (Meinhard *et al.*, 2001, 2002), but whether H_2O_2 acts via the oxidation of specific Cys residues is not known. The activity of the tyrosine phosphatase AtPTP1 in plants may also be inhibited by H_2O_2, and this might be subsequently involved in the control of MAPK activity (Gupta & Luan, 2003). However, the *in vivo* effects of H_2O_2, and the exact mechanism of H_2O_2 action are yet to be determined. As all tyrosine phosphatases contain a conserved cysteine in their catalytic site, a mechanism of H_2O_2 inhibition similar to that reported with PP1B might be involved.

Oxidation of proteins must be reversed to terminate the signal. With the protein phosphatase PP1B, GSH might play this role, but other reducing peptides could also be involved. For example, such a peptide might be thioredoxin (Trx), an ubiquitous disulphide oxidoreductase which contains two active cysteine groups within a conserved domain structure (Buchanan et al., 2002). Different groups of Trx exist in plants, with different isoforms being located in different sub-cellular locations (see Chapter 2). For example, Trx-h is a cytosolic form, whilst Trx-f and Trx-m are located in the chloroplast (Laloi et al., 2001; Meyer et al., 2002). Trx catalyses reduction reactions that include the conversion of –S–S– bridges to –SH groups. One of the targets of Trx is the enzyme peroxiredoxin (Prx), which, in its reduced form, can then react with and remove H_2O_2. Furthermore, the Cys-sulphinic acid of yeast Prx, Tsa1, is reduced by the enzyme sulphiredoxin (Srx) in a reaction requiring ATP and magnesium. This catalysis involves the transient formation of a disulphide bridge and the subsequent formation of the sulphenic acid group (Biteau et al., 2003). This reduction mechanism is important, as it shows that the formation of the sulphinic acid group is not necessarily an irreversible step in protein modification by ROS.

Recent studies have identified thioredoxin-interacting proteins in plants (Yano et al., 2001). A very recent report, using thioredoxin-affinity chromatography in association with mass spectrometry, identified a wide range of proteins, including those involved in antioxidant activity, e.g. ascorbate peroxidase; proteins involved in protein biosynthesis and degradation, and several metabolic proteins including glyceraldehyde 3-phosphate dehydrogenase and malate dehydrogenase (Yamazaki et al., 2004). It is possible that these proteins can react with H_2O_2 and then be re-reduced by thioredoxin for further rounds of signalling. For example, a role for thioredoxin in plant defence signalling has recently been demonstrated. Thioredoxin interacts with the resistance protein Cf-9, a membrane-located protein involved in pathogen recognition (Rivas et al., 2004). Prevention of Trx expression by gene silencing led to an accelerated hypersensitive response, accompanied by elevated accumulation of ROS. Furthermore, defence-related genes were found to be induced, and protein kinase activity was modulated. This is the first report of thioredoxin controlling disease resistance in plants, and shows a link between redox changes and physiological effects. Redox changes could also be initiated via indirect effects of ROS, e.g. via the release of active signalling molecules – oxidation of lipids results in the formation of phytoprostanes which leads to the activation of protein kinases, defence gene expression and phytoalexin formation (Thoma et al., 2003).

As well as changes to specific proteins, ROS might also alter the redox poise of the cytoplasm, termed by Schafer and Buettner (2001) as the redox environment in the cell. The cytoplasm is maintained at a very reducing redox state, which would potentially become moderately oxidised in the presence of ROS. Some redox proteins, such as cytochrome c which is released from the mitochondria, could be influenced by such a changing redox state, potentially leading to signalling events (Hancock et al., 2001). However, some caution does need to be exercised here, as not all ROS will have the same effect. Superoxide, for example, is relatively reducing, and would reduce cytochrome c, while H_2O_2 is relatively oxidising.

Furthermore, no robust methods for measuring the intracellular redox state exist, the ratios and concentrations of GSH and its oxidised form GSSG often being used (Hancock *et al.*, 2004). Modified forms of green fluorescent protein have been proposed as a new tool, but have yet to see widespread use (Ostergaard *et al.*, 2001).

7.4   Regulators of ROS signalling

7.4.1   ROS, calcium and ion channels

Calcium is a ubiquitous second messenger in plants. Increased cytosolic calcium concentrations activate various signalling proteins such as protein kinases and phosphatases and subsequent downstream responses (Scrase-Field & Knight, 2003) (Figure 7.4). Moreover, ROS effects seem to be closely linked with calcium fluxes (Mori & Schroeder, 2004). In guard cells, H_2O_2 activates plasma membrane calcium-permeable cation channels leading to an influx of calcium and thus stomatal closure (Pei *et al.*, 2000). In similar work, Kohler *et al.* (2003) have shown that H_2O_2 regulates the activity of both calcium and K^+ channels located in the plasma membrane, both of which are essential for stomatal closure. ROS regulation of calcium levels is not unique to guard cells. In roots, application of hydroxyl radicals causes increases in cytosolic calcium concentrations, which regulate cell growth and elongation (Schopfer *et al.*, 2002; Demidchek *et al.*, 2003; Foreman *et al.*, 2003). Mutants defective in the *Arabidopsis* NADPH oxidase isoform *AtrbohC* are unable to display calcium oscillations and have shorter root hairs (Foreman *et al.*, 2003). In other work, allelopathic chemicals exuded from an invasive plant triggered ROS production in *Arabidopsis* root hairs, leading to elevations in cytosolic calcium (Bais *et al.*, 2003). During pathogen challenge, ROS cause an increase in the levels of cytosolic calcium, but calcium has also been shown to increase levels of ROS (see Bowler & Fluhr, 2000). Therefore, calcium and ROS must interact very closely to regulate ROS homeostasis (Yang & Poovaiah, 2002).

H_2O_2 can also regulate the activities of K^+ channels and plasma membrane H^+-ATPases. Using various *Arabidopsis* mutants, Suhita *et al.* (2004) have demonstrated recently that ABA-induced H_2O_2 production (via NADPH oxidase) leads to the activation of outward rectifying K^+ channels in guard cells. However, H_2O_2 also inhibits the inward rectifying K^+ channels in guard cells (Kohler *et al.*, 2003). This latter response was found to be irreversible, suggesting that ROS targets these channels directly, via irreversible oxidation of reactive amino acid residues. In related work, it has been shown that K^+ uptake (or lack of) also regulates ROS levels in plants – Shin and Schachtman (2004) show that K^+ deficiency leads to enhanced ROS production in roots, which subsequently modulates gene expression and kinetics of K^+ uptake. pH changes in the cell also regulates H_2O_2 synthesis and action, at least in guard cells: Suhita *et al.* (2004) suggest a scheme whereby ABA activation of calcium channels leads to cytosolic alkalinisation, subsequently leading to the production of H_2O_2 through NADPH oxidase. However, H_2O_2 has also been

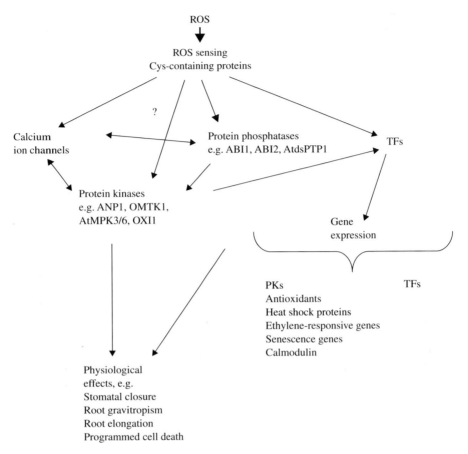

Figure 7.4 Downstream signalling from ROS. Intracellular signalling initiated by ROS involves many signal transduction components, including ion channels, protein kinases (PKs), protein phosphatases (PPs) and transcription factors (TFs), often leading to their sequential activation or inhibition in cascades. Such signalling can lead to direct physiological effects or gene expression.

shown to cause cytosolic alkalinisation leading to stomatal closure (Zhang *et al.*, 2001). Therefore, H_2O_2 can regulate as well as be regulated by changes in ion concentrations within cells: again, the micro-location of these molecules will determine their effects on cellular responses.

7.4.2 *Reversible protein phosphorylation*

Reversible protein phosphorylation is a key cell signalling event, and ROS such as H_2O_2 activate various protein kinases (PKs), including those belonging to the mitogen activated protein kinase (MAPK) family. MAPKs are a highly conserved family of PKs in eukaryotes that form part of a three-tier phosphorylation signalling cascade. Upon perceiving a stimulus, an upstream MAPK kinase kinase phosphorylates a MAPKK, which subsequently phosphorylates a MAPK on conserved threonine and tyrosine residues. Activation of a MAPK can lead to the phosphorylation of transcription factors, which ultimately regulates gene expression (Figure 7.4). In *Arabidopsis*, H_2O_2 challenge leads to the activation of the MAPKs, AtMPK3 and AtMPK6 (Kovtun *et al.*, 2000; Desikan *et al.*, 1999, 2001b), this process being essential for mediating tolerance to various stresses (Kovtun *et al.*, 2000). Activation of AtMPK3 and AtMPK6 is regulated by ANP1, an *Arabidopsis* MAPKKK that is also activated following H_2O_2 treatment (Kovtun *et al.*, 2000). In other work, H_2O_2 has also been shown to cause activation of OMTK1, a MAPKKK in alfalfa, which subsequently regulates the activities of the downstream MAPK kinase MMK3. Because OMTK1 activates MMK3 in response to H_2O_2 but not in response to elicitor or ethylene challenge, OMTK1 has been suggested to act as a scaffold protein, determining signalling specificity (Nakagami *et al.*, 2004). However, whether H_2O_2 interacts with ANP1/NPK1 or OMTK1 via direct modification remains to be determined. There is also evidence that H_2O_2-activation of MAPKs occurs via the action of other protein kinases. The AGC kinase OXI1 has been identified as a kinase that is activated following addition of H_2O_2, and whose function is required for AtMPK3 and AtMPK6 activation (Rentel *et al.*, 2004). Together, these observations suggest that H_2O_2 does not activate AtMPK3 and AtMPK6 directly; rather, interaction of these MAPKs with other H_2O_2-activated PKs regulates their activities. It is possible that specific PKs that are activated by ROS are tethered in certain cellular microdomains, and that the formation of specific macromolecular complexes via, e.g. scaffold proteins or other protein–protein interactions, results in the targeting of specific downstream signalling proteins, which subsequently determines specificity to a particular stimulus.

Protein phosphorylation is usually accompanied by dephosphorylation, and, therefore, the role of phosphatases in H_2O_2 signalling needs to be considered. H_2O_2 regulation of PP activity has been reported. Usually, H_2O_2 is found to inhibit phosphatase activity, thereby enhancing the steady-state-phosphorylation levels in a cell. Therefore, to effect signalling, H_2O_2 may increase kinase activity, or decrease phosphatase activity, the resulting phosphorylation in the cell possibly being the same. In reality, H_2O_2 probably does both, giving a level of subtlety to the signalling pathways involving ROS.

As mentioned above, H_2O_2 has been shown to directly modify thiol residues on PPs. Although there is no direct evidence to show that such modification alters signalling responses in cells, there is some circumstantial evidence that suggests it does. For example, guard cells of the PP2C mutant *abi1-1* are unable to synthesise H_2O_2, whilst those of *abi2-1* are unable to respond to H_2O_2 (Murata *et al.*, 2001); both ABI1 and ABI2 are targets of direct modification by H_2O_2 *in vitro* (Meinhard *et al.*, 2001, 2002). Moreover, it is not known whether ABI1/2 interact with the MAPK cascades regulated by H_2O_2. The *Arabidopsis* phosphatase AtPTP1 is also a target for direct modification by H_2O_2 *in vitro* (Gupta & Luan, 2003). AtPTP1 interacts with AtMPK6, possibly forming a 'signalosome'. This, together with the possible regulation of AtMPK6 by other kinases (see earlier), suggests that formation of macromolecular complexes in specific microdomains of the cell can regulate activation of the MAPK pathway in response to distinct stimuli.

7.4.3   ROS regulation of gene expression

Gene expression might occur both as a direct result of ROS signalling through modification of TF activity (as mentioned earlier), or indirectly, through a signal transduction cascade such as the MAPK cascade. Either way, ROS regulation of gene expression is evident from numerous studies that have demonstrated either direct regulation by ROS, a requirement for ROS via its removal, either through pharmacological or genetic studies, or regulation of antioxidant activity. Although most studies have implicated H_2O_2 as the primary ROS altering gene expression, there are others demonstrating different ROS having similar effects. The *Arabidopsis* runaway cell death mutant *lsd1* produces uncontrolled levels of superoxide, leading to changes in defence gene expression and cell death lesions (Jabs *et al.*, 1996). Although superoxide is the ROS implicated here, it seems rather surprising that H_2O_2 did not have a similar effect, given that superoxide does not easily traverse membranes, and spontaneous and catalysed dismutation of superoxide leads to hydrogen peroxide. It is possible that different stimuli cause an excess of superoxide production in chloroplasts, resulting in subsequent alterations in gene expression. Methyl viologen (MV), a redox-active compound that enhances chloroplast-derived superoxide production, caused changes in the expression of a number of genes in tobacco. Moreover, acclimation of tobacco plants by pre-treating with MV caused concerted activation of different sets of defence-related genes that contribute toward oxidative stress tolerance (Vranova *et al.*, 2002a). A recent study used a genetic approach to delineate changes in gene expression regulated either via singlet oxygen or superoxide. The conditional *fluorescent (flu)* mutant of *Arabidopsis* selectively produces singlet oxygen within chloroplasts when returned to the light after a short dark exposure (Op den Camp *et al.*, 2003). Analysis of gene expression of these mutants exposed to paraquat (a superoxide-generating compound) revealed that the expression of only a small number of genes was affected by paraquat treatment, which differed from those whose expression was affected by singlet oxygen. For example, a gene encoding ferritin, a protein that maintains iron levels in plants, was found to be upregulated by paraquat but not by singlet oxygen, whereas signal transduction genes such as MAP kinase kinase 4, the protein phosphatase ABI1 and several protein kinases, were up-regulated specifically by singlet oxygen (Op den Camp *et al.*, 2003). Because of the short half-life of singlet oxygen and superoxide, it is likely that these ROS are not having a direct effect on nuclear gene expression, but rather that these molecules interact with components that are more closely associated with their site of origin from within chloroplasts, such as membrane lipids (Op den Camp *et al.*, 2003). Indeed, in other studies, it has been shown that ROS oxidation reactions generate cycloprostanes that trigger defence gene expression, MAPK activation and formation of phytoalexins (Thoma *et al.*, 2003). Experimental strategies using mutants that lack specific ROS-generating enzymes (e.g. *rboh*) can reveal important mechanisms by which different ROS can selectively effect changes in cell function. For example, it will be interesting to note whether a lack of production of one type of ROS (e.g. superoxide, via Rboh) will affect the activities of other enzymes that produce different types of ROS.

Global changes in gene expression in response to treatment with exogenous H_2O_2 has identified a number of genes whose function can be assigned to signal transduction, defence, transcription, metabolism as well as cell structure (Desikan *et al.*, 2001a; Vandenabeele *et al.*, 2003). Signal transduction genes included those encoding protein kinases, calmodulin, histidine kinase and a tyrosine phosphatase, and genes encoding TFs were also induced, suggesting that cells might adapt for potentially greater or faster responses in the future, or for acclimation tolerance toward other stresses. Other SA, NO, JA and ethylene-related genes were also regulated by H_2O_2, indicating potential cross-talk (Desikan *et al.*, 2001a; Vranova *et al.*, 2002a). Further analyses of a small subset of H_2O_2-regulated genes showed additional regulation by drought, UV-B and elicitor treatments. The expression of some of these required H_2O_2, whilst others did not, indicating both specificity and cross-talk between signalling pathways (Desikan *et al.*, 2001a). It is not yet clear whether specific sub-classes of genes are regulated by specific ROS, i.e. if each type of ROS has a gene expression fingerprint – a careful global bioinformatic analysis of gene expression will be required to elucidate this. It is likely, though, that even if various genes are commonly regulated by different ROS, the capacity of the cells to cope with each ROS could also be selective, as observed recently for yeast (Thorpe *et al.*, 2004).

Removal of ROS via regulation of antioxidant activities is another approach to study the requirement for ROS in regulating gene expression. Tobacco plants with antisense suppression of either catalase, ascorbate peroxidase, or both, showed enhanced expression of other antioxidant genes, indicating a compensatory mechanism to cope with elevated levels of ROS (Rizhsky *et al.*, 2002). Knock-out plants that lacked *APX* not only showed altered growth responses, but also exhibited altered levels of expression of specific antioxidant genes and heat shock proteins during light stress, reflecting the plant's ability to cope with excess ROS generated during light stress (Pnueli *et al.*, 2003).

Changes in gene expression may be regulated by ROS either directly or indirectly. This could occur via direct modification of TFs by ROS, as described above, or via activation of another signalling protein (e.g. protein kinase) that subsequently activates a TF causing transcriptional changes (Figure 7.4). TFs sensitive to specific ROS have not yet been identified in plants, although there is some speculation in other systems that superoxide and H_2O_2-sensitive TFs might be unique. In one recent example of redox regulation of TF activity in plants, the disease resistance protein NPR1 was found to confer redox regulation of binding of the TF TGA1 to specific SA-regulatory sequences, leading to subsequent changes in gene expression

(Despres *et al.*, 2003). Several H_2O_2-regulated genes also possess *cis*-elements that might be sensitive to H_2O_2 (Desikan *et al.*, 2001a). However, there are likely to be different *cis*-elements unique to each type of ROS (Garreton *et al.*, 2002), further indicating specific regulation of signalling pathways. It is also possible that signalling pathways upstream of TFs are common to various stimuli (and different ROS), but that specificity and cross-talk occur at the level of TF regulation, i.e. there could be specific superoxide/hydrogen peroxide-sensitive TFs, which could act as a focal point for signalling pathways.

Using a proteomic approach, Sweetlove *et al.* (2002) demonstrated that the expression of various mitochondrial proteins was up- or down-regulated following exposure to H_2O_2. The up-regulated proteins included antioxidant defence proteins such as peroxiredoxins and protein disulphide isomerase, whereas those associated with the TCA cycle were down-regulated. However, H_2O_2 can affect the proteome not just by increasing or decreasing the levels of proteins indirectly, but also by direct modification of protein activities, as described earlier. A detailed analysis of the protein profile of various organelles following exposure to different ROS is required, to decode specific signalling pathways in response to various stimuli.

7.5   Regulation of ROS production

The exact site of production of ROS determines its action, whether deep inside the cell, or as intra- or inter-cellular signals. Localisation of ROS production and action to specific microlocales, and the action of efficient antioxidant mechanisms that fine-tunes the concentrations of ROS at particular parts of the cell could limit responses to localised 'hot-spots' (Neill *et al.*, 2002). ROS are generated within the cell at low levels via non-enzymatic routes, during photosynthesis and respiration, in chloroplasts and mitochondria, respectively. However, increased levels of ROS could be produced via these routes following exposure to different stimuli, subsequently activating signalling processes. ROS such as H_2O_2 can also be produced via 'dedicated' enzymes (Desikan *et al.*, 2004b), such as NADPH oxidase or peroxidase, that have the capacity to reduce oxygen to yield ROS in the apoplast. Other ROS-generating enzymes include xanthine oxidase, oxalate oxidase and amine oxidase (Figure 7.1).

Much work has focused on the NADPH oxidase-like enzymes, one of the major enzymatic routes of synthesis of ROS in plant cells (termed rboh, *r*espiratory *b*urst *o*xidase *h*omologue). This is discussed in more detail in Chapter 8. One of the NADPH oxidase homologues cloned from plants has been shown to be a plasma-membrane-bound enzyme (Keller *et al.*, 1998), and shows striking similarity to the large subunit of the mammalian NADPH oxidase complex, gp91-*phox*, although the plant protein contains an additional EF-hand region that binds calcium. Several isoforms of Rboh have been identified in plants such as tomato, tobacco, pea and *Arabidopsis*; at least ten different isoforms have been identified in *Arabidopsis*, with some of their expression profiles varying in different tissues (Torres *et al.*, 1998).

Like its mammalian counterparts, the plant oxidase protein contains consensus sequences in domains that potentially bind to redox prosthetic groups, i.e. flavins and haems, which could result in the formation of superoxide from the reduction of molecular oxygen. However, although several gp91-*phox* homologues have been identified in plants, there is only one study to date that demonstrates the superoxide-generating capacity of this protein *in vitro* (Sagi & Fluhr, 2001). Nevertheless, of the various isoforms of this protein that have been cloned from *Arabidopsis*, reverse genetic studies have identified knock-out mutants of three Rboh proteins (D, E and F) that fail to produce ROS in response to pathogens and also exhibit altered cell death (Torres *et al.*, 2002). As the Rboh proteins are thought to be plasma membrane associated, generation of superoxide by Rboh proteins suggests that their products and resultant action will be outside the cell, or that the signal is somehow transduced (possibly via dismutation to H_2O_2) to within the cell, a phenomenon which remains unanswered.

Differential expression of the different isoforms of Rboh in different tissues indicates that they have distinct functions in *Arabidopsis* (Torres *et al.*, 1998), suggesting signalling specificity. For example, *atrbohD, atrbohE* and *atrbohF* single, double and triple mutants have compromised responses to pathogen attack and guard cell responses to abscisic acid (Torres *et al.*, 2002; Kwak *et al.*, 2003); *atrbohC* mutants have defects in root hair development (Foreman *et al.*, 2003), and ABA-inhibition of root elongation is reduced in *atrbohD/F* double and *atrbohF* single mutants. As well as activation of the superoxide-generating activity of the plant oxidases, some of the isoforms are also induced at the mRNA level by various stimuli (Desikan *et al.*, 1998; Kwak *et al.*, 2003), indicating another level of complexity in the regulation of ROS production. Therefore, it is likely that these Rboh proteins have different activation profiles and that varying levels of ROS can be generated by Rboh proteins in different cell types and/or in different parts of the cell in response to different stimuli, thereby regulating the biological effects of ROS at that particular site of action. More studies are required here to discern whether the lack of expression of a particular *rboh* isoform in a specific tissue will affect ROS production not just in that tissue, but in other cell types as well – e.g. if the roots and leaves of *atrbohC* and *atrbohD* plants, respectively, lack ROS production, it will be interesting to determine whether *atrbohC* leaves will respond to ABA, and whether *atrbohD* has defective root hairs.

Peroxidases are another distinct enzymatic source of ROS. They are located in cell walls, and also have a role in ROS production during pathogen challenge (Bolwell *et al.*, 2002). Therefore, it is possible that the same stimulus induces ROS production via multiple enzymatic routes, but at different levels, all of which are required for a full signalling response, and that the lack of one functionality will be compensated for by another. A thorough analysis of the *rboh* and peroxidase knock-out mutants in response to distinct stimuli will elucidate the roles and degeneracy associated with such enzymes.

Regulation of the synthesis of ROS through NADPH oxidase occurs via the activity of other signalling molecules, such as G proteins, inositol phosphates, sphingolipids and protein kinases/phosphatases. During hypoxia, H_2O_2 production in *Arabidopsis* seedlings is regulated by the small GTPase proteins known as AtROPs (Baxter-Burrell *et al.*, 2002). During pathogen defence in rice, the blast fungus *Magnoporthe grisea* or a sphingolipid elicitor both transmit signals to the Gα subunit of a heterotrimeric G protein (Suharsono *et al.*, 2002). Accumulation of Gα is induced by pathogen or elicitor signals, and Gα in turn regulates the activity of a small GTPase, OsRac1. OsRac1 regulates the activity of NADPH oxidase, leading to ROS production during pathogen defence. However, the situation is complex, as recent work has also shown that OsRac1 down-regulates the expression of a metallothionein gene that acts as a ROS scavenger (Wong *et al.*, 2004). Therefore, the small GTPase protein could act both as an inducer of ROS synthesis as well as a suppressor of ROS scavenging. The recent discovery of a G protein coupled receptor binding to a Gα subunit which regulates stomatal function of guard cells in response to ABA and sphingolipids (Pandey & Assmann, 2004), opens up new horizons to explore ROS signalling in this situation.

In guard cells, pharmacological experiments suggested that inositol phosphate action is required for H_2O_2 production following ABA treatment (Park *et al.*, 2003). Calcium is also known to regulate both H_2O_2 synthesis and action – Sagi and Fluhr (2001) showed that the superoxide-generating activity of the tobacco Rboh protein is calcium-dependent, which fits in with the observation that calcium actually binds to the EF hand of the Rboh protein (Keller *et al.*, 1998). Yang and Poovaiah (2002) showed that calcium/calmodulin binds and inactivates catalases *in vitro*, leading to a decrease in H_2O_2 levels. In guard cells, H_2O_2 stimulates the activity of plasma membrane-located calcium channels (Pei *et al.*, 2000); therefore calcium has both positive and negative effects in regulating homoeostasis of ROS.

Protein phosphorylation also plays a role in regulating H_2O_2 synthesis. A protein kinase, OST1, has been shown to regulate ABA-induced H_2O_2 synthesis in guard cells. Mutants in this kinase show reduced generation of H_2O_2 in response to ABA, that results in increased water loss through plants (Mustilli *et al.*, 2002), which is expected as H_2O_2 is a stimulus causing stomatal closure. Moon *et al.* (2003) showed that H_2O_2 induced the expression of a NDP kinase that interacts with AtMPK3 and AtMPK6, and that over-expression of NDPK2 caused the down-regulation of H_2O_2 production, which in turn enhanced tolerance to various abiotic stresses. Therefore, it appears that kinases regulate both H_2O_2 production and action, emphasising their importance as key signalling proteins determining specificity. However, whether any interaction between these various components occurs in all cell types or is unique to a specific tissue is not known. Moreover, it is quite likely that some of the molecules mentioned above, such as calcium-binding proteins, PKs, G proteins and NADPH oxidase interact to form macromolecular complexes, which in turn regulate ROS synthesis and action. Such regulation could also be compartmentalised to specific locations within cells, again emphasising how precise co-ordination of events is required for a final response. For example, there is some evidence that calcium-dependent protein kinases regulate ROS production during pathogen defence (Romeis *et al.*, 2000), and that calcium levels and NADPH oxidase activity regulate one another (Sagi & Fluhr, 2001; Foreman *et al.*, 2003). Different NADPH oxidases could also be regulated in different ways, adding another layer of specificity in the production of ROS in specific cell types. Macromolecular complexes formed between heterotrimeric G proteins, small GTPases and metallothioneins offer more complexity in ROS metabolism, and MAPK signalling modules determine precisely what events should follow-on from a particular stimulus.

7.5.1 ROS removal

The responsiveness of different tissues to ROS could result not only from the presence and the function of distinct ROS-generating proteins as described above, but also via the lack or presence of antioxidants in the cell. In barley aleurone cells, e.g. gibberellin treatment causes a decrease in the activities of catalase, SOD and ascorbate peroxidase, resulting in increased levels of H_2O_2 and thus cell death (Fath *et al.*, 2001). Different antioxidants also have different affinities for ROS which could further determine the efficiency of ROS removal – e.g. catalase has a high affinity for H_2O_2 whereas ascorbate peroxidase has a low affinity for H_2O_2 (Mittler, 2002). The location and action of these enzymes could thus affect how ROS levels are fine-tuned. Manipulation of levels of antioxidants within cells either via over-expression or knock-out studies has also provided previously unidentified links between antioxidants, ROS and physiological responses. For example, Chen and Gallie (2004) showed that manipulating the levels of ascorbate in guard cells altered stomatal apertures, providing new links between the redox state of guard cells and the signalling role of H_2O_2 in this system. Furthermore, plants over-expressing APX showed altered stomatal responses and growth and development, suggesting its important regulatory roles (Pnueli *et al.*, 2003). Manipulation of antioxidant levels has also shown that subsequent exposure of the plant to oxidative stress situations can be tolerated better. Plants lacking both catalase and APX were less sensitive to oxidative stress, and contained reduced amounts of ROS (Rizhsky *et al.*, 2002). Moreover, *Arabidopsis* plants expressing a transgenic APX exhibited systemic resistance to high light intensity (Karpinski *et al.*, 1999). Therefore, the specificity and sensitivity of different cells to ROS are associated intimately with the antioxidant status of a given cell at a given time.

7.6 Cross-talk with other signalling molecules/pathways

The signalling modules and complexes mentioned above also offer a focal point for convergence of signalling cross-talk. In reality, the plant faces multiple threats from the environment; there is never exposure to one single stimulus. It is now apparent that ROS are generated following exposure of cells to multiple stimuli. The multi-tude of biological responses that are mediated by ROS also means that the signalling pathways that are involved are precisely fine-tuned. ROS signalling can occur in series, whereby a particular stimulus activates a ROS generating enzyme resulting in ROS production; this could, subsequently, activate a subset of signalling proteins leading to a physiological response. However, regulation of the synthesis of ROS could also occur via these signalling proteins. Cross-talk occurs when a second stimulus activates the same or another ROS-generating enzyme, or the same signalling protein is activated by more than one stimulus, either via or independent of ROS (Figure 7.5). NADPH oxidase activity is affected by various abiotic and biotic stimuli; therefore, this offers an explanation why acclimation and cross-talk occurs – when plants are exposed to mild levels of one stress, they are acclimated to subsequent exposure of the same stress. Signalling targets of ROS such as PKs, PPases, calcium etc. could also act as targets (indirect) of other signals, which might offer further cross-talk with other signalling pathways. Interaction of ROS with other signals, either locally or distally, also requires precise co-ordination of events, such that a favourable outcome is attained. In summary, ROS can be viewed as one sub-section of an orchestra, whose co-ordinated actions are required for a coherent outcome. Therefore, it is important to consider some of these other orchestral partners that interact with ROS.

Guard cells represent one of the most complex, yet unique and specialised, plant cells, in which almost every signalling event can be studied in detail. Several studies,

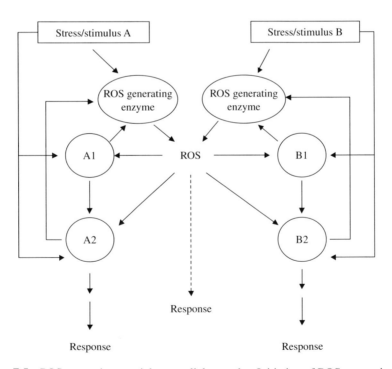

Figure 7.5 ROS can activate serial or parallel cascades. Initiation of ROS generation may lead to signal serial signal transduction cascades, but, more likely, more than one parallel cascade is involved, with probable cross-talk between them.

as indicated earlier, have demonstrated a role for ROS in regulating stomatal movements. Specifically, the plant hormone abscisic acid (ABA), synthesised in response to drought stress, induces the generation of H_2O_2, resulting in stomatal closure. There have also been some studies implicating ROS as signals mediating the synthesis of ABA in other cell types (Zhao *et al.*, 2001). Whether ABA and ROS cause stomatal closure via similar downstream mechanisms is not yet very clear. Both ABA and H_2O_2 cause activation of plasma membrane calcium channels (Pei *et al.*, 2000; Kohler *et al.*, 2003), whilst they seem to have opposite effects on K^+ channels (Kohler *et al.*, 2003), suggesting a branched pathway. Modification of PP2C activities of ABI1 and ABI2 by H_2O_2 occurs *in vitro*, and these proteins are essential signalling intermediates in the ABA-closure signalling pathway, again implicating some point of convergence of ABA and H_2O_2 signalling (Meinhard *et al.*, 2001, 2002). However, it is not yet known whether these interactions are unique to guard cells, or might be present in other ABA-sensitive tissues as well, such as roots and seeds. We have shown recently that darkness-induced stomatal closure also requires the generation of H_2O_2 in both pea (Desikan *et al.*, 2004) and *Arabidopsis* (Desikan, unpublished). Pharmacological data indicate that NADPH oxidase is the source of this ROS, but further genetic analyses are required.

SA is a plant hormone associated with plant defence responses following attack by pathogens. There have been some contradictory reports over the years regarding the relationship between ROS and SA. Whilst some have indicated that ROS leads to elevated levels of SA, SA accumulation was also proposed to potentiate ROS generation (Van Camp *et al.*, 1998). However, using SA-deficient plants it has been shown that redox signalling occurs independent of SA (Grant *et al.*, 2000), suggesting differences in ROS–SA interactions. In an interesting study, Kawano and Muto (2000) have shown that extracellularly secreted peroxidases catalyse SA-dependent formation of superoxide and H_2O_2 in tobacco cells; however, SA was found to suppress the production of hydroxyl radicals in this system. This reveals complex links between different types of ROS and their interaction with other molecules.

Wounding induces the generation of ROS via release of the plant hormone jasmonic acid (JA), these ROS then acting as signals to cause the expression of various defence-related genes (Orozco-Cardenas *et al.*, 1999, 2001). In guard cells, the JA derivative, Me-JA, has been shown to cause cytoplasmic alkalinisation, followed by NADPH oxidase-dependent accumulation of ROS, activation of K^+ channels and stomatal closure (Suhita *et al.*, 2004). Treatment with ABA also caused these events; therefore, cross-talk is evident between JA, ABA and ROS, with common downstream targets. During pathogen-induced defence responses, ROS generation and MAPK activation occurred independent of JA (Grant *et al.*, 2000). However, JA accumulation requires the activation of the wound-inducible MAPK, WIPK (Seo *et al.*, 1999), which is also activated by ROS (Samuel *et al.*, 2001), again suggesting signalling complexities.

Ethylene, a stress hormone involved in many aspects of plant growth and development, has been suggested to act as a positive regulator of ROS production and cell death (Overmyer *et al.*, 2003). During the monitoring of changes in gene expression in response to H_2O_2, several ethylene-related genes were found to be up-regulated (Desikan *et al.*, 2001a; Vandenabeele *et al.*, 2003), including genes involved in ethylene biosynthesis as well as ethylene-response genes, suggesting that ROS can interact with ethylene signalling at various steps.

Auxin association with ROS has also been noted. Auxin induces cell elongation in roots via the release of hydroxyl radicals, derived from superoxide and H_2O_2 in a Haber–Weiss reaction occurring in the cell wall (Schopfer *et al.*, 2002). Auxin signalling occurs via the regulation of MAPK activity, and Kovtun *et al.* (2000) have proposed a link between auxin and ROS signalling at the level of MAPKs. ROS cause activation of ANP1, a MAPKKK which subsequently leads to the activation of downstream MAPKs, AtMPK3 and AtMPK6. Kovtun *et al.* (2000) have shown that activation of these MAPKs leads to repression of auxin-inducible genes, which implies that H_2O_2, directly or indirectly, negatively modulates auxin responses. However, in maize roots, asymmetric application of auxin, or gravity, both led to increased generation of H_2O_2 (Joo *et al.*, 2001). The differing effects and interaction of ROS and auxin could be due to tissue specificity, but could also be dependent on cross-talk with other signalling molecules that are unique in certain cells.

Reactive nitrogen species (RNS) such as nitric oxide (NO) have chemical properties very similar to those of ROS. Furthermore, they often appear to be made temporally and spatially together (Desikan *et al.*, 2004a) and, therefore, close interplay between these two signals is very likely. During pathogen-induced defence responses, both H_2O_2 and NO synthesis occurs, and a critical ratio between the two determines whether or not the defence response of programmed cell death occurs (Delledonne *et al.*, 2001). In guard cells, ABA treatment causes the generation of both H_2O_2 and NO, both being essential for stomatal closure. Exposure to UV is linked to both H_2O_2 and NO production (Neill *et al.*, 2002). However, the downstream signalling mechanisms by which H_2O_2 and NO mediate these responses are not very clear. It is quite likely that H_2O_2 and NO have both common and distinct targets, determining specificity in response to a particular stimulus. Both H_2O_2 and NO activate MAPKs, but whether they are the same MAPKs is not yet known. Both H_2O_2 and NO also signal through PPases, at least in guard cells (Murata *et al.*, 2001; Desikan *et al.*, 2002). However, whilst NO acts via the second messenger cyclic GMP, H_2O_2 does not (Desikan, unpublished), indicating unique pathways. In addition to its signalling interactions with ROS, NO also has antioxidant properties and can react with, and thus remove, any ROS (Lamattina *et al.*, 2003). Thus, the outcome of many processes probably depends on the localisation and the rate of NO generation relative to ROS.

7.7   Systemic signalling

Although H_2O_2 interacts with, potentially activates, inhibits or induces proteins in cells in the immediate location of its production, it is also possible that ROS act as transmissible signals, mediating long-distance effects. Alvarez *et al.* (1998) showed that in *Arabidopsis* plants challenged with incompatible pathogens, a primary generation of ROS in infected leaves leads to a secondary generation of ROS in distal leaves, expression of defence-related genes and systemic immunity, suggesting that some long-distance transmissible signal mediates the plant immune response. H_2O_2 has also been shown to act as a systemic signal during abiotic stresses – exposure to high light intensity induced the expression of APX2-LUC and the systemic release of H_2O_2 leading to an acclimatory response in untreated parts of the plant (Karpinski *et al.*, 1999). During wounding of tomato plants, the local and systemic release of H_2O_2 causes the expression of defence-related genes, resulting in a systemic defence response (Orozco-Cardenas *et al.*, 2001). Application of H_2O_2 to *Arabidopsis* roots activated MAP kinase activity in the leaves. The same pattern of MAP kinase was caused by direct injection of H_2O_2 to the leaves, suggesting long-distance transmission of a H_2O_2-derived signal (Capone *et al.*, 2004). It is possible that in all these situations, ROS is not the primary signal, but other signalling intermediates, such as lipid derivatives or SA, could be involved.

Thus, H_2O_2 may act as a distal as well as a local signal, mediating either similar or dissimilar signalling pathways in the different target cells that result in particular physiological responses. It would be interesting to determine whether ROS generated in specific cells such as guard cells following ABA treatment, will lead to alterations in cellular responses to further stimulation with the same, or a new, stimulus, or whether other cells and tissues, such as roots are affected too.

7.8 Conclusion

In just a few years, the view of ROS has changed from one that saw them as toxic metabolic side products that required antioxidant mechanisms in order to protect cells from their effects, to one that places them at the centre of plant biology as key signals regulating growth and development and coordinating responses to abiotic and biotic stress. Continued application of large-scale post-genomic analyses will elucidate the gene, protein and metabolite networks that are ROS-responsive. Biochemical characterisation of the various sources of ROS in cells, along with the mechanisms by which ROS generation and removal are regulated and by which ROS signalling is transduced, will facilitate further functional analyses to determine those biological process in which ROS play a role. There can be no doubt that there are many new developments and surprises waiting to be uncovered.

References

Alvarez, M.E., Pennell, R.I., Meijer, P.-J., Ishikawa, A., Dixon, R.A. and Lamb, C. (1998) 'Reactive oxygen intermediates mediate a systemic signal network in the establishment of plant immunity', *Cell* **92**, 773–784.

Bais, H., Vepachedu, R., Gilroy, S., Callaway, R.M. and Vivanco, J.M. (2003) 'Allelopathy and exotic plant invasion: from molecules and genes to species interactions', *Science* **301**, 1377–1380.

Baxter-Burrell, A., Yang, Z., Springer, P.S. and Bailey-Serres, J. (2002) 'RopGAP4-dependent Rop GTPase rheostat control of *Arabidopsis* oxygen deprivation tolerance', *Science* **296**, 2026–2028.

Biteau, B., Labarre, J. and Toledano, M.B. (2003) 'ATP-dependent reduction of cysteine-sulphinic acid by *S. cerevisiae* sulphiredoxin', *Nature* **425**, 980–984.

Bolwell, G.P., Bindschedler, L.V., Blee, K.A. *et al.* (2002) 'The apoplastic oxidative burst in response to biotic stress in plants: a three-component system', *Journal of Experimental Botany* **53**, 1367–1376.

Bowler, C. and Fluhr, R. (2000) 'The role of calcium and activated oxygens as signals for controlling cross-tolerance', *Trends in Plant Science* **5**, 241–245.

Buchanan, B.B., Schürmann, P., Wolosiuk, R.A. and Jacquot, J.-P. (2002) 'The ferredoxin/thioredoxin system: from discovery to molecular structures and beyond', *Photosynthetic Research* **73**, 215–222.

Buck, V., Quinn, J., Pino, T.S. *et al.* (2001) 'Peroxide sensors for the fission yeast stress-activated mitogen-activated protein kinase pathway', *Molecular Biology of the Cell* **12**, 407–419.

Capone, O.R., Tiwari, B.S. and Levine, A. (2004) 'Rapid transmission of oxidative and nitrosative stress signals from roots to shoots in *Arabidopsis*', *Plant Physiology and Biochemistry* **42**, 425–428.

Chen, Z. and Gallie, D.R. (2004) 'The ascorbic acid redox state controls guard cell signalling and stomatal movement', *The Plant Cell* **16**, 1143–1162.

Cooper, C.E., Patel, R.P., Brookes, P.S. and Darley-Usmar, V.M. (2002) 'Nanotransducers in cellular redox signalling: modification of thiols by reactive oxygen and nitrogen species', *Trends in Biochemical Sciences* **27**, 489–492.

Danon, A. (2002) 'Redox reactions of regulatory proteins: do kinetics promote specificity?', *Trends in Biochemical Sciences* **27**, 197–203.

Delaunay, A., Pflieger, D., Barrault, M.-B., Vinh, J. and Toledano, M.B. (2002) 'A thiol peroxidase is an H_2O_2 receptor and redox-transducer in gene activation', *Cell* **111**, 471–481.

Delledonne, M., Murgia, I., Ederle, D. *et al.* (2002) 'Reactive oxygen intermediates modulate nitric oxide signalling in the plant hypersensitive disease-resistance response', *Plant Physiology and Biochemistry* **40**, 605–610.

Delledonne, M., Zeier, J., Marocco, A. and Lamb, C. (2001) 'Signal interactions between nitric oxide and reactive oxygen intermediates in the plant hypersensitive disease resistance response', *Proceedings of the National Academy of Sciences USA* **98**, 13454–13459.

Demidchek, V., Shabala, S.N., Coutts, K.B., Tester, M.A. and Davies, J.M. (2003) 'Free oxygen radicals regulate plasma membrane Ca^{2+}- and K^+-permeable channels in plant root cells', *Journal of Cell Science* **116**, 81–88.

Desikan, R., A.-H. Mackerness, S., Hancock, J.T. and Neill, S.J. (2001a) 'Regulation of the *Arabidopsis* transcriptome by oxidative stress', *Plant Physiology* **127**, 159–172.

Desikan, R., Burnett, E., Hancock, J.T. and Neill, S.J. (1998) 'Harpin and hydrogen peroxide induce the expression of a homologue of gp91-*phox* in *Arabidopsis thaliana* suspension cultures', *Journal of Experimental Botany* **49**, 1767–1771.

Desikan, R., Cheung, M.-K., Bright, J., Henson, D., Hancock, J.T. and Neill, S.J. (2004a) 'ABA, hydrogen peroxide and nitric oxide signalling in stomatal guard cells', *Journal of Experimental Botany* **55**, 205–212.

Desikan, R., Cheung, M.-K., Clarke, A. *et al.* (2004) 'Hydrogen peroxide is a common signal for darkness and ABA-induced stomatal closure in *Pisum sativum*', *Functional Plant Biology* **31**, 913–920.

Desikan, R., Clarke, A., Hancock, J.T. and Neill, S.J. (1999) 'H_2O_2 activates a MAP kinase-like enzyme in *Arabidopsis thaliana* suspension cultures', *Journal of Experimental Botany* **50**, 1863–1866.

Desikan, R., Griffiths, R., Hancock, J. and Neill, S. (2002) 'A new role for an old enzyme: nitrate reductase-mediated nitric oxide generation is required for abscisic acid-induced stomatal closure in *Arabidopsis thaliana*', *Proceedings of the National Academy of Sciences USA* **99**, 16319–16324.

Desikan, R., Hancock, J.T. and Neill, S.J. (2004b) Oxidative stress signalling, in *Plant Responses to Abiotic Stress* (ed. H. Hirt and K. Shinozaki). *Topics in Current Genetics* **4**, 121–148.

Desikan, R., Hancock, J.T., Ichimura, K., Shinozaki, K. and Neill, S.J. (2001b) 'Harpin induces activation of the *Arabidopsis* mitogen-activated protein kinases AtMPK4 and AtMPK6', *Plant Physiology* **126**, 1579–1587.

Despres, C., Chubak, C., Rochon, A. *et al.* (2003) 'The *Arabidopsis* NPR1 disease resistance protein is a novel cofactor that confers redox regulation of DNA binding activity to the basic domain/leucine zipper transcription factor TGA1', *The Plant Cell* **15**, 2181–2191.

Fath, A., Bethke, P.C. and Jones, R.L. (2001) 'Enzymes that scavenge reactive oxygen species are down-regulated prior to gibberellic acid-induced programmed cell death in barley aleurone', *Plant Physiology* **126**, 156–166.

Foreman, J., Demidchik, V., Bothwell, J.H.F. *et al.* (2003) 'Reactive oxygen species produced by NADPH oxidase regulate plant cell growth', *Nature* **422**, 442–446.

Garreton, V., Carpinelli, J., Jordana, X. and Holuigue, L. (2002) 'The *as-1* promoter element is an oxidative stress-responsive element and salicylic acid activates it via oxidative species', *Plant Physiology* **130**, 1516–1526.

Grant, J.J., Yun, B.-W. and Loake, G.J. (2000) 'Oxidative burst and cognate redox signalling reported by luciferase imaging: identification of a signal network that functions independently of ethylene, SA and Me-JA but is dependent on MAPKK activity', *The Plant Journal* **24**, 569–582.

Gupta, R. and Luan, S. (2003) 'Redox control of protein tyrosine phosphatases and mitogen activated protein kinases in plants', *Plant Physiology* **132**, 1149–1152.

Hancock, J.T., Desikan, R. and Neill, S.J. (2001) 'Does the redox status of cytochrome c act as a fail-safe mechanism in the regulation of programmed cell death?', *Free Radical Biology and Medicine* **31**, 697–703.

Hancock, J.T., Desikan, R., Neill, S.J. and Cross, A.R. (2004) 'New equations for redox and nano signal transduction', *Journal of Theoretical Biology* **226**, 65–68.

Hwang, I., Chen, H.-C. and Sheen, J. (2002) 'Two-component signal transduction pathways in *Arabidopsis*', *Plant Physiology* **129**, 500–515.

Jabs, T. (1999) 'Reactive oxygen intermediates as mediators of programmed cell death in plants and animals', *Biochemical Pharmacology* **57**, 231–245.

Jabs, T., Dietrich, R.A. and Dangl, J.L. (1996) 'Initiation of runaway cell death in an *Arabidopsis* mutant by extracellular superoxide', *Science* **273**, 1853–1856.

Joo, J.H., Bae, Y.S. and Lee, J.S. (2001) 'Role of auxin-induced reactive oxygen species in root gravitropism', *Plant Physiology* **126**, 1055–1060.

Karpinski, S., Reynolds, H., Karpinska, B., Wingsle, G., Creissen, G. and Mullineaux, P. (1999) 'Systemic signaling and acclimation in response to excess excitation energy in *Arabidopsis*', *Science* **284**, 654–657.

Kawano, T. and Muto, S. (2000) 'Mechanism of peroxidase actions for salicylic acid-induced generation of active oxygen species and an increase in cytosolic calcium in tobacco cell suspension culture', *Journal of Experimental Botany* **51**, 685–693.

Keller, T., Damude, H.G., Werner, D., Doerner, P., Dixon, R.A. and Lamb, C. (1998) 'A plant homolog of the neutrophil NADPH oxidase gp91-*phox* subunit gene encodes a plasma membrane protein with Ca^{2+} binding motifs', *The Plant Cell* **10**, 255–266.

Kohler, B., Hills, A. and Blatt, M.R. (2003) 'Control of guard cell ion channels by hydrogen peroxide and abscisic acid indicates their action through alternate signalling pathways', *Plant Physiology* **131**, 385–388.

Kovtun, Y., Chiu, W.-L., Tena, G. and Sheen, J. (2000) 'Functional analysis of oxidative stress-activated mitogen-activated protein kinase cascade in plants', *Proceedings of the National Academy of Sciences USA* **97**, 2940–2945.

Kwak, J.M., Mori, I.C., Pei, Z.-M. *et al.* (2003) 'NADPH oxidase *AtrbohD* and *AtrbohF* genes function in ROS-dependent ABA signaling in *Arabidopsis*', *EMBO Journal* **22**, 2623–2633.

Laloi, C., Apel, K. and Danon, A. (2004) 'Reactive oxygen signalling: the latest news', *Current Opinion in Plant Biology* **7**, 323–328.

Laloi, C., Rayapuram, N., Chartier, Y., Grienenberger, J.M., Bonnard, G. and Meyer, Y. (2001) 'Identification and characterisation of a mitochondrial thioredoxin system in plants', *Proceedings of the National Academy of Sciences USA* **98**, 14144–14149.

Lamattina, L., Garcia-Mata, C., Grazano, M. and Pagnussat, G. (2003) 'Nitric oxide: the versatility of an extensive signal molecule', *Annual Review of Plant Biology* **54**, 109–136.

Meinhard, M. and Grill, E. (2001) 'Hydrogen peroxide is a regulator of ABI1, a protein phosphatase 2C from *Arabidopsis*', *FEBS Letters* **508**, 443–446.

Meinhard, M., Rodriguez, P.L. and Grill, E. (2002) 'The sensitivity of ABI2 to hydrogen peroxide links the abscisic acid-response regulator to redox signalling', *Planta* **214**, 775–782.

Meyer, Y., Vignols, F. and Reichleld, J.P. (2002) 'Classification of plant thioredoxins by sequence similarity and intron position', *Methods in Enzymology* **347**, 394–402.

Mittler, R. (2002) 'Oxidative stress, antioxidants and stress tolerance', *Trends in Plant Science* **7**, 405–410.

Moon, H., Lee, B., Choi, G. *et al.* (2003) 'NDP kinase 2 interacts with two oxidative stress-activated MAPKs to regulate cellular redox state and enhances multiple stress tolerance in transgenic plants', *Proceedings of the National Academy of Sciences USA* **100**, 358–363.

Mori, I.C. and Schroeder, J.I. (2004) 'Reactive oxygen species activation of plant Ca^{2+} channels. A signalling mechanism in polar growth, hormone transduction, stress signalling, and hypothetically mechanotransduction', *Plant Physiology* **135**, 702–708.

Mou, Z., Fan, W.H. and Dong, X.N. (2003) 'Inducers of plant systemic acquired resistance regulate NPR1 function through redox changes', *Cell* **113**, 935–944.

Murata, Y., Pei, Z.-M., Mori, I.C. and Schroeder, J. (2001) 'Abscisic acid activation of plasma membrane Ca^{2+} channels in guard cells requires cytosolic NAD(P)H and is differentially disrupted upstream and downstream of reactive oxygen species production in *abi1-1* and *abi2-1* protein phosphatase 2C mutants', *The Plant Cell* **13**, 2513–2523.

Mustilli, A.-C., Merlot, S., Vavasseur, A., Fenzi, F. and Giraudat, J. (2002) '*Arabidopsis* OST1 protein kinase mediates the regulation of stomatal aperture by abscisic acid and acts upstream of reactive oxygen species production', *The Plant Cell* **14**, 3089–3099.

Nakagami, H., Kiegerl, S. and Hirt, H. (2004) 'OMTK1, a novel MAPKKK, channels oxidative stress signalling through direct MAPK interaction', *Journal of Biological Chemistry* **279**, 26959–26966.

Neill, S., Desikan, R. and Hancock, J. (2002) 'Hydrogen peroxide signalling', *Current Opinion in Plant Biology* **5**, 388–395.

Op den Camp, R.G.L., Przybyla, D., Ochsenbein, C. *et al.* (2003) 'Rapid induction of distinct stress responses after the release of singlet oxygen in *Arabidopsis*', *The Plant Cell* **15**, 2320–2332.

Orozco-Cardenas, M.L. and Ryan, C. (1999) 'Hydrogen peroxide is generated systemically in plant leaves by wounding and systemin via the octadecanoid pathway', *Proceedings of the National Academy of Sciences USA* **96**, 6553–6557.

Orozco-Cardenas, M.L., Narvaez-Vasquez, J. and Ryan, C.A. (2001) 'Hydrogen peroxide acts as a second messenger for the induction of defense genes in tomato plants in response to wounding, systemin, and methyl jasmonate', *The Plant Cell* **13**, 179–191.

Ostergaard, H., Henriksen, A., Hansen, F.G. and Einther, J.R. (2001) 'Shedding light on disulfide bond formation: engineering a redox switch in green fluorescent protein', *EMBO Journal* **21**, 5853–5862.

Overmyer, K., Brosche, M. and Kangasjarvi, J. (2003) 'Reactive oxygen species and hormonal control of cell death', *Trends in Plant Science* **8**, 335–342.

Pandey, S. and Assmann, S.M. (2004) 'The *Arabidopsis* putative G protein-coupled receptor GCR1 interacts with the G protein alpha subunit GPA1 and regulates abscisic acid signalling', *The Plant Cell* **16**, 1616–1632.

Park, K.-Y., Jung, J.-Y., Park, J. *et al.* (2003) 'A role for phosphatidylinositol 3-phosphate in abscisic acid-induced reactive oxygen species generation in guard cells', *Plant Physiology* **132**, 92–98.

Pei, Z.-M., Murata, Y., Benning, G. *et al.* (2000) 'Calcium channels activated by hydrogen peroxide mediate abscisic acid signalling in guard cells', *Nature* **406**, 731–734.

Perrone, S.T., McDonald, K.L., Sutherland, M.W. and Guest, D.I. (2003) 'Superoxide release is necessary for phytoalexin accumulation in *Nicotiana tabacum* cells during the expression of cultivar-race ad non-host resistance towards *Phytophthora* spp', *Physiological and Molecular Plant Pathology* **62**, 127–135.

Pnueli, L., Liang, H., Rozenberg, M. and Mittler, R. (2003) 'Growth suppression, altered stomatal responses, and augmented induction of heat shock proteins in cytosolic ascorbate peroxidase (*Apx1*)-deficient *Arabidopsis* plants', *The Plant Journal* **34**, 187–203.

Rentel, M.C., Lecorieux, D., Ouaked, F. *et al.* (2004) 'OXI1 kinase is necessary for oxidative burst-mediated signalling in *Arabidopsis*', *Nature* **427**, 858–861.

Rhee, S.G., Bae, Y.S., Lee, S.-R. and Kwon, J. (2000) 'Hydrogen peroxide: a key messenger that modulates protein phosphorylation through cysteine oxidation', *Science's STKE* http://stke.sciencemag.org/cgi/content/full/OC_sigtrans;2000/53/pe1

Rivas, S., Rougon-Cardoso, A., Smoker, M., Schauser, L., Yoshioka, H. and Jones, J.D.G. (2004) 'CITRX thioredoxin interacts with the tomato Cf-9 resistance protein and negatively regulates defence', *EMBO Journal* **23**, 2156–2165.

Rizhsky, L., Hallak-Herr, E., Van Breusegem, F. *et al.* (2002) 'Double antisense plants lacking ascorbate peroxidase and catalase are less sensitive to oxidative stress than single antisense plants lacking ascorbate peroxidase or catalase', *The Plant Journal* **32**, 329–342.

Rodriguez, A.A., Grunberg, K.A. and Taleisnik, E.L. (2002) 'Reactive oxygen species in the elongation zone of maize leaves are necessary for leaf extension', *Plant Physiology* **129**, 1627–1632.

Romeis, T., Piedras, P. and Jones, J.D.G. (2000) 'Resistance gene-dependent activation of a calcium-dependent protein kinase in the plant defense response', *The Plant Cell* **12**, 803–815.

Sagi, M. and Fluhr, R. (2001) 'Superoxide production by plant homologues of the gp91^{phox} NADPH oxidase. Modulation of activity by calcium and by tobacco mosaic virus infection', *Plant Physiology* **126**, 1281–1290.

Salmeen, A., Andersen, J.N., Myers, M.P. *et al.* (2003) 'Redox regulation of protein tyrosine phosphatase 1B involves a sulphenyl-amide intermediate', *Nature* **423**, 769–773.

Samuel, M.A., Miles, G.P. and Ellis, B.E. (2000) 'Ozone treatment rapidly activates MAP kinase signalling in plants', *The Plant Journal* **22**, 367–376.

Schafer, F.Q. and Buettner, G.R. (2001) 'Redox environment of the cell as viewed through the redox state of the glutathione disulfide/glutathione couple', *Free Radical Biology Medicine* **30**, 1191–1212.

Schopfer, P., Liszkay, A., Bechtold, M., Frahry, G. and Wagner, A. (2002) 'Evidence that hydroxyl radicals mediate auxin-induced extension growth', *Planta* **214**, 821–828.

Scrase-Field, S.A.M.G. and Knight, M.R. (2003) 'Calcium: just a chemical switch?', *Current Opinion in Plant Biology* **6**, 500–506.

Seo, S., Sano, H. and Ohashi, Y. (1999) 'Jasmonate-based wound signal transduction requires activation of WIPK, a tobacco mitogen-activated protein kinase', *The Plant Cell* **11**, 289–298.

Shin, R. and Schachtman, D.P. (2004) 'Hydrogen peroxide mediates plant root cell response to nutrient deprivation', *Proceedings of the National Academy of Sciences USA* **101**, 8827–8832.

Singh, K.K. (2000) 'The *Sacccharomyces cerevisiae* SLN1P-SSK1P two-component system mediates response to oxidative stress and in an oxidant-specific fashion', *Free Radical Biology and Medicine* **29**, 1043–1050.

Suharsono, U., Fujisawa, Y., Kwasaki, T., Iwaski, Y., Satoh, H. and Shimamoto, K. (2002) 'The heterotrimeric G protein α subunit acts upstream of the small GTPase Rac in disease resistance of rice', *Proceedings of the National Academy of Sciences USA* **99**, 13307–13312.

Suhita, D., Raghavendra, A.S., Kwak, J.M. and Vavasseur, A. (2004) 'Cytosolic alkalinisation precedes reactive oxygen species production during methyl jasmonate- and abscisic acid-induced stomatal closure', *Plant Physiology* **134**, 1536–1545.

Sweetlove, L.J., Heazlewood, J.L., Herald, V. *et al.* (2002) 'The impact of oxidative stress on *Arabidopsis* mitochondria', *The Plant Journal* **32**, 891–904.

Thoma, I., Loeffler, C., Sinha, A.K. *et al.* (2003) 'Cyclopentenone isoprostanes induced by reactive oxygen species trigger defense gene activation and phytoalexin accumulation in plants', *The Plant Journal* **34**, 363–375.

Thorpe, G.W., Fong, C.S., Alic, N., Higgins, V.J. and Dawes, I.W. (2004) 'Cells have distinct mechanisms to maintain protection against reactive oxygen species: oxidative-stress-responsive genes', *Proceedings of the National Academy of Sciences USA* **101**, 6564–6569.

Torres, M.A., Dangl, J.L. and Jones, J.D.G. (2002) '*Arabidopsis* gp91^{phox} homologues *AtrbohD* and *AtrbohF* are required for accumulation of reactive oxygen intermediates in the plant defense response', *Proceedings of the National Academy of Sciences USA* **99**, 517–522.

Torres, M.A., Onouchi, H., Hamada, S., Machida, C., Hammond-Kossack, K.E. and Jones, J.D.G. (1998) 'Six *Arabidopsis thaliana* homologues of the human respiratory burst oxidase (gp91-phox)', *The Plant Journal* **14**, 365–370.

Van Camp, W., Montagu, M.V. and Inze, D. (1998) 'H_2O_2 and NO: redox signals in disease resistance', *Trends in Plant Science* **3**, 330–334.

Vandenabeele, S., Der Kelen, K.V., Dat, J. *et al.* (2003) 'A comprehensive analysis of hydrogen peroxide-induced gene expression in tobacco', *Proceedings of the National Academy of Sciences USA* **100**, 16113–16118.

Van Montfort, R.L.M., Congrave, M., Tisi, D., Carr, R. and Jhoti, H. (2003) 'Oxidation state of the active-site cysteine in protein tyrosine phosphatase 1B', *Nature* **423**, 773–777.

Vranova, E., Atichartpongkul, S., Villarroel, R., Van Montagu, M., Inze, D. and Van Camp, W. (2002a) 'Comprehensive analysis of gene expression in *Nicotiana tabacum* leaves acclimated to oxidative stress', *Proceedings of the National Academy of Sciences USA* **99**, 10870–10875.

Vranova, E., Inze, D. and Van Breusegem, F. (2002b) 'Signal transduction during oxidative stress', *Journal of Experimental Botany* **53**, 1227–1236.

Wong, H.L., Sakamoto, T., Kwasaki, T., Umemura, K. and Shimamoto, K. (2004) 'Down-regulation of metallothionein, a reactive oxygen scavenger, by the small GTPase OsRac1 in rice', *Plant Physiology* **135**, 1447–1456.

Yamazaki, D., Motohashi, K., Kasama, T., Hara, Y. and Hisabori, T. (2004) 'Target proteins of the cytosolic thioredoxins in *Arabidopsis thaliana*', *Plant and Cell Physiology* **45**, 18–27.

Yang, T. and Poovaiah, B.W. (2002) 'Hydrogen peroxide homeostasis: activation of plant catalase by calcium/calmodulin', *Proceedings of the National Academy of Sciences USA* **99**, 4097–4102.

Yano, H., Wong, J.H., Lee, Y.M., Cho, M.-J. and Buchanan, B.B. (2001) 'A strategy for the identification of proteins targeted by thioredoxin', *Proceedings of the National Academy of Sciences USA* **98**, 4794–4799.

Zhang, X., Dong, F.C., Cao, J.F. and Song, C.P. (2001) 'Hydrogen peroxide-induced changes in intracellular pH of guard cells precede stomatal closure', *Cell Research* **11**, 37–43.

Zhao, Z., Chen, G. and Zhang, C. (2001) 'Interaction between reactive oxygen species and nitric oxide in drought-induced abscisic acid synthesis in root tips of wheat seedlings', *Australian Journal of Plant Physiology* **28**, 1055–1061.

8 Reactive oxygen species in plant development and pathogen defence

Mark A. Jones and Nicholas Smirnoff

8.1 Introduction

This chapter covers aspects of the roles of reactive oxygen species (ROS) in plant growth and development and in defence against pathogens. Plants, like animals, produce ROS via oxidase enzymes when attacked by pathogens (the oxidative burst). This can have a direct role in defence (e.g. by damaging the pathogen and enabling peroxidative cross-linking of cell wall proteins and polysaccharides). ROS also initiate signal transduction cascades that activate defence-related gene expression along with varied input from other signalling molecules, such as salicylic acid (SA), nitric oxide (NO) and jasmonic acid (JA). ROS have a role in initiating the hypersensitive response, a form of programmed cell death (PCD) induced by incompatible (avirulent) pathogens. ROS are also involved in the PCD that occurs as part of normal development. The cell cycle is sensitive to redox state (Reichheld et al., 1999), presumably to protect against the potentially mutagenic action of ROS during DNA replication. The role of the antioxidants ascorbate and glutathione in redox state and cell division is discussed in Chapters 1 and 3. Hydrogen peroxide (H_2O_2) inhibits cell expansion by peroxidative cross-linking, but hydroxyl radicals generated in the wall can cause scission of polysaccharides and can loosen or soften the wall. This has been proposed to contribute to auxin-induced growth (Schopfer, 2001) and fruit ripening (Dumville & Fry, 2003) and is discussed in Chapter 9. More recently, it has become apparent that ROS are required for the growth of at least some cell types, and for abscisic acid (ABA) signalling leading to stomatal closure. Plasma membrane localised nicotinamide adenine dinucleotide phosphate (NADPH) oxidase (NOX) enzymes that generate extracellular superoxide are implicated in many of these processes and are the focus of this chapter. In particular, the role of NOX-derived ROS as signals that control development will be considered. The effects of ROS on growth and development resulting from more direct alterations to cell wall structure are covered in Chapter 9. The perception of ROS and redox state and the signal transduction pathways leading to changes in gene expression are also covered in Chapter 7.

8.2 The roles of ROS in plant development

There is growing evidence that ROS have important roles as signalling molecules in mammalian cell growth and development (Sauer et al., 2001). The major source of these signalling ROS is the superoxide anion ($O_2^{.-}$) generating plasma membrane (PM) NOX complex (Babior, 1999). $O_2^{.-}$ readily gives rise to other ROS including H_2O_2 and the hydroxyl radical ($OH^{.}$), the former reaction being catalysed by superoxide dismutases (Halliwell & Gutteridge, 1999). In contrast to other ROS, H_2O_2 is a relatively stable molecule (Halliwell & Gutteridge, 1999) which may act as a long distance signalling molecule (see Section 8.3.2). The kinetics of NOX activation and inactivation are ideal for an enzyme involved in signalling pathways (Sauer et al., 2001). gp91^{phox} is the catalytic subunit of the archetypal mammalian NOX complex in phagocytes (Babior, 1999). One of the first links established between ROS formation and cell growth control was the effect of *Mox1* (*Nox1* a *gp91^{phox}* homologue) overexpression on increased $O_2^{.-}$ generation and cell growth in mouse fibroblasts (Suh et al., 1999). It was subsequently shown that H_2O_2 mediated the effects of *Nox1* overexpression on cell growth, transformation and related gene expression, suggesting that H_2O_2 functions as an intracellular signalling molecule in mammalian cells (Arnold et al., 2001). Human NOXs have also been implicated in a number of other developmental processes including angiogenesis (Arbiser et al., 2002).

The recent discoveries that NOX-derived ROS are required for sexual development in the filamentous fungus *Aspergillus nidulans* (Lara-Ortiz et al., 2003) and root-hair development in *Arabidopsis thaliana* (Foreman et al., 2003) have raised the question of whether ROS have a fundamental role in eukaryotic cell development. This possibility is strengthened by an analysis of the occurrence of NOX homologues, which shows that they are consistently absent in unicellular eukaryotic organisms, regardless of phylogenetic position, but are consistently present in multicellular eukaryotic organisms (Lalucque & Silar, 2003). This distribution of NOX suggests a primary role for NOX in multicellular development rather than in pathogen defence. Like mammalian cells, plant cells produce ROS as a by-product of their normal aerobic metabolism (Mittler, 2002) and as part of their defences against pathogen attack (Bolwell, 1999). However, consistent with the emerging role of NOXs in mammalian cell development, plant NOXs, specifically, appear to be important for plant cell development, despite the multiple sources of ROS in plants (Mittler, 2002). Like their mammalian homologues, plant NOXs are intrinsic PM proteins (Keller et al., 1998) that catalyse the formation of $O_2^{.-}$ from molecular oxygen using reduced NADPH as an electron donor (Sagi & Fluhr, 2001). *Arabidopsis* has a family of 10 NOX genes, termed *Atrboh* (*A. thaliana r*espiratory *b*urst *o*xidase *h*omologues) that are homologous to *gp91^{phox}* (Keller et al., 1998; Torres et al., 1998; Foreman et al., 2003).

Several studies have identified roles for ROS in plant development without identifying the source of ROS formation. $O_2^{.-}$ was both found to be present during, and essential for, morphogenesis of etiolated wheat seedlings. Furthermore, seedlings incubated in antioxidants had reduced $O_2^{.-}$ and distorted development (Shorning et al., 2000). Low doses of exogenous H_2O_2 have been reported to promote seed germination and plant growth (Korystov & Narimanov, 1997). In another study,

H_2O_2 formation was coincident with the emergence of bud primordia during plant regeneration from strawberry callus culture (Tian *et al.*, 2003). ROS also have also been implicated in auxin-mediated root gravitropism (Joo *et al.*, 2001). Upon gravi-stimulation, ROS formation was increased in the lower cortex of horizontally oriented primary roots. Furthermore, exogenous topically applied H_2O_2 resulted in gravity-independent root curvature whereas ROS scavengers inhibited root gravitropism (Joo *et al.*, 2001). Although exogenous auxin applied to one side of the root tip stimulated localized ROS formation, the auxin transport inhibitor N-(1-naphthyl)phthalamic acid (NPA) did not prevent exogenous H_2O_2-induced gravitropism, suggesting that ROS formation acts later than auxin in root gravitropism (Joo *et al.*, 2001). OH· scavengers such as dimethylsulfoxide inhibited root and shoot elongation (Schopfer, 2001), suggesting that ROS are required for growth generally as well as for root gravitropism. One proposed mechanism for ROS-induced growth involves the generation of OH· in the cell wall by a Fenton reaction, mediated by Cu^{2+} and ascorbate or by peroxidase activity. The OH· cause scission of polysaccharides and thereby loosen the wall. This process could be involved in the regulation of cell expansion and fruit ripening. The formation and function of ROS in the cell wall is discussed in more detail in Chapter 9.

Still further studies have implicated NOX activity in developmental processes through the indirect evidence provided by the flavin-containing oxi-dase inhibitor diphenyleneiodonium (DPI). As DPI not only inhibits NOX, but also the H_2O_2-generating activity of peroxidase (Frahry & Schopfer, 1998), the precise source of DPI-sensitive ROS is unclear in the absence of genetic evi-dence. DPI-sensitive ROS formation coincided with the start of secondary wall deposition in cotton fibres (Potikha *et al.*, 1999). Furthermore, both DPI and ROS scavengers inhibited wall differentiation, and exogenous H_2O_2 induced early secondary wall formation in these highly differentiated cells (Potikha *et al.*, 1999). It appears that a ROS, probably H_2O_2, may act as a developmental signal in the differentiation of secondary walls in cotton fibres. DPI-sensitive ROS formation was also detected in the expanding zone of maize leaf blades (Rodriguez *et al.*, 2002). Exogenous H_2O_2 partially rescued the DPI-induced inhibition of leaf segment elongation.

Genetic evidence for the involvement of NOX genes in plant growth and devel-opment has recently come from tomato plants expressing plant *NOX* RNAi trans-genes (Sagi *et al.*, 2004). These plants had reduced leaf ROS formation and highly pleiotropic mutant phenotypes with both vegetative and reproductive structures being disrupted. These results suggest that NOX-derived ROS mediate multiple developmental signalling processes in plants. Interestingly, NOX protein levels were responsive to a range of different plant hormones (Sagi *et al.*, 2004) suggest-ing that NOXs may function as signal transducers in a number of different hormone signalling networks. Root elongation is also reduced in complete loss-of-function (LOF) mutants of three *Arabidopsis* NOX genes, *AtrbohC/RHD2* (Foreman *et al.*, 2003), *AtrbohD* and *AtrbohF* (Kwak *et al.*, 2003).

8.3   NADPH oxidase and ROS in plant cell morphogenesis

8.3.1   NADPH oxidases in polarised plant cells

Recently, NOX-derived ROS have been implicated in root-hair development (Foreman *et al.*, 2003). Root hairs are highly polarised outgrowths from single root epidermal cells. ROS formation is lower in the tip-growth defective root hairs of *rhd2* plants, which carry premature stop codons in the NOX gene *AtrbohC/RHD2*, than in wild-type (WT) hairs (Foreman *et al.*, 2003). This shows that AtrbohC/RHD2 produces the $O_2^{\cdot-}$ required for root-hair tip growth. Furthermore, WT hairs treated with DPI phenocopy *rhd2* hairs (Foreman *et al.*, 2003). These ROS appear to activate PM Ca^{2+} channels in root hairs (see Section 8.3.4). Additionally, a LOF mutant of the serine/threonine protein kinase OXI1, which is induced by H_2O_2 and activates MAPK3 and MAPK6, has short root hairs (Rentel *et al.*, 2004). This provides further evidence that ROS have a signalling role in root-hair development. ROS are also formed in embryonic cells of the alga *Fucus serratus* in response to hyperosmotic stress (Coelho *et al.*, 2002). The initial extracellular ROS formation is required for a Ca^{2+} wave in the apical cytoplasm. A nonselective cation channel that is stimulated by H_2O_2 may be involved in this increase in cytoplasmic $[Ca^{2+}]_{cyt}$ (Coelho *et al.*, 2002; see later). These results suggest that a common role for ROS in polarised plant cell growth may be the activation of PM Ca^{2+} channels. There is not yet any published evidence to suggest whether NOXs might also play a role in pollen tube tip growth. Interestingly, however, the reactive nitrogen species NO does have a role in pollen tube growth and reorientation (Prado *et al.*, 2004) because exogenous NO leads to growth reorientation. Endogenous NO was localised to peroxisomes but both NO and peroxisomes were absent from the cytoplasm at the growing tip of pollen tubes (Prado *et al.*, 2004).

8.3.2   NADPH-oxidase-mediated effects on gene expression

ROS accumulated in K^+-deficient *Arabidopsis* roots and modulated up-regulation of genes induced by K^+ deficiency (Shin & Schachtman, 2004). Exogenous H_2O_2 restored the expression of K^+-deficiency-induced genes in *rhd2* plants, expression of which is not normally induced in this root-hairless, short-rooted mutant. Expression of *NOX* RNAi transgenes in tomato plants also resulted in changes in the expression level of a number of transcription factors, including ectopic up-regulation in leaves of homeotic-type genes normally only expressed in reproductive structures (Sagi *et al.*, 2004). Interestingly, H_2O_2 induced expression of a NOX gene in *Arabidopsis* suspension cultured cells (Desikan *et al.*, 1998), and in cultured tomato cells $O_2^{\cdot-}$ and H_2O_2 induced the expression of different subsets of extensin genes (Wisniewski *et al.*, 1999). H_2O_2 can readily cross the PM through water channels (Henzler & Steudle, 2000). Once inside the plant cell, it can subsequently act as an intracellular (Kovtun *et al.*, 2000) and intercellular (Allan & Fluhr, 1997) signalling molecule. It appears that NOX-derived ROS, and possibly ROS from other sources, are involved in signalling cascades that lead to changes in gene expression, probably mediated by MAP kinase cascades (Kovtun *et al.*, 2000; see Chapter 7).

It is well established that under conditions of water deficit ABA induces stomatal closure by reducing the turgor of guard cells. The first indication that ROS might be involved in this process came when it was discovered that H_2O_2 was both required for ABA-induced stomatal closure and for ABA-induced activation of Ca^{2+} channels in the PM of *Arabidopsis* guard cells (Pei *et al.*, 2000). The inhibitory effect of exogenous H_2O_2 on ABA-induced stomatal closure was reversed by the antioxidant ascorbic acid, exogenous catalase and DPI (Zhang *et al.*, 2001). Furthermore, the level of endogenous H_2O_2 formation was dependent upon the dose of exogenous ABA, and H_2O_2 formation was found to precede stomatal closure (Zhang *et al.*, 2001). The positive effect of elicitors on stomatal closure was suppressed by exogenous catalase or AA, suggesting that H_2O_2 also mediates pathogen-induced stomatal closure (Lee *et al.*, 1999). It appears that ROS are involved in a fundamental aspect of stomatal closure.

Two NOX genes, which are expressed in guard cells, *AtrbohD* and *AtrbohF*, appear to be functionally redundant in ABA signalling as the *atrbohD/F* double mutant disrupts ABA-induced stomatal closure and ABA-dependent ROS formation (Kwak *et al.*, 2003). Exogenous H_2O_2 rescues the defective stomatal closure in this double mutant. The OST1 ABA-activated protein kinase appears to act between ABA perception and ROS formation in guard cells as *ost1* guard cells are defective in ABA-induced ROS formation and exogenous H_2O_2 rescues stomatal closure in *ost1* (Mustilli *et al.*, 2002). Similarly, the JAR1 protein, which adenylates jasmonic acid (JA) (Staswick *et al.*, 2002), probably acts upstream of NOX-dependent ROS formation as methyl jasmonate increases DPI-sensitive ROS formation in WT guard cells but not in guard cells of the NOX double mutant *atrbohD/F* (Suhita *et al.*, 2004). Both ABI1 and ABI2, protein serine/threonine phosphatases 2C (PP2C) that negatively regulate ABA signalling, are inactivated *in vitro* by H_2O_2 (Meinhard & Grill, 2001; Meinhard *et al.*, 2002). It appears that ROS may feedback to negatively regulate key negative regulators of guard cell ABA signalling.

In the mammalian phagocyte, NOX-dependent ROS formation is activated by phosphatidylinositol 3-phosphate (PI3P) (Ellson *et al.*, 2001), a phosphoinositide generated by phosphatidylinositol 3-kinase (PI3K) activity. The PI3K inhibitors wortmannin or LY294002, and overexpression of a PI3P-binding protein, inhibited both ABA-induced ROS generation and ABA-induced stomatal closure in guard cells (Park *et al.*, 2003). Exogenous H_2O_2 partially rescued these inhibitory effects. It appears that a regulatory role for PI3P in NOX activation might be conserved between plants and animals. Dehydroascorbate reductase (DHAR) catalyses the reduction of dehydroascorbate to AA helping to maintain cellular redox balance (Noctor & Foyer, 1998; Chapter 3). Transgenic plants overexpressing DHAR had lowered H_2O_2 levels in guard cells and reduced stomatal closure (Chen & Gallie, 2004). Furthermore, guard cells in these plants were less sensitive to exogenous H_2O_2 or ABA. These results implicate AA as an important endogenous regulator of ROS in guard cells.

8.3.3   The regulatory effect of ROS on calcium channels

H_2O_2-activated Ca^{2+} channels in the guard cell PM (Pei et al., 2000) and ABA-induced activation of guard cell Ca^{2+} channels was found to be NADPH-dependent (Murata et al., 2001) suggesting the involvement of a NADPH-requiring enzyme such as NOX. ABA-induced ROS formation and Ca^{2+} channel activation were both defective in abi1-1, but H_2O_2-induced activation of Ca^{2+} channels and stomatal closure were not, suggesting that ABI1 acts between ABA perception and ROS formation. Whereas ABA-induced ROS formation was not defective in abi2-1, ABA-induced Ca^{2+} channel activation and H_2O_2-induced activation of Ca^{2+} channels and stomatal closure were, suggesting that ABI2 acts between ROS formation and Ca^{2+} channel activation (Murata et al., 2001). Genetic evidence for the involvement of NOX in calcium channel activation in guard cells came from the atrbohD/F NOX double mutant, which disrupted ABA-induced increases in $[Ca^{2+}]_{cyt}$ and ABA-induced activation of PM Ca^{2+} channels in guard cells (Kwak et al., 2003). Exogenous H_2O_2 rescued Ca^{2+} channel activation in atrbohD/F guard cells.

Root-hair elongation requires a tip-high $[Ca^{2+}]_{cyt}$ gradient and root hairs show a tip-localised influx of extracellular Ca^{2+} (Bibikova et al., 1997; Wymer et al., 1997). Exogenous H_2O_2 inhibited root-hair elongation and subsequently lead to increased $[Ca^{2+}]_{cyt}$ (Jones et al., 1998). Patch-clamp experiments have identified a hyperpolarization-activated Ca^{2+} channel (HACC) in the hair tip (Very & Davies, 2000) and in the PM of root cell protoplasts (Demidchik et al., 2003). rhd2 roots treated with exogenous $OH^•$ had ballooned hair-like structures (Foreman et al., 2003) indicating that the effect of this LOF mutation on root-hair morphogenesis could be partially rescued by exogenous $OH^•$. Furthermore, PM HACCs in rhd2 protoplasts were stimulated by exogenous $OH^•$ but not by H_2O_2 (Foreman et al., 2003). This effect of $OH^•$ on the root-hair HACC is in contrast to the H_2O_2-activated nonselective cation channels (NSCCs) identified in the Fucus embryo (Coelho et al., 2002) and in guard cells (Pei et al., 2000).

In addition to NOXs, other sources of ROS may regulate Ca^{2+} channels during cell development (Mittler, 2002). For instance, blue light, but not red light, activated voltage dependent Ca^{2+} channels in the PM of mesophyll cells (Stoelzle et al., 2003). This activation was not DPI-sensitive. Similarly, a non-ROS regulated type of Ca^{2+} channel may be present in root hairs (Very & Davies, 2000; Demidchik et al., 2003).

8.3.4   The role of calcium in regulating plant NOXs

Similar to a divergent human gp91^{phox} homologue, NOX5 (Banfi et al., 2001), but unlike the phagocyte NOX, plant gp91^{phox} proteins contain one or two EF-hand motifs in an N-terminal extension (Keller et al., 1998; Torres et al., 1998). Plant NOXs bind Ca^{2+} in vitro (Keller et al., 1998), and NOX activity is up-regulated directly by Ca^{2+} in vitro (Sagi & Fluhr, 2001). Furthermore, DPI-sensitive ROS formation in maize leaf homogenates is Ca^{2+}-sensitive (Jiang & Zhang, 2003).

These results suggest that Ca^{2+} may be an important factor in regulating NOX-mediated O_2^- production in plants.

8.3.5   The role of ROP GTPases in regulating plant NOXs

The activity of mammalian Nox1, which affects cell growth and transformation, is regulated by a Ras small GTPase molecular switch (Mitsushita et al., 2004). In the phagocyte, a RHO GTPase of the RAS superfamily (Rac) regulates the activity of gp91^{phox} (Diekmann et al., 1994). NOX activation requires the assembly of a multi-protein complex at the PM. Rac interacts directly with the p67^{phox} (Diekmann et al., 1994; Diebold & Bokoch, 2001) and gp91^{phox} (Diebold & Bokoch, 2001) subunits of this complex. ROP GTPases (11 genes in Arabidopsis) form a plant-specific sub-family of RHO, within the RAS superfamily (Yang, 2002). ROPs are strong candidates to regulate plant NOXs as maize ROP activates ROS formation in mammalian cells (Hassanain et al., 2000) and human Rac can activate DPI-sensitive ROS formation in plant cells (Park et al., 2000). Furthermore, ROPs regulate DPI-sensitive ROS formation in planta (Yang, 2002; Agrawal et al., 2003). A rice ROP, OsRac1, regulates both ROS-mediated plant cell death (Kawasaki et al., 1999) and ROS formation during plant disease resistance (Ono et al., 2001). A cotton ROP gene, Rac13, induced during the onset of secondary wall differentiation, activated ROS formation in transformed soybean and Arabidopsis cells (as a constitutively active mutant). In contrast, antisense and dominant negative rac13 reduced ROS formation (Potikha et al., 1999). Significantly, ROP2 regulates DPI-sensitive ROS formation in response to hypoxia in Arabidopsis roots (Baxter-Burrell et al., 2002) and constitutively active ROP2 was found to regulate O_2^- production in Arabidopsis leaf extracts (Park et al., 2004). Both ROP GTPases and NOX-generated ROS have, separately, been shown to be involved in ABA-induced stomatal closure, although, as yet, no functional relationship has been established between any members of these two gene families (Pei et al., 2000; Lemichez et al., 2001; Zhang et al., 2001; Zheng et al., 2002; Kwak et al., 2003). Figure 8.1 is a summary diagram illustrating the current understanding of the proposed regulation of NOX activity by ROP GTPases, Ca^{2+}, phosphatidic acid and PI3P.

8.4   Programmed cell death and senescence

Roles for ROS have been proposed in programmed cell death (PCD) and in senescence. PCD is a process in which cells die in an organised manner as a part of normal development or in response to stimuli. Examples include xylogenesis, gametogenesis, incompatibility between pollen and stigma, death of aleurone cells, formation of holes in leaves (e.g. Monstera) and aerenchyma formation. PCD is also triggered by stresses such as heat shock, UV radiation, ozone, H_2O_2 and singlet oxygen (Beers & McDowell, 2001). This suggests that ROS are involved in PCD. PCD is an active process distinct from necrotic cell death caused by more extreme

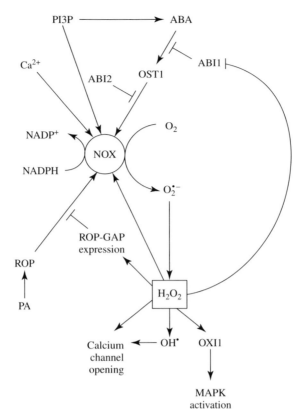

Figure 8.1 Proposed mechanisms for the regulation of plasma membrane NADPH oxidase (NOX) activity. This schematic diagram is based on evidence from ABA-induced stomatal closure in guard cells, the induction of alcohol dehydrogenase gene expression by hypoxia, root-hair development and wounding/pathogen responses as described in the text. Some of the factors activating NADPH oxidase are shown operating in parallel but it is not known which are independent pathways. NADPH oxidase has six transmembrane domains and has predicted N- and C-terminal cytoplasmic domains. The N-terminus has one or two Ca^{2+}-binding EF-hand motifs and the enzyme is activated by Ca^{2+} *in vitro*. The C-terminus has the FAD- and NADPH-binding sites. Oxygen reduction occurs on the apoplastic side. H_2O_2 can activate (e.g. during programmed cell death) or inactivate (e.g. during the hypoxic response) NOX in a feedback mechanism. ROP GTPases in the active GTP-bound form are proposed to activate NOX and this is reversed by ROP-GTPase activating proteins (ROP-GAPs). ABA-induced ROS formation in guard cells requires the OST1 protein kinase and is antagonised by the ABI1 and ABI2 protein phosphatase 2C proteins. ABI1 and ABI2 are inhibited by H_2O_2 producing a possible feed-forward loop for ABA-induced H_2O_2 formation. ABA-induced ROS formation is decreased by phosphatidylinositol 3-kinase (PI3K) inhibitors, which prevent synthesis of phosphatidylinositol 3-phosphate (PI3P), and by PI3P binding proteins. Phosphatidic acid (PA) stimulates ROS formation and programmed cell death in a ROP-dependent manner. The OXI1 serine/threonine protein kinase is induced by H_2O_2, indicating a signalling role for NOX-derived H_2O_2, and is required for normal root-hair development. The potentially complex control over NOX activity may differ according to cell type or environmental conditions. Along with the occurrence of 10 NADPH oxidase homologues that have tissue specific expression, this regulatory network could provide tight control over ROS formation. The arrows indicate activation but do not imply that it is direct and some interactions are still speculative.

stress: low H_2O_2 concentration triggers PCD in cell cultures, while high concentrations cause necrotic cell death. The hypersensitive response (HR) to avirulent pathogens is also a PCD process involving ROS and is considered in Section 8.6. The mechanism of PCD differs in detail between species and cell type and may or may not have resemblances to mammalian apoptosis (Beers & McDowell, 2001). As in apoptosis, release of cytochrome c from mitochondria could play a role in initiating cell death in response, e.g. to H_2O_2 or salicylic acid (Tiwari *et al.*, 2002). This process is associated with redox changes in the mitochondria. Plant mitochondria have a redox-controlled alternative oxidase (AOX) that diverts electrons from ubiquinone and may, therefore, prevent over-reduction of the electron transport chain and consequent O_2^- formation (Moller, 2001). Transgenic tobacco cell cultures with decreased AOX activity are more susceptible to H_2O_2 or SA-induced PCD suggesting its role is in preventing oxidative stress in mitochondria, thereby decreasing cytochrome c release (Robson & Vanlerberghe, 2002). The protein phosphatase inhibitor cantharidin induces PCD in tobacco cell cultures by a cytochrome-c-independent mechanism but sensitivity to this compound is also greater in low AOX cells, suggesting a role for mitochondria in this case as well (Robson & Vanlerberghe, 2002). Cytoplasmic male sterile tobacco plants that lack functional complex I illustrate the role of mitochondria more generally in controlling the antioxidant system. They have increased AOX expression, altered expression of a range of antioxidant genes and enhanced tolerance to ozone and tobacco mosaic virus (Dutilleul *et al.*, 2003). The hormone-mediated death of aleurone cells during the germination of cereal seeds is a part of normal development and provides a clear example of a role for ROS in PCD (Bethke & Jones, 2001; Fath *et al.*, 2001a,b; Beligni *et al.*, 2002). In this system, gibberellic acid (GA) stimulates cell death while ABA inhibits it. ABA increases the resistance of the cells to UV and H_2O_2 and increases catalase activity, while GA decreases the activity of catalase, ascorbate peroxidase and superoxide dismutase (SOD) and causes a decrease in UV and H_2O_2 resistance. It is suggested that the reduced ability to scavenge ROS in GA-treated cells causes cell death as a result of peroxidative damage to vacuolar membranes followed by release of hydrolytic enzymes. PCD induced by heat shock in tobacco BY2 cells is mediated by O_2^- and H_2O_2 formation and can be prevented by ascorbate and SOD as well as transcription and translation inhibitors (Vacca *et al.*, 2004). In a manner similar to the aleurone system, cell death is associated with reduced expression of cytosolic ascorbate peroxidase (Vacca *et al.*, 2004). NO may also be required for ROS-induced PCD, e.g. in the HR response (Delledonne *et al.*, 2001) and in tobacco BY2 cells (de Pinto *et al.*, 2002). O_2^- and NO react rapidly to form $ONOO^-$, which could be the actual trigger for PCD (de Pinto *et al.*, 2002). However, $ONOO^-$ may not itself cause cell death, instead, if enough H_2O_2 accumulates, this may co-operate with NO to induce cell death (Torres *et al.*, 2002). In contrast, NO reduces ROS-induced cell death in the barley aleurone system (Beligni *et al.*, 2002) possibly by reacting with O_2^-. Transgenic tobacco plants with 10% of WT catalase activity accumulate H_2O_2 in high light as a result of photorespiration. These plants show PCD on exposure to high light, including disruption of mitochondrial ultrastructure and release of cytochrome c, as well as induction of pathogen response gene expression. PCD is accompanied by an oxidative burst, which is inhibited by DPI, and induction of a NOX homologue, suggesting that the initial accumulation of H_2O_2 triggers a NOX-dependent oxidative burst that leads to cell death (Dat et al., 2003). Ozone also induces an oxidative burst and PCD in a similar manner (see Chapter 11). There is, therefore, strong evidence that ROS and some other conditions, such as heat shock and elicitors (see later), induce an oxidative burst, which triggers PCD. Depending on the specific system, this may or may not be NO-dependent.

Senescence is the controlled process by which some organs, e.g. leaves and petals, die at the end of their lives. Many of the processes involved in senescence are distinct from PCD, although PCD may be the final step in some cases. Roles for ROS in senescence have been suggested but are not clear-cut. For example, senescence is associated with the appearance of oxidative stress symptoms such as lipid peroxidation (Spiteller, 2003) and with changes in activity and expression of antioxidant enzymes which variously increase or decrease (del Rio et al., 1998; Jimenez et al., 1998). Oxidative stresses also cause accelerated senescence (Trippi et al., 1989). In leaves, the visible signs of senescence are loss of chlorophyll and unmasking or accumulation of the carotenoids and anthocyanins that provide the characteristic yellow and red colours. These pigments are themselves effective antioxidants and protectants against photooxidation, and have been proposed to have an antioxidant role (see Chapter 6). Leaf senescence involves extensive and reversible reorganisation of metabolism and ultrastructure, particularly in chloroplasts and peroxisomes associated with loss of photosynthetic enzymes and the breakdown of proteins and lipids with their eventual conversion to sugars and amino acids which are exported from the leaf. This reorganisation can result in increased potential for ROS formation, e.g. by photosensitisation of singlet oxygen formation from chlorophyll and its coloured catabolites (Apel & Hirt, 2004) and from H_2O_2 formed by β-oxidation of fatty acids in peroxisomes/glyoxysomes. However, at the same time, ROS formation from photosynthesis and photorespiration decreases in line with loss of photosynthetic capacity. Release of redox active metals such as Fe, Cu and Mn from proteins could also result in increased potential for OH˙ formation via the Fenton reaction. Increased expression of metal-binding proteins such as metallothionein occurs during senescence (Buchanan-Wollaston et al., 2003). Despite a possibly increased risk of oxidative stress, glutathione and ascorbate, the main soluble antioxidants in leaves, both decrease in parallel with photosynthetic capacity (Dertinger et al., 2003). Ascorbate biosynthethic capacity is lost (Bartoli et al., 2000), so this cannot simply be accounted for by its oxidation. In contrast, the α-tocopherol (vitamin E) content of senescing leaves is consistently reported to increase (Dertinger et al., 2003). α-Tocopherol is a membrane-localised antioxidant synthesised in plastids. This, along with carotenoids, is implicated in protecting against lipid peroxidation caused by photosynthesis (Chapter 3). Increased tocopherol synthesis is likely to be facilitated by release of the phytyl tail of chlorophyll that is used in its synthesis, and expression of some of its biosynthetic enzymes increases during senescence (see Chapter 3). Lipophilic antioxidants may, therefore, have roles during the dismantling of chloroplasts and in fatty acid oxidation during leaf senescence. The transcriptome of senescing leaves includes a range of antioxidant enzymes (Page *et al.*, 2001; Buchanan-Wollaston *et al.*, 2003), including catalase. It is apparent that ROS formation does not itself initiate natural senescence (Dertinger *et al.*, 2003), but there are changes in the antioxidant system consistent with a need to protect different cellular components: e.g. metallothionein and ferritin maintain low concentration of redox active metals. However, faster dark-induced senescence and increased expression of some senescence-associated genes (SAGs) in a low ascorbate *A. thaliana* mutant (*vtc1*), suggest that the relationship between antioxidants and senescence could be more subtle (Barth *et al.*, 2004). Furthermore, increasing the ascorbate pool by feeding a biosynthetic precursor delays senescence of oat leaves (Borraccino *et al.*, 1994). Since many of the genes induced during senescence are also related to pathogen defence and may be under redox and SA control (Morris *et al.*, 2000), and are also more highly expressed in *vtc1* (Pastori *et al.*, 2003; Barth *et al.*, 2004), decreased ascorbate content could be required to allow expression of these genes during senescence. The possible modulating role of ascorbate in leaf redox state is discussed in Chapter 3. The metabolic processes associated with nitrogen fixing legume root nodules render them liable to ROS production and oxidative stress and they are susceptible to rapid senescence, induced by many environmental factors (Matamoros *et al.*, 2003). Ascorbate and glutathione, and associated ascorbate–glutathione cycle enzymes are present at high concentrations in nodules until senescence (Dalton *et al.*, 1993; Becana *et al.*, 2000). Infusion of ascorbate into soybean nodules prolonged and increased the rate of nitrogen fixation (Dalton *et al.*, 1993; Bashor & Dalton, 1999; Ross *et al.*, 1999).

8.5 ROS and antioxidants in response to pathogens and wounding

Plants respond to pathogen infection and mechanical or herbivore-induced wounding (Yahraus *et al.*, 1995; Orozco-Cardenas *et al.*, 2001), or even insect footsteps (Bown *et al.*, 2002) by localised production of ROS, a phenomenon often referred to as the 'oxidative burst'. The key aspects are reviewed by Bolwell (1999), Grant and Loake (2000) and Lamb and Dixon (1997). Abiotic elicitors such as cell wall fragments produced by wounding or wall degrading enzymes can induce the oxidative burst. An analogous defence-related oxidative burst involving NOX occurs in mammalian phagocytes as described in Section 8.3.2. The most persuasive evidence for the role of NOX in the oxidative burst is the decrease in ROS formation in knockout mutants of *AtrbohD* and *AtrbohF* in response to incompatible bacterial (*Pseudomonas syringae*) and fungal (*Peronospora parasitica*) infection (Torres *et al.*, 2002). A knockout of *AtrbohD* was most effective, but in the case of *P. parasitica*, the *AtrbohF* knockout produced smaller but dense deposits of oxidised 3,3′-diaminobenzidine (DAB), the peroxidase substrate used to detect H_2O_2. This suggests that control over the localisation of NOX activity is complex and may differ between members of the family. As discussed in Section 8.3.5, there is some evidence that NOX activity may be activated by ROP GTPases during response to pathogens (Kawasaki *et al.*, 1999; Park *et al.*, 2000, 2004; Ono *et al.*, 2001; Agrawal *et al.*, 2003).

While NOXes are likely to be involved in the oxidative burst, other cell wall oxidases that produce ROS include oxalate oxidase (germin), amine oxidases and peroxidase. Oxalate oxidase is a wall-localised enzyme that generates H_2O_2 during oxidation of oxalic acid. It has been associated with pathogen defence (Zhang *et al.*, 1995; Caliskan & Cuming, 1998; Schweizer *et al.*, 1999; Donaldson *et al.*, 2001; Hu *et al.*, 2003; Tabuchi *et al.*, 2003; Caliskan *et al.*, 2004; Le Deunff *et al.*, 2004). Amine oxidases occur in the cell wall and can generate H_2O_2 using substrates such as spermidine and putrescine. Increased amine oxidase expression or activity has been associated with pathogen response, xylem vessel lignification and PCD (Moller & McPherson, 1998; Rea *et al.*, 2002, 2004; Cona *et al.*, 2003). Cell wall peroxidases can generate O_2^- if supplied with a reductant (e.g. NAD[P]H, cysteine or palmitate) and oxygen (Bolwell *et al.*, 1998, 2002; Bolwell, 1999; Chen & Schopfer, 1999; Schweikert *et al.*, 2000, 2002; Blee *et al.*, 2001, 2003; Liszkay *et al.*, 2003). The role of peroxidase in H_2O_2 generation during the oxidative burst is strengthened by cloning of a French bean cell wall peroxidase (FBP1) that can generate H_2O_2 in the presence of cysteine at pH 7.2 (Blee *et al.*, 2001; Bolwell *et al.*, 2002). Since wall alkalinisation is an early response to pathogen attack, this is a potential source of ROS, possibly in the early burst. The nature of the actual reductant is unknown, but the possible role of FBP1 in the defence response is suggested by antisense suppression of its homologue(s) in *A. thaliana*. The plants are more susceptible to bacterial and fungal pathogens (Bolwell *et al.*, 2002).

The ROS formed in these reactions are involved in defence responses (e.g. direct effects on the pathogen, increased oxidative cross-linking of cell wall polymers, induction of pathogen response gene expression). The complex of signalling molecules involved in these pathogen responses (Rusterucci *et al.*, 2001; Aviv *et al.*, 2002; Epple *et al.*, 2003; Shah, 2003) is beyond the scope of this chapter but is covered in Chapters 7 and 11.

References

Agrawal, G.K., Rakwal, R. and Agrawal, V.P. (2003) 'Small GTPase Rop: molecular switch for plant defence responses', *FEBS Letters* **546**, 173–180.

Allan, A.C. and Fluhr, R. (1997) 'Two distinct sources of elicited reactive oxygen species in tobacco epidermal cells', *The Plant Cell* **9**, 1559–1572.

Apel, K. and Hirt, H. (2004) 'Reactive oxygen species: metabolism, oxidative stress, and signal transduction', *Annual Review of Plant Biology* **55**, 373–399.

Arbiser, J.L., Petros, J., Klafter, R. *et al.* (2002) 'Reactive oxygen generated by Nox1 triggers the angiogenic switch', *Proceedings of the National Academy of Sciences USA* **99**, 715–720.

Arnold, R.S., Shi, J., Murad, E. *et al.* (2001) 'Hydrogen peroxide mediates the cell growth and transformation caused by the mitogenic oxidase Nox1', *Proceedings of the National Academy of Sciences USA* **98**, 5550–5555.

Aviv, D.H., Rusterucci, C., Holt, B.F., Dietrich, R.A., Parker, J.E. and Dangl, J.L. (2002) 'Runaway cell death, but not basal disease resistance, in *Isd1* is SA- and NIM1/NPR1-dependent', *The Plant Journal* **29**, 381–391.

Babior, B.M. (1999) 'NADPH oxidase: an update', *Blood* **93**, 1464–1476.

Banfi, B., Molnar, G., Maturana, A. *et al.* (2001) 'A Ca^{2+}-activated NADPH oxidase in testis, spleen, and lymph nodes', *Journal of Biological Chemistry* **276**, 37594–37601.

Barth, C., Moeder, W., Klessig, D.F. and Conklin, P.L. (2004) 'The timing of senescence and response to pathogens is altered in the ascorbate-deficient *Arabidopsis* mutant vitamin *vtc1*', *Plant Physiology* **134**, 1784–1792.

Bartoli, C.G., Pastori, G.M. and Foyer, C.H. (2000) 'Ascorbate biosynthesis in mitochondria is linked to the electron transport chain between complexes III and IV', *Plant Physiology* **123**, 335–343.

Bashor, C.J. and Dalton, D.A. (1999) 'Effects of exogenous application and stem infusion of ascorbate on soybean (*Glycine max*) root nodules', *New Phytologist* **142**, 19–26.

Baxter-Burrell, A., Yang, Z.B., Springer, P.S. and Bailey-Serres, J. (2002) 'RopGAP4-dependent Rop GTPase rheostat control of *Arabidopsis* oxygen deprivation tolerance', *Science* **296**, 2026–2028.

Becana, M., Dalton, D.A., Moran, J.F., Iturbe-Ormaetxe, I., Matamoros, M.A. and Rubio, M.C. (2000) 'Reactive oxygen species and antioxidants in legume nodules', *Physiologia Plantarum* **109**, 372–381.

Beers, E.P. and McDowell, J.M. (2001) 'Regulation and execution of programmed cell death in response to pathogens, stress and developmental cues', *Current Opinion in Plant Biology* **4**, 561–567.

Beligni, M.V., Fath, A., Bethke, P.C., Lamattina, L. and Jones, R.L. (2002) 'Nitric oxide acts as an antioxidant and delays programmed cell death in barley aleurone layers', *Plant Physiology* **129**, 1642–1650.

Bethke, P.C. and Jones, R.L. (2001) 'Cell death of barley aleurone protoplasts is mediated by reactive oxygen species', *The Plant Journal* **25**, 19–29.

Bibikova, T.N., Zhigilei, A. and Gilroy, S. (1997) 'Root hair growth in *Arabidopsis thaliana* is directed by calcium and an endogenous polarity', *Planta* **203**, 495–505.

Blee, K.A., Choi, J.W., O'Connell, A.P., Schuch, W., Lewis, N.G. and Bolwell, G.P. (2003) 'A lignin-specific peroxidase in tobacco whose antisense suppression leads to vascular tissue modification', *Phytochemistry* **64**, 163–176.

Blee, K.A., Jupe, S.C., Richard, G., Zimmerlin, A., Davies, D.R. and Bolwell, G.P. (2001) 'Molecular identification and expression of the peroxidase responsible for the oxidative burst in French bean (*Phaseolus vulgaris* L.) and related members of the gene family', *Plant Molecular Biology* **47**, 607–620.

Bolwell, G.P. (1999) 'Role of active oxygen species and NO in plant defence responses', *Current Opinion in Plant Biology* **2**, 287–294.

Bolwell, G.P., Bindschedler, L.V., Blee, K.A. *et al.* (2002) 'The apoplastic oxidative burst in response to biotic stress in plants: a three-component system', *Journal of Experimental Botany* **53**, 1367–1376.

Bolwell, G.P., Davies, D.R., Gerrish, C., Auh, C.K. and Murphy, T.M. (1998) 'Comparative biochemistry of the oxidative burst produced by rose and French bean cells reveals two distinct mechanisms', *Plant Physiology* **116**, 1379–1385.

Borraccino, G., Mastropasqua, L., DeLeonardis, S. and Dipierro, S. (1994) 'The role of the ascorbic acid system in delaying the senescence of oat (*Avena sativa* L.) leaf segments', *Journal of Plant Physiology* **144**, 161–166.

Bown, A.W., Hall, D.E. and MacGregor, K.B. (2002) 'Insect footsteps on leaves stimulate the accumulation of 4-aminobutyrate and can be visualized through increased chlorophyll fluorescence and superoxide production', *Plant Physiology* **129**, 1430–1434.

Buchanan-Wollaston, V., Earl, S., Harrison, E. *et al.* (2003) 'The molecular analysis of leaf senescence – a genomics approach', *Plant Biotechnology Journal* **1**, 3–22.

Caliskan, M. and Cuming, A.C. (1998) 'Spatial specificity of H$_2$O$_2$-generating oxalate oxidase gene expression during wheat embryo germination', *The Plant Journal* **15**, 165–171.

Caliskan, M., Turet, M. and Cuming, A.C. (2004) 'Formation of wheat (*Triticum aestivum* L.) embryogenic callus involves peroxide-generating germin-like oxalate oxidase', *Planta* **219**, 132–140.

Chen, S.X. and Schopfer, P. (1999) 'Hydroxyl radical production in physiological reactions – a novel function of peroxidase', *European Journal of Biochemistry* **260**, 726–735.

Chen, Z. and Gallie, D.R. (2004) 'The ascorbic acid redox state controls guard cell signaling and stomatal movement', *The Plant Cell* **16**, 1143–1162.

Coelho, S.M., Taylor, A.R., Ryan, K.P., Sousa-Pinto, I., Brown, M.T. and Brownlee, C. (2002) 'Spatiotemporal patterning of reactive oxygen production and Ca^{2+} wave propagation in *Fucus* rhizoid cells', *The Plant Cell* **14**, 2369–2381.

Cona, A., Cenci, F., Cervelli, M. *et al.* (2003) 'Polyamine oxidase, a hydrogen peroxide-producing enzyme, is up-regulated by light and down-regulated by auxin in the outer tissues of the maize mesocotyl', *Plant Physiology* **131**, 803–813.

Dalton, D.A., Langeberg, L. and Treneman, N.C. (1993) 'Correlations between the ascorbate-glutathione pathway and effectiveness in legume root nodules', *Physiologia Plantarum* **87**, 365–370.

Dat, J.F., Pellinen, R., Beeckman, T. *et al.* (2003) 'Changes in hydrogen peroxide homeostasis trigger an active cell death process in tobacco', *The Plant Journal* **33**, 621–632.

de Pinto, M.C., Tommasi, F. and De Gara, L. (2002) 'Changes in the antioxidant systems as part of the signaling pathway responsible for the programmed cell death activated by nitric oxide and reactive oxygen species in tobacco bright-yellow 2 cells', *Plant Physiology* **130**, 698–708.

del Rio, L.A., Pastori, G.M., Palma, J.M. *et al.* (1998) 'The activated oxygen role of peroxisomes in senescence', *Plant Physiology* **116**, 1195–1200.

Demidchik, V., Shabala, S.N., Coutts, K.B., Tester, M.A. and Davies, J.M. (2003) 'Free oxygen radicals regulate plasma membrane Ca^{2+} and K^+– permeable channels in plant root cells', *Journal of Cell Science* **116**, 81–88.

Dertinger, U., Schaz, U. and Schulze, E.D. (2003) 'Age-dependence of the antioxidative system in tobacco with enhanced glutathione reductase activity or senescence-induced production of cytokinins', *Physiologia Plantarum* **119**, 19–29.

Desikan, R., Reynolds, A., Hancock, J.T. and Neill, S.J. (1998) 'Harpin and hydrogen peroxide both initiate programmed cell death but have differential effects on defence gene expression in *Arabidopsis* suspension cultures', *Biochemical Journal* **330**, 115–120.

Diebold, B.A. and Bokoch, G.M. (2001) 'Molecular basis for Rac2 regulation of phagocyte NADPH oxidase', *Nature Immunology* **2**, 211–215.

Diekmann, D., Abo, A., Johnston, C., Segal, A.W. and Hall, A. (1994) 'Interaction of Rac with p67[(Phox)] and regulation of phagocytic NADPH oxidase activity', *Science* **265**, 531–533.

Donaldson, P.A., Anderson, T., Lane, B.G., Davidson, A.L. and Simmonds, D.H. (2001) 'Soybean plants expressing an active oligomeric oxalate oxidase from the wheat GF-2.8 (germin) gene are resistant to the oxalate-secreting pathogen *Sclerotina sclerotiorum*', *Physiological and Molecular Plant Pathology* **59**, 297–307.

Dumville, J.C. and Fry, S.C. (2003) 'Solubilisation of tomato fruit pectins by ascorbate: a possible non-enzymic mechanism of fruit softening', *Planta* **217**, 951–961.

Dutilleul, C., Garmier, M., Noctor, G. *et al.* (2003) 'Leaf mitochondria modulate whole cell redox homeostasis, set antioxidant capacity, and determine stress resistance through altered signaling and diurnal regulation', *The Plant Cell* **15**, 1212–1226.

Ellson, C.D., Gobert-Gosse, S., Anderson, K.E. *et al.* (2001) 'PtdIns(3)P regulates the neutrophil oxidase complex by binding to the PX domain of p40(phox)', *Nature Cell Biology* **3**, 679–682.

Epple, P., Mack, A.A., Morris, V.R.F. and Dangl, J.L. (2003) 'Antagonistic control of oxidative stress-induced cell death in *Arabidopsis* by two related, plant-specific zinc finger proteins', *Proceedings of the National Academy of Sciences USA* **100**, 6831–6836.

Fath, A., Bethke, P.C., Belligni, M.V., Spiegel, Y.N. and Jones, R.L. (2001a) 'Signalling in the cereal aleurone: hormones, reactive oxygen and cell death', *New Phytologist* **151**, 99–107.

Fath, A., Bethke, P.C. and Jones, R.L. (2001b) 'Enzymes that scavenge reactive oxygen species are down-regulation prior to gibberellic acid-induced programmed cell death in barley aleurone', *Plant Physiology* **126**, 156–166.

Foreman, J., Demidchik, V., Bothwell, J.H.F. *et al.* (2003) 'Reactive oxygen species produced by NADPH oxidase regulate plant cell growth', *Nature* **422**, 442–446.

Frahry, G. and Schopfer, P. (1998) 'Inhibition of O_2 reducing activity of horseradish peroxidase by diphenyleneiodonium', *Phytochemistry* **48**, 223–227.

Grant, J.J. and Loake, G.J. (2000) 'Role of reactive oxygen intermediates and cognate redox signaling in disease resistance', *Plant Physiology* **124**, 21–29.

Halliwell, B. and Gutteridge, J.M.C. (1999) *Free Radicals in Biology and Medicine*, Oxford University Press, Oxford.

Hassanain, H.H., Sharma, Y.K., Moldovan, L. *et al.* (2000) 'Plant Rac proteins induce superoxide production in mammalian cells', *Biochemical and Biophysical Research Communications* **272**, 783–788.

Henzler, T. and Steudle, E. (2000) 'Transport and metabolic degradation of hydrogen peroxide in *Chara corallina*: model calculations and measurements with the pressure probe suggest transport of H_2O_2 across water channels', *Journal of Experimental Botany* **51**, 2053–2066.

Hu, X., Bidney, D.L., Yalpani, N. *et al.* (2003) 'Overexpression of a gene encoding hydrogen peroxide-generating oxalate oxidase evokes defense responses in sunflower', *Plant Physiology* **133**, 170–181.

Jiang, M. and Zhang, J. (2003) 'Cross-talk between calcium and reactive oxygen species originated from NADPH oxidase in abscisic acid-induced antioxidant defence in leaves of maize seedlings', *Plant Cell and Environment* **26**, 929–939.

Jimenez, A., Hernandez, J.A., Pastori, G., del Rio, L.A. and Sevilla, F. (1998) 'Role of the ascorbate–glutathione cycle of mitochondria and peroxisomes in the senescence of pea leaves', *Plant Physiology* **118**, 1327–1335.

Jones, D.L., Gilroy, S., Larsen, P.B., Howell, S.H. and Kochian, L.V. (1998) 'Effect of aluminum on cytoplasmic Ca^{2+} homeostasis in root hairs of *Arabidopsis thaliana* (L.)', *Planta* **206**, 378–387.

Joo, J.H., Bae, Y.S. and Lee, J.S. (2001) 'Role of auxin-induced reactive oxygen species in root gravitropism', *Plant Physiology* **126**, 1055–1060.

Kawasaki, T., Henmi, K., Ono, E. *et al.* (1999) 'The small GTP-binding protein Rac is a regulator of cell death in plants', *Proceedings of the National Academy of Sciences USA* **96**, 10922–10926.

Keller, T., Damude, H.G., Werner, D., Doerner, P., Dixon, R.A. and Lamb, C. (1998) 'A plant homolog of the neutrophil NADPH oxidase gp91(phox) subunit gene encodes a plasma membrane protein with Ca^{2+} binding motifs', *The Plant Cell* **10**, 255–266.

Korystov, Y.N. and Narimanov, A.A. (1997) 'Low doses of ionizing radiation and hydrogen peroxide stimulate plant growth', *Biologia* **52**, 121–124.

Kovtun, Y., Chiu, W.L., Tena, G. and Sheen, J. (2000) 'Functional analysis of oxidative stress-activated mitogen-activated protein kinase cascade in plants', *Proceedings of the National Academy of Sciences USA* **97**, 2940–2945.

Kwak, J.M., Mori, I.C., and Pei, Z.M. (2003) 'NADPH oxidase *AtrbohD* and *AtrbohF* genes function in ROS-dependent ABA signaling in *Arabidopsis*', *EMBO Journal* **22**, 2623–2633.

Lalucque, H. and Silar, P. (2003) 'NADPH oxidase: an enzyme for multicellularity?', *Trends in Microbiology* **11**, 9–12.

Lamb, C. and Dixon, R.A. (1997) 'The oxidative burst in plant disease resistance', *Annual Review of Plant Physiology and Plant Molecular Biology* **48**, 251–275.

Lara-Ortiz, T., Riveros-Rosas, H. and Aguirre, J. (2003) 'Reactive oxygen species generated by microbial NADPH oxidase NoxA regulate sexual development in *Aspergillus nidulans*', *Molecular Microbiology* **50**, 1241–1255.

Le Deunff, E., Davoine, C., Le Dantec, C., Billard, J.P. and Huault, C. (2004) 'Oxidative burst and expression of germin/oxo genes during wounding of ryegrass leaf blades: comparison with senescence of leaf sheaths', *The Plant Journal* **38**, 421–431.

Lee, S., Choi, H., Suh, S. *et al.* (1999) 'Oligogalacturonic acid and chitosan reduce stomatal aperture by inducing the evolution of reactive oxygen species from guard cells of tomato and *Commelina communis*', *Plant Physiology* **121**, 147–152.

Lemichez, E., Wu, Y., Sanchez, J.P., Mettouchi, A., Mathur, J. and Chua, N.H. (2001) 'Inactivation of AtRac1 by abscisic acid is essential for stomatal closure', *Genes & Development* **15**, 1808–1816.

Liszkay, A., Kenk, B. and Schopfer, P. (2003) 'Evidence for the involvement of cell wall peroxidase in the generation of hydroxyl radicals mediating extension growth', *Planta* **217**, 658–667.

Matamoros, M.A., Dalton, D.A., Ramos, J., Clemente, M.R., Rubio, M.C. and Becana, M. (2003) 'Biochemistry and molecular biology of antioxidants in the rhizobia–legume symbiosis', *Plant Physiology* **133**, 499–509.

Meinhard, M. and Grill, E. (2001) 'Hydrogen peroxide is a regulator of ABI1, a protein phosphatase 2C from *Arabidopsis*', *FEBS Letters* **508**, 443–446.

Meinhard, M., Rodriguez, P.L. and Grill, E. (2002) 'The sensitivity of ABI2 to hydrogen peroxide links the abscisic acid-response regulator to redox signalling', *Planta* **214**, 775–782.

Mitsushita, J., Lambeth, J.D. and Kamata, T. (2004) 'The superoxide-generating oxidase Nox1 is functionally required for Ras oncogene transformation', *Cancer Research* **64**, 3580–3585.

Mittler, R. (2002) 'Oxidative stress, antioxidants and stress tolerance', *Trends in Plant Science* **7**, 405–410.

Moller, I.M. (2001) 'Plant mitochondria and oxidative stress: electron transport, NADPH turnover, and metabolism of reactive oxygen species', *Annual Review of Plant Physiology and Plant Molecular Biology* **52**, 561–591.

Moller, S.G. and McPherson, M.J. (1998) 'Developmental expression and biochemical analysis of the *Arabidopsis* atao1 gene encoding an H_2O_2-generating diamine oxidase', *The Plant Journal* **13**, 781–791.

Morris, K., Mackerness, S.A.H., Page, T. *et al.* (2000) 'Salicylic acid has a role in regulating gene expression during leaf senescence', *The Plant Journal* **23**, 677–685.

Murata, Y., Pei, Z.M., Mori, I.C. and Schroeder, J. (2001) 'Abscisic acid activation of plasma membrane Ca^{2+} channels in guard cells requires cytosolic NAD(P)H and is differentially disrupted upstream and downstream of reactive oxygen species production in *abi1-1* and *abi2-1* protein phosphatase 2C mutants', *The Plant Cell* **13**, 2513–2523.

Mustilli, A.C., Merlot, S., Vavasseur, A., Fenzi, F. and Giraudat, J. (2002) '*Arabidopsis* OST1 protein kinase mediates the regulation of stomatal aperture by abscisic acid and acts upstream of reactive oxygen species production', *The Plant Cell* **14**, 3089–3099.

Noctor, G. and Foyer, C.H. (1998) 'Ascorbate and glutathione: keeping active oxygen under control', *Annual Review of Plant Physiology and Plant Molecular Biology* **49**, 249–279.

Ono, E., Wong, H.L., Kawasaki, T., Hasegawa, M., Kodama, O. and Shimamoto, K. (2001) 'Essential role of the small GTPase Rac in disease resistance of rice', *Proceedings of the National Academy of Sciences USA* **98**, 759–764.

Orozco-Cardenas, M.L., Narvaez-Vasquez, J. and Ryan, C.A. (2001) 'Hydrogen peroxide acts as a second messenger for the induction of defense genes in tomato plants in response to wounding, systemin, and methyl jasmonate', *The Plant Cell* **13**, 179–191.

Page, T., Griffiths, G. and Buchanan-Wollaston, V. (2001) 'Molecular and biochemical characterization of postharvest senescence in broccoli', *Plant Physiology* **125**, 718–727.

Park, J., Choi, H.T., Lee, S., Lee, T., Yang, Z.B. and Lee, Y. (2000) 'Rac-related GTP-binding protein in elicitor-induced reactive oxygen generation by suspension-cultured soybean cells', *Plant Physiology* **124**, 725–732.

Park, J., Gu, Y., Lee, Y., Yang, Z.B. and Lee, Y. (2004) 'Phosphatidic acid induces leaf cell death in *Arabidopsis* by activating the Rho-related small G protein GTPase-mediated pathway of reactive oxygen species generation', *Plant Physiology* **134**, 129–136.

Park, K.Y., Jung, J.Y., Park, J. *et al.* (2003) 'A role for phosphatidylinositol 3-phosphate in abscisic acid-induced reactive oxygen species generation in guard cells', *Plant Physiology* **132**, 92–98.

Pastori, G.M., Kiddle, G., Antoniw, J. *et al.* (2003) 'Leaf vitamin C contents modulate plant defense transcripts and regulate genes that control development through hormone signaling', *The Plant Cell* **15**, 939–951.

Pei, Z.M., Murata, Y., Benning, G. *et al.* (2000) 'Calcium channels activated by hydrogen peroxide mediate abscisic acid signalling in guard cells', *Nature* **406**, 731–734.

Potikha, T.S., Collins, C.C., Johnson, D.I., Delmer, D.P. and Levine, A. (1999) 'The involvement of hydrogen peroxide in the differentiation of secondary walls in cotton fibers', *Plant Physiology* **119**, 849–858.

Prado, A.M., Porterfield, D.M. and Feijo, J.A. (2004) 'Nitric oxide is involved in growth regulation and re-orientation of pollen tubes', *Development* **131**, 2707–2714.

Rea, G., de Pinto, M.C., Tavazza, R. *et al.* (2004) 'Ectopic expression of maize polyamine oxidase and pea copper amine oxidase in the cell wall of tobacco plants', *Plant Physiology* **134**, 1414–1426.

Rea, G., Metoui, O., Infantino, A., Federico, R. and Angelini, R. (2002) 'Copper amine oxidase expression in defense responses to wounding and *Ascochyta rabiei* invasion', *Plant Physiology* **128**, 865–875.

Reichheld, J.P., Vernoux, T., Lardon, F., Van Montagu, M. and Inze, D. (1999) 'Specific checkpoints regulate plant cell cycle progression in response to oxidative stress', *The Plant Journal* **17**, 647–656.

Rentel, M.C., Lecourieux, D., Ouaked, F. *et al.* (2004) 'OXI1 kinase is necessary for oxidative burst-mediated signalling in *Arabidopsis*', *Nature* **427**, 858–861.

Robson, C.A. and Vanlerberghe, G.C. (2002) 'Transgenic plant cells lacking mitochondrial alternative oxidase have increased susceptibility to mitochondria-dependent and -independent pathways of programmed cell death', *Plant Physiology* **129**, 1908–1920.

Rodriguez, A.A., Grunberg, K.A. and Taleisnik, E.L. (2002) 'Reactive oxygen species in the elongation zone of maize leaves are necessary for leaf extension', *Plant Physiology* **129**, 1627–1632.

Ross, E.J.H., Kramer, S.B. and Dalton, D.A. (1999) 'Effectiveness of ascorbate and ascorbate peroxidase in promoting nitrogen fixation in model systems', *Phytochemistry* **52**, 1203–1210.

Rusterucci, C., Aviv, D.H., Holt, B.F., Dangl, J.L. and Parker, J.E. (2001) 'The disease resistance signaling components EDS1 and PAD4 are essential regulators of the cell death pathway controlled by LSD1 in *Arabidopsis*', *The Plant Cell* **13**, 2211–2224.

Sagi, M. and Fluhr, R. (2001) 'Superoxide production by plant homologues of the gp91$^{(phox)}$ NADPH oxidase. Modulation of activity by calcium and by tobacco mosaic virus infection', *Plant Physiology* **126**, 1281–1290.

Sagi, M., Davydov, O., Orazova, S. *et al.* (2004) 'Plant respiratory burst oxidase homologs impinge on wound responsiveness and development in *Lycopersicon esculentum*', *The Plant Cell* **16**, 616–628.

Sauer, H., Wartenberg, M. and Hescheler, J. (2001) 'Reactive oxygen species as intracellular messengers during cell growth and differentiation', *Cellular Physiology and Biochemistry* **11**, 173–186.

Schopfer, P. (2001) 'Hydroxyl radical-induced cell-wall loosening *in vitro* and *in vivo*: implications for the control of elongation growth', *The Plant Journal* **28**, 679–688.

Schweikert, C., Liszkay, A. and Schopfer, P. (2000) 'Scission of polysaccharides by peroxidase-generated hydroxyl radicals', *Phytochemistry* **53**, 565–570.

Schweikert, C., Liszkay, A. and Schopfer, P. (2002) 'Polysaccharide degradation by Fenton reaction- or peroxidase-generated hydroxyl radicals in isolated plant cell walls', *Phytochemistry* **61**, 31–35.

Schweizer, P., Christoffel, A. and Dudler, R. (1999) 'Transient expression of members of the germin-like gene family in epidermal cells of wheat confers disease resistance', *The Plant Journal* **20**, 540–552.

Shah, J. (2003) 'The salicylic acid loop in plant defense', *Current Opinion in Plant Biology* **6**, 365–371.

Shin, R. and Schachtman, D.P. (2004) 'Hydrogen peroxide mediates plant root cell response to nutrient deprivation', *Proceedings of the National Academy of Sciences USA* **101**, 8827–8832.

Shorning, B.Y., Smirnova, E.G., Yaguzhinsky, L.S. and Vanyushin, B.F. (2000) 'Necessity of superoxide production for development of etiolated wheat seedlings', *Biochemistry – Moscow* **65**, 1357–1361.

Spiteller, G. (2003) 'The relationship between changes in the cell wall, lipid peroxidation, proliferation, senescence and cell death', *Physiologia Plantarum* **119**, 5–18.

Staswick, P.E., Tiryaki, I. and Rowe, M.L. (2002) 'Jasmonate response locus JARL and several related *Arabidopsis* genes encode enzymes of the firefly luciferase superfamily that show activity on jasmonic, salicylic, and indole-3-acetic acids in an assay for adenylation', *The Plant Cell* **14**, 1405–1415.

Stoelzle, S., Kagawa, T., Wada, M., Hedrich, R. and Dietrich, P. (2003) 'Blue light activates calcium-permeable channels in *Arabidopsis* mesophyll cells *via* the phototropin signaling pathway', *Proceedings of the National Academy of Sciences USA* **100**, 1456–1461.

Suh, Y.A., Arnold, R.S., Lassegue, B. *et al.* (1999) 'Cell transformation by the superoxide-generating oxidase Mox1', *Nature* **401**, 79–82.

Suhita, D., Raghavendra, A.S., Kwak, J.M. and Vavasseur, A. (2004) 'Cytoplasmic alkalization precedes reactive oxygen species production during methyl jasmonate- and abscisic acid-induced stomatal closure', *Plant Physiology* **134**, 1536–1545.

Tabuchi, T., Kumon, T., Azuma, T., Nanmori, T. and Yasuda, T. (2003) 'The expression of a germin-like protein with superoxide dismutase activity in the halophyte *Atriplex lentiformis* is differentially regulated by wounding and abscisic acid', *Physiologia Plantarum* **118**, 523–531.

Tian, M., Gu, Q. and Zhu, M.Y. (2003) 'The involvement of hydrogen peroxide and antioxidant enzymes in the process of shoot organogenesis of strawberry callus', *Plant Science* **165**, 701–707.

Tiwari, B.S., Belenghi, B. and Levine, A. (2002) 'Oxidative stress increased respiration and generation of reactive oxygen species, resulting in ATP depletion, opening of mitochondrial permeability transition, and programmed cell death', *Plant Physiology* **128**, 1271–1281.

Torres, M.A., Dangl, J.L. and Jones, J.D.G. (2002) '*Arabidopsis* gp91[(phox)] homologues *AtrbohD* and *AtrbohF* are required for accumulation of reactive oxygen intermediates in the plant defense response', *Proceedings of the National Academy of Sciences USA* **99**, 517–522.

Torres, M.A., Onouchi, H., Hamada, S., Machida, C., Hammond-Kosack, K.E. and Jones, J.D.G. (1998) 'Six *Arabidopsis thaliana* homologues of the human respiratory burst oxidase (gp91[(phox)])', *The Plant Journal* **14**, 365–370.

Trippi, V.S., Gidrol, X. and Pradet, A. (1989) 'Effects of oxidative stress caused by oxygen and hydrogen peroxide on energy metabolism and senescence in oat leaves', *Plant and Cell Physiology* **30**, 157–162.

Vacca, R.A., de Pinto, M.C., Valenti, D., Passarella, S., Marra, E. and De Gara, L. (2004) 'Production of reactive oxygen species, alteration of cytosolic ascorbate peroxidase, and impairment of mitochondrial metabolism are early events in heat shock-induced programmed cell death in tobacco bright-yellow 2 cells', *Plant Physiology* **134**, 1100–1112.

Very, A.A. and Davies, J.M. (2000) 'Hyperpolarization-activated calcium channels at the tip of *Arabidopsis* root hairs', *Proceedings of the National Academy of Sciences USA* **97**, 9801–9806.

Wisniewski, J.P., Cornicle, P. and Montillet, J.L. (1999) 'The extensin multigene family responds differentially to superoxide or hydrogen peroxide in tomato cell cultures', *FEBS Letters* **447**, 264–268.

Wymer, C.L., Bibikova, T.N. and Gilroy, S. (1997) 'Cytoplasmic free calcium distributions during the development of root hairs of *Arabidopsis thaliana*', *The Plant Journal* **12**, 427–439.

Yahraus, T., Chandra, S., Legendre, L. and Low, P.S. (1995) 'Evidence for a mechanically induced oxidative burst', *Plant Physiology* **109**, 1259–1266.

Yang, Z.B. (2002) 'Small GTPases: versatile signaling switches in plants', *The Plant Cell* **14**, S375–S388.

Zhang, X., Zhang, L., Dong, F.C., Gao, J.F., Galbraith, D.W. and Song, C.P. (2001) 'Hydrogen peroxide is involved in abscisic acid-induced stomatal closure in *Vicia faba*', *Plant Physiology* **126**, 1438–1448.

Zhang, Z.G., Collinge, D.B. and Thordalchristensen, H. (1995) 'Germin-like oxalate oxidase, a H_2O_2 producing enzyme, accumulates in barley attacked by the powdery mildew fungus', *The Plant Journal* **8**, 139–145.

Zheng, Z.L., Nafisi, M., Tam, A. *et al.* (2002) 'Plasma membrane-associated ROP10 small GTPase is a specific negative regulator of abscisic acid responses in *Arabidopsis*', *The Plant Cell* **14**, 2787–2797.

9 Reactive oxygen species in cell walls

Robert A.M. Vreeburg and Stephen C. Fry

9.1 The cell wall and the apoplast

The cell wall is the outermost layer of the plant cell. Its thickness varies from about 0.1 to 10 µm depending on the cell type and the maturity of the cell. Cell walls serve numerous special roles associated with their position as a frontier zone between the plasma membrane and the environment. They are physically strong (capable of withstanding high turgor pressures, resisting mechanical tissue damage and helping to prevent the ingress of micro-organisms) but may simultaneously be capable of the controlled plastic extension that defines plant growth. They are capable of binding and harmlessly sequestering certain heavy metals, e.g. copper, which would be toxic to the protoplasts if present in excess; yet, cell walls are also permeable to most low-M_r solutes. The wall can thus permit the passage of nutrients and signalling molecules from cell to cell; indeed, the wall may itself also act as a source of signalling molecules (oligosaccharins).

Cell walls are composed mainly of polysaccharides, with smaller amounts of glycoproteins and, in certain specialised cells, also other polymers such as lignin, cutin, suberin and silica. Bonds exist between the plasma membrane and the wall, possibly contributed by membrane-anchored arabinogalactan-proteins, but the membrane, nevertheless, quickly and reversibly withdraws from contact with the wall when the cell is plasmolysed by bathing in sugar or salt solutions. The wall is normally considered to be *part of* the cell, and it would thus be incorrect to refer to it as the 'extracellular matrix'. Here, the cell excluding the wall is considered 'a protoplast', and the wall is described as 'extraprotoplasmic'. All the walls of a plant (plus the aqueous sap that permeates them) together constitute the apoplast. The apoplastic fluid located within the volume occupied by the wall forms a continuum with any extracellular fluid, such as mucilaginous secretions. All the plant's protoplasts (interconnected by plasmodesmata) together constitute the symplast. Physically, the cell wall is a composite material, composed of at least two phases: a skeleton of rigid, inextensible, cellulosic microfibrils and a matrix of non-cellulosic polymers. The matrix is more highly hydrated and exhibits greater molecular mobility than microfibrils.

The cell wall may be composed of relatively discrete layers, deposited at different times during cell development and serving different biological roles. A *primary* cell wall layer is one whose microfibrils were being deposited at a time when at least the particular area of wall under consideration was capable of irreversible expansion (growth). Primary walls are often approximately 0.1 μm thick, but some are much thicker, e.g. those of collenchyma. Some primary walls later become altered (cross-linked or lignified), preventing further growth in area, although lignification per se does not stop the wall layer from being considered primary. The primary walls of neighbouring cells adhere to each other via a pectin-rich middle lamella. A *secondary* cell wall layer is one whose microfibrils were being deposited after the wall area under consideration had lost the ability to expand. It is located between the primary wall and the plasma membrane. Secondary walls may continue to thicken for considerable periods, often reaching approximately 10 μm of thickness, e.g. in sclerenchyma. Primary and secondary wall layers of the same cell often differ markedly in polysaccharide and glycoprotein composition. Lignification, if any, usually begins in the primary wall and may then spread into the secondary wall.

9.2   Reactive oxygen species

In this chapter, we consider the three reactive oxygen species (ROS) which can be regarded formally as being produced by the progressive addition of four single [H] atoms (each equivalent to one free electron) to O_2 as it is reduced, step-wise, to two H_2O (Figure 9.1). These three ROS (also known as 'reactive oxygen intermediates') are, respectively, the hydroperoxyl radical ($HO_2^{\cdot}$, better known as its ionised form, the superoxide radical, $O_2^{\cdot-}$), hydrogen peroxide (H_2O_2), and the hydroxyl radical ($^{\cdot}OH$). They are known as ROS because they tend to be highly reactive with many biologically relevant molecules. We will discuss them in order of increasing reactivity, $H_2O_2 < O_2^{\cdot-} < {}^{\cdot}OH$.

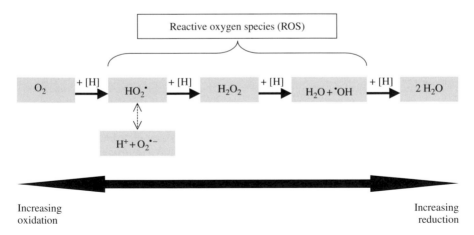

Figure 9.1   The formal inter-relationship between O_2, H_2O and the three intermediary ROS discussed in this chapter.

9.3 H₂O₂ in plant cell walls

9.3.1 Evidence for the presence of H₂O₂ in the wall/apoplast of the living plant cell

Chemical assays (see Chapter 3) for H_2O_2 in the aqueous media of plant-cell-suspension cultures give clear evidence of apoplastic H_2O_2 production, since the medium is a continuum of the apoplast. Apoplastic H_2O_2 production within tissues can be localised at the EM level by observation of deposits of cerium per-hydroxide (Bestwick *et al.*, 1998; Rodríguez *et al.*, 2002), an insoluble product formed by reaction of H_2O_2 with soluble cerium salts. The same chemistry is also applicable to light microscopy (Liu *et al.*, 1995). Added catalase or KI decreases these deposits, supporting the view that they indicate H_2O_2. A second non-enzymic method used to localise H_2O_2 in tissues is the potassium iodide/starch stain used by Olson and Varner (1993). H_2O_2 oxidises I^- to I_2, which is then complexed by the starch to form a blue–purple colour. These authors warn of the possibility that I^- can be oxidised by other electron acceptors. To test the specificity of the method for H_2O_2, control experiments with the H_2O_2-scavenging enzyme, catalase, are advised.

An enzymic approach consists of bathing the botanical specimen in a solution of peroxidase plus an artificial substrate of this enzyme: at sites of H_2O_2 production, the characteristic coloured or fluorescent product is generated. For optimal spatial resolution, this product should be insoluble so that it accumulates at, or very near, the site of H_2O_2 generation. Useful substrates include syringaldazine, 'Amplex red' (Molecular Probes Inc.), and 3,3'-diaminobenzidine (Le Deunff *et al.*, 2004). This approach has been used to detect sites at which polyamines can cause H_2O_2 production: Angelini *et al.* (1990) used a method involving the addition of polyamines plus peroxidase and a chromogenic substrate. The endogenous polyamine oxidase activity generated H_2O_2, which was then detected by its ability to drive the formation of a coloured peroxidase product.

9.3.2 External and internal factors that trigger apoplastic H₂O₂ formation

Phytohormones. Auxin has been reported to suppress the accumulation of the H_2O_2-generating enzyme polyamine oxidase in maize mesocotyl epidermis, whereas light (which inhibits the growth of this organ) increases the enzyme's expression (Cona *et al.*, 2003). Thus, in this case, an H_2O_2-generating enzyme correlates negatively with growth rate, suggesting that the H_2O_2 plays a role in wall tightening. However, Federico and Angelini (1988) showed that copper amine oxidase activity is higher in etiolated tissue than after exposure to light. Thus, a positive role in growth can also be proposed for an H_2O_2-generating system.

Developmental programme. H_2O_2 production can be observed at various distances along the developing root, indicating developmentally regulated H_2O_2 synthesis. In the recent past, H_2O_2 was thought of solely as an agent that would 'tighten' the cell wall, e.g. by enabling the oxidative coupling of wall-polymer-bound phenolic groups (ferulate, p-coumarate, tyrosine). In the light of the relation between H_2O_2 production and growth cessation, it is interesting to note that the degradation of wall-associated proteins by endopeptidases is enhanced by H_2O_2 (Gómez et al., 1995). H_2O_2 levels in apoplastic fluid of bean hypocotyls increased when hypocotyls reached their maximal length. At this time, proteins extracted from the cell wall showed an electrophoretic pattern resembling that of a protein extract from growing hypocotyls treated with endopeptidases. Gómez et al. proposed a model in which the final stages of elongation are associated with an increase in apoplastic H_2O_2 content that triggers oxidative denaturation and endopeptidase-catalysed hydrolysis of wall proteins. The role of H_2O_2 in the cessation of elongation might, therefore, be two-fold: tightening the cell wall by forming more cross-links, and inactivating wall proteins that might be involved in catalysing turgor-driven cell extension.

It has been rather surprising to find that H_2O_2 production sometimes peaks in the regions of rapid cell expansion. The possible role of this H_2O_2 in $^{\cdot}OH$ production and hence wall-loosening, is discussed more fully later. There is also evidence for the developmentally programmed production of H_2O_2 at sites of lignin synthesis, i.e. in differentiating vessel elements and tracheids (Section 9.3.3).

Microbial elicitors and wounding. Numerous studies have shown that certain chemical components of microbial origin (elicitors) are able to trigger the production of H_2O_2 by plant cells and tissues. This phenomenon is an integral part of the 'oxidative burst' and is thought to contribute to the plant's arsenal of active defence mechanisms which enable it to resist some potential pathogens (Wojtaszek, 1997; Grant & Loake 2000; Grant et al., 2000; Bindschedler et al., 2001). For example, certain fungal oligosaccharins (biologically active oligosaccharides derived from chitin, chitosan, fungal β-$(1 \rightarrow 3),(1 \rightarrow 6)$-glucans or fungal glycoproteins) are able to elicit a range of defence responses including phytoalexin synthesis and H_2O_2 production. In addition to the effects of microbial elicitors, simple mechanical wounding may promote the formation of H_2O_2, e.g. by induction of oxalate oxidase activity (Le Deunff et al., 2004).

Salinity and aluminium stress. High salt, which inhibits growth, e.g. in rice roots, was found to be associated with increased H_2O_2 production (Lin & Kao, 2001). NaCl caused an increase in levels of ionically wall-bound peroxidase. Furthermore, added H_2O_2 inhibited the growth of rice roots. These observations support a cross-linking role for H_2O_2 in tightening the cell wall when peroxidase levels are high. Conversely, near-toxic concentrations of Al^{3+} promoted H_2O_2 production in NADH-bathed barley roots, possibly providing $^{\cdot}OH$ and thus a mechanism to favour continuation of cell expansion in the face of growth-inhibiting concentrations of Al^{3+} (Tamás et al., 2004). However, natural H_2O_2 production (without exogenous NADH) remains to be tested in this system.

9.3.3 Proposed roles for apoplastic H_2O_2 in vivo

Lignification. Lignin is a hydrophobic, phenolic polymer that endows cell walls with waterproofness and resistance to enzymic digestion. It occurs constitutively in water-conducting xylem cells, whose walls need to be relatively waterproof, and in certain fibres, which provide structural support and physical protection. Lignin may also be induced in cells that are not normally lignified, e.g. the epidermal cells of wheat leaves after attempted infection by *Puccinia graminis*. Lignification is a final modification of certain cells, usually rapidly followed by cell death.

The building blocks of lignin are coniferyl alcohol, syringyl alcohol and *p*-coumaryl alcohol. These monolignols become oxidatively coupled to generate dimers and ultimately insoluble polymers; this process of polymer biosynthesis, unlike the biosynthesis of wall polysaccharides, occurs in the wall (and middle lamella) itself. There is controversy as to whether the oxidative coupling of monolignols *in vivo* involves peroxidase with H_2O_2, or laccase (phenol oxidase) with O_2, or both (Boudet, 2003). *In vitro*, artificial 'lignification' (the production of dehydrogenation polymers) can be achieved by either of these enzymic systems. If peroxidase contributes to lignification, apoplastic H_2O_2 will be required, and, as discussed earlier, there is evidence that this resource is indeed available.

Ferulate dimerisation, trimerisation, etc. Other oxidative phenolic coupling reactions occurring in the plant cell wall include the dimerisation of polysaccharide-esterified feruloyl residues to form dehydrodiferuloyl residues (often simply called diferulates). This reaction would potentially cross-link a pair of feruloylated polysaccharide molecules and thereby contribute to cell wall assembly and/or to the tightening of wall material already present. However, it remains to be determined whether the dehydrodiferulates are inter-polymeric (i.e. cross-linking two formerly separate polysaccharide chains) or intra-polymeric (forming a molecular loop). The feruloylated polysaccharides which could undergo such reactions include arabinoxylans in gramineous monocots and pectic arabinogalactan moieties in certain dicots (especially the Centrospermae, e.g. spinach and beet).

Hatfield and Ralph (1999) attempted to model the oxidative coupling of feruloylated arabinoxylans *in silico*, based on the predicted behaviour of an oligo-β-(1 $\rightarrow$ 4)-xylan (Xyl_{16}) backbone with two feruloyl-arabinose (Fer-Ara) groups variously spaced along this backbone. They concluded that at all spacings tested (namely, with the second Fer-Ara group sited one to five Xyl residues from the first), most of the known dehydrodiferulates failed, for steric reasons, to act as intra-chain bonds. (The single exception was when the two Fer-Ara groups were on the nth and $(n + 3)$rd Xyl residues, and the dimer was 5,5'-dehydrodiferulate.) However, the widest spacing tested was with the Fer-Ara groups on the nth and $(n + 5)$th Xyl residues, a very short span of the backbone chain in comparison with real cell wall arabinoxylans, which often have M_r approximately equal to 10^6 (Kerr & Fry, 2003, 2004), corresponding to a backbone of roughly 4000 Xyl residues and thus a total chain length of about 2 μm. This polysaccharide length is more than an order of magnitude greater than the diameter of a typical Golgi vesicle (~0.1 μm), which presumably contained the arabinoxylan molecule prior to secretion. It follows that, *in vivo*, the polysaccharide chain must be capable of coiling, such that very widely spaced Fer-Ara residues (on, say, the nth and $(n + 2000)$th Xyl residues) could come into close contact – a situation which would certainly enable intra-polymeric looping via dehydrodiferulate bridges. Notwithstanding these arguments, inter-polymeric cross-linking would also seem feasible, albeit yet to be demonstrated.

The oxidative coupling does not stop at the dehydro*di*ferulate stage, but proceeds to form trimers and larger products, including some that are chromatographically immobile on TLC (Fry *et al.*, 2000; Rouau *et al.*, 2003).

As with lignification, it cannot yet be proven whether the oxidative coupling of feruloyl residues is catalysed by peroxidase (and thus requiring H_2O_2) or by phenol oxidase (and, therefore, able to proceed with only O_2). Resolving this question is an interesting challenge for the future.

A variable proportion of the dehydrodiferulate formation in cultured maize cells takes place within the protoplast, presumably in the Golgi vesicles, prior to secretion of the feruloyl-polysaccharides (Fry *et al.*, 2000). It again remains to be determined whether this process operates with vesicle-associated peroxidase and H_2O_2.

Apoplastic oxidative coupling of feruloyl-[³H]polysaccharides (naturally occurring soluble extracellular polysaccharides, SEPs) and [¹⁴C]feruloyl-oligosaccharides can conveniently be studied in cell cultures. Cell-suspension cultures of maize released feruloylated SEPs into their medium. Pulse-labelling with [³H]arabinose was used to monitor changes in the SEPs' M_r (estimated by gel-permeation chromatography) with time after synthesis. Newly released ³H-SEPs were 1.3–1.6 MDa, but a certain time (e.g. 3 days) after radiolabelling, the ³H-SEPs abruptly increased to more than 17 MDa, indicating extensive cross-linking. Cross-linking of ³H-SEPs *in vivo* was delayed (up to ~7 days after radiolabelling) by exogenous sinapate, chlorogenate or rutin – agents predicted to compete with feruloyl-polysaccharides for the available H_2O_2 that is likely to be involved in the SEP cross-linking. The ability of certain phenolics to prevent the cross-linking of ³H-SEPs supports the idea that the cross-linking involved phenolic oxidation (Kerr & Fry, 2004).

The cultured maize cells were also tested for their ability to bind an exogenous low-M_r model substrate, [¹⁴C]feruloyl-[³H]Ara-Xyl-Xyl (FAXX) (A. Encina & S.C. Fry, unpublished). In preliminary experiments, binding was inconsistent (0–40% within 6 h) but increased notably if the cells were re-suspended in fresh medium. During the binding, radioactivity from both [*feruloyl*-¹⁴C]FAXX and [*pentosyl*-³H]FAXX was equally bonded to the cell walls, indicating that the FAXX became linked to the walls as an intact unit, and, therefore, by oxidative coupling rather than after hydrolysis of the ester bond to release free [¹⁴C]ferulic acid. This was confirmed by detection of [¹⁴C]diferulate and larger oxidation products after alkaline hydrolysis. The binding was suppressed by inhibitors of peroxidase action, e.g. KI or ascorbate. Exogenous H_2O_2 had no effect, suggesting that endogenous apoplastic H_2O_2 was sufficient. The reason for the difference between washed and unwashed cultures was traced to the accumulation in spent medium of a low-M_r, heat-labile, hydrophilic extracellular inhibitor that reversibly prevented FAXX binding (A. Encina & S.C. Fry, unpublished). The identity of this naturally occurring apoplastic anti-oxidant will be of interest.

Isodityrosine, pulcherosine and di-isodityrosine production. The tyrosine residues of cell wall glycoproteins are also susceptible to oxidative phenolic coupling to form a dimer (isodityrosine), a trimer (pulcherosine) and a tetramer (di-isodityrosine). There is evidence that some of the isodityrosine residues are intra-polymeric (Epstein & Lamport, 1984), although others may be inter-polymeric. It has been shown by molecular modelling that the trimer and tetramer are very likely to be inter-polymeric (Brady *et al.*, 1997). The sub-cellular location of the oxidative coupling of Tyr has not been established, but is widely assumed to be apoplastic; and the nature of the oxidant used has not been proven, although it is widely assumed to be H_2O_2.

Oxidative burst toxic for micro-organisms. Although outside the scope of the present chapter, it is relevant to mention that apoplastic H_2O_2 may be produced during the 'oxidative burst' elicited by some invading micro-organisms, and is thought to provide an oxidising environment inimical to the proliferation of the microbial cells (see Chapters 7 and 8; Grant & Loake 2000; Bolwell *et al.*, 2002). In plants that do not normally exhibit an H_2O_2-generating oxalate oxidase activity (see below) in their cell walls, e.g. soya, the introduction of oxalate oxidase activity by genetic manipulation has been shown to confer resistance to oxalate-secreting pathogens (Donaldon *et al.*, 2001). This may point to a physiological role in disease resistance for apoplastic H_2O_2.

Growth inhibition. The proposal that H_2O_2 leads to growth inhibition can potentially be tested by application of dimethylthiourea, which scavenges H_2O_2. Putrescine-treated maize roots, which exhibited decreased elongation, had their growth partially restored by dimethylthiourea (de Agazio & Zacchini, 2001). However, dimethylthiourea is also a good scavenger of ˙OH (see later; Figure 9.4), and the evidence is thus ambiguous. Another effective scavenger of H_2O_2, which has not been extensively used in plant studies, is pyruvate, which reacts non-enzymically as follows

$$CH_3\text{--}CO\text{--}COOH + H_2O_2 \rightarrow CH_3\text{--}COOH + H_2O + CO_2 \qquad (9.1)$$

and which has a relatively low reactivity with ˙OH (Figure 9.4).

9.3.4 Proposed mechanisms of H_2O_2 formation

Oxidases. Numerous oxidases are known which use molecular O_2 as electron acceptor and generate H_2O_2 as a product (e.g. Enzyme Commission catagories 1.1.3.–, 1.2.3.–, 1.3.3.–, 1.4.3.–, 1.5.3.– and 1.8.3.–), and which could, in principle, contribute to H_2O_2 production. Some of these activities have been reported in the plant cell wall, including oxalate oxidase, flavin-containing polyamine oxidase (e.g. acting on spermidine) and copper-containing amine oxidases that can act on diamines (e.g. putrescine) and polyamines.

Oxalate oxidase is found, sometimes in large quantities, in some cell walls (especially those of cereals), but is not universal in the plant kingdom. A protein of the 'germin' class, oxalate oxidase is a homopentameric glycoprotein that catalyses the reaction

$$HOOC–COOH + O_2 \rightarrow 2CO_2 + H_2O_2 \tag{9.2}$$

Since the production oxalate oxidase may increase sharply in the early stages of germination, e.g. in wheat grains (Lane et al., 1993), it has been suggested that it contributes in some way to the 'explosive' cell expansion that causes germination. Possibly the H_2O_2 formed enables production of wall-loosening ˙OH (see later). In wheat coleoptiles, the germin protein as well as oxalate oxidase activity were reported to be concentrated in the epidermis (Caliskan et al., 2004) – a tissue that may require H_2O_2 production either for wall loosening during growth or for cross-linking the aromatic moieties of cutin.

Some germin-like proteins found in cell walls lack oxalate oxidase activity (Kim et al., 2004); some have superoxide dismutase (SOD) activity (Woo et al., 2000) or ADP-glucose pyrophosphatase/phosphodiesterase (Rodriguez-Lopez et al., 2001) – a surprising activity for a wall protein, which would not be expected to have access to sugar-nucleotides in vivo.

Diamine oxidase and polyamine oxidase are distinct enzymes, differentially expressed e.g. in tissues of the barley grain during filling (Asthir et al., 2002). Like oxalate oxidase, they are also found in some, but not all, plant cell walls. The reaction catalysed can be illustrated by that involving putrescine:

$$H_2N–(CH_2)_4–NH_2 + O_2 + H_2O \rightarrow H_2N–(CH_2)_3–CHO + H_2O_2 + NH_3 \tag{9.3}$$

The amino-aldehyde produced cyclises non-enzymatically to Δ^1-pyrroline.

Copper-containing amine oxidases are weakly wall-bound homodimers, each subunit possessing one Cu atom and a post-translationally modified Tyr residue (2,4,5-trihydroxyphenylalanine quinone) which participates in the reaction. Accumulation of mRNA for a copper-containing diamine oxidase in Arabidopsis was found principally in the tracheary elements and lateral root cap cells, support-ing a role in lignification and programmed cell death (Møller & McPherson, 1998).

Infection of tobacco with tobacco mosaic virus (TMV), which elicits a hyper-sensitive response (localised host cell death), was found to increase the biosynthesis of polyamines and their export into the apoplast and also increase polyamine oxidase activity in the cell wall. Furthermore, guazatine, a polyamine oxidase inhibitor, reduced the hypersensitive response (Yoda et al., 2003). It was proposed that the apoplastic H_2O_2 generated by polyamine oxidase contributed to the localised cell death.

Diamine oxidase activity was localised (by detection of the H_2O_2 formed in the presence of putrescine) in the cortex of *Pisum* stems, but it was not detected in the lignifying vascular tissue (Liu *et al.*, 1995). In *Rhizobium* infection threads in *Pisum*, diamine oxidase and peroxidase were both implicated in the insolubilisation of a specific plant glycoprotein, supporting a role for putrescine-generated H_2O_2 in wall cross-linking (Wisniewski *et al.*, 2000). Overexpression of diamine and polyamine oxidases in the apoplast of transgenic plants did not cause any increase in apoplastic H_2O_2 production unless the relevant amines were also added (Rea *et al.*, 2004). This suggests that the secretion of amines rather than the activity of the oxidases is rate-limiting for H_2O_2 production.

Peroxidases. Grant *et al.* (2000) found evidence for both NAD(P)H oxidase- and peroxidase-mediated H_2O_2 generation during the oxidative burst. Peroxidases are extremely versatile enzymes. Although best known for their ability to utilise H_2O_2 in the oxidation of phenols (e.g. forming dehydrodiferulate and isodityrosine), peroxidases can also contribute to H_2O_2 *production*. Acting in the latter mode, they require an electron donor (i.e. reductant) plus O_2 as electron acceptor. *In vitro*, the electron donor can be NAD(P)H, as demonstrated in the cell walls of liverworts (Ishida *et al.*, 1987) and numerous other plants (Bolwell *et al.*, 2001). The production of an unidentified endogenous reductant, needed for wall-bound peroxidase to generate H_2O_2 *in vivo*, has been proposed (Bindschedler *et al.*, 2001). The nature of the endogenous reductant(s) involved is unclear, but exogenous palmitate can fulfil the role experimentally (Bolwell *et al.*, 2002).

9.4 $O_2^{.-}$ and $HO_2^{.}$ in plant cell walls

9.4.1 *Properties of $O_2^{.-}$ and $HO_2^{.}$*

$O_2^{.-}$ is a moderately reactive free radical. It is the ionised form of the very highly reactive $HO_2^{.}$, these two species rapidly and non-enzymically interconverting:

$$HO_2^{.} \leftrightarrow O_2^{.-} + H^+ \qquad (9.4)$$

The pK_a of this ionisation is approximately 4.8, and thus, in the (relatively acidic) apoplast, both radical species can be important. This is in contrast to the situation within the protoplasm, where the pH is usually greater than 7, and $O_2^{.-}$ greatly exceeds $HO_2^{.}$. For simplicity, we will generally refer to $O_2^{.-}$, even though $HO_2^{.}$ may be at least equally important in this chapter.

Experimentally, $O_2^{.-}$ can be generated by addition of xanthine plus xanthine oxidase, although this enzyme has a relatively high pH optimum and may thus fail to operate satisfactorily at typical apoplastic pH values. Alternatively, $O_2^{.-}$ can be directly added in the form of the commercially available crystalline (but relatively impure) salt, potassium superoxide (KO_2).

9.4.2   Detection of $O_2^{\cdot-}$

Specific detection of endogenous $O_2^{\cdot-}$ is of critical importance for the study of $O_2^{\cdot-}$ production. The requirement that the method is specific for $O_2^{\cdot-}$, and not for other ROS, is often tested by use of the $O_2^{\cdot-}$-scavenging enzyme, SOD, and the H_2O_2-scavenging enzyme, catalase. For studies in the cell wall, low-M_r scavengers are also used, such as Mn-desferrioxamine (desferal), since enzymes may not diffuse through the wall matrix very easily. The signal of an $O_2^{\cdot-}$-specific detection method should decrease only when an $O_2^{\cdot-}$-scavenging agent, like SOD, is added, and not when an H_2O_2-scavenging agent, such as catalase, is added. When a system contains a high endogenous SOD activity, hardly any $O_2^{\cdot-}$ production might be found, since the $O_2^{\cdot-}$ is rapidly converted to H_2O_2. This problem might be overcome by inhibiting SOD action with, e.g., DIECA (N,N-diethyldithiocarbonate).

A sensitive method for measuring $O_2^{\cdot-}$ is the chemiluminescence of lucigenin (bis-N-methylacrodinium). Before reacting with $O_2^{\cdot-}$, lucigenin has to be reduced to a lucigenin cation radical, which might be done by the same biological system that produces the $O_2^{\cdot-}$. When a lucigenin radical reacts with $O_2^{\cdot-}$, lucigenin dioxetane is formed, which will decompose to two molecules of N-methylacridone. One of the two molecules will decompose to an excited state and will, upon relaxation to the ground state, emit a photon that can be captured and counted. This method has been applied with great success to membrane vesicles. When lucigenin is used in too high a concentration, auto-oxidation may occur, resulting in the production of $O_2^{\cdot-}$, which will then give a signal. Since this artefact involves $O_2^{\cdot-}$, scavenging of $O_2^{\cdot-}$ with SOD will result in a decreased signal and the lucigenin signal might be interpreted as originating from plant-generated $O_2^{\cdot-}$. The concentration of lucigenin must be chosen carefully to prevent such false positives, and tested with, e.g., concomitant measurements of oxygen consumption (Li et al., 1998).

A second method used to determine apoplastic $O_2^{\cdot-}$ production utilises tetrazolium salts. Upon reduction by $O_2^{\cdot-}$, tetrazolium salts change colour, which can be measured with a spectrophotometer. Nitro blue tetrazolium (NBT) gives a blue precipitate after reduction by $O_2^{\cdot-}$, a property used to localise apoplastic production sites in plant tissues. 3'-[1-[(Phenylamino)-carbonyl]-3,4-tetrazolium]-bis(4-methoxy-6-nitro)benzenesulphonic acid hydrate (XTT) and 2-[4-iodophenyl]-3-[4-nitrophenyl]-5-[2,4-disulphophenyl]-2H-tetrazolium (WST-1) are more recently introduced tetrazolium salts (Sutherland & Leamonth, 1997; Berridge & Tan, 1998) that remain in solution upon reduction by $O_2^{\cdot-}$. WST-1 has been shown not to cross the plasma membrane, and to remain apoplastic after addition to the tissue (Berridge & Tan, 1998). Both XTT and WST-1 gave similar results when used to study $O_2^{\cdot-}$ production in the cell walls of germinating radish seeds (Schopfer et al., 2001). In contrast to lucigenin, which can give false positive results, XTT can give false negative results. NADH oxidation by maize coleoptile segments was inhibited 50% by 500 μM XTT, a similar percentage inhibition to that observed on addition of 1 mM KCN (Frahry & Schopfer, 2001). Superoxide production by an XTT-sensitive mechanism might thus be underestimated by this method.

SOD-inhibitable reduction of cytochrome c (cyt c) is a third method applied to study apoplastic $O_2^{\cdot-}$ production, and the specificity of this method is sometimes used to verify the results obtained with other techniques. This method is mainly used with vesicles, and cyt c may not penetrate the cell wall. The use of cyt c reduction as a measure of $O_2^{\cdot-}$ production in plasma membrane vesicles might be seriously hampered by the presence of cyt c reductases, giving an overestimate or false positive results (Murphy et al., 1998).

9.4.3   Formation of $O_2^{\cdot-}$ in plant cell walls

Apoplastic $O_2^{\cdot-}$ can originate from NAD(P)H oxidases that are present in the cell wall or bound to the plasma membrane. To act as an NADH oxidase, wall-bound peroxidase has first to be activated by H_2O_2 to transform the peroxidase (with Fe^{3+}) to compound I, in which the iron is oxidised to Fe^{4+} and an oxygen atom is bound to the enzyme. Compound I can then oxidise NADH to an NAD$^{\cdot}$ radical, transforming itself to compound II (with the iron still as Fe^{4+}, but with an extra hydrogen atom bonded to the enzyme). Compound II can oxidise another NADH to NAD$^{\cdot}$, and this reaction will transform the peroxidase back into its Fe^{3+} ground state. The NAD$^{\cdot}$ that is formed in these reactions will react with O_2 to form NAD^+ and $O_2^{\cdot-}$ (Halliwell, 1978).

Both wall-bound peroxidase and plasma-membrane-bound NAD(P)H oxidases, with the NAD(P)H oxidising site facing the apoplastic side, would require NAD(P)H in the apoplast, since NAD(P)H is membrane-impermeant (Lin, 1982). The reducing equivalents can, however, be delivered to the apoplast by a shuttle mechanism, using another substrate to carry the reducing power across the plasma membrane. A proposed shuttle mechanism relies on a malate–oxaloacetate cycle. Malate can get oxidised to oxaloacetate in the cell wall by a wall-bound malate dehydrogenase, thereby reducing NAD^+ to NADH (Gross, 1977; Gross et al., 1977). The oxaloacetate that is produced in the cell wall can be reduced in the cytoplasm to malate by a membrane-bound malate dehydrogenase, using NADH as reducing substrate (Córdoba-Pedregosa et al., 1998). Both malate and NAD^+ (but not NADH) have been claimed to be present in the apoplast (Shinkle et al., 1992; Otter & Polle, 1997; Gabriel & Kesselmeier, 1999), and malate transporters have been found in the plasma membrane of plants, but the proposed malate–oxaloacetate shuttle remains hypothetical.

Membrane-bound NAD(P)H oxidases can also oxidise NAD(P)H on the symplastic side of the plasma membrane, transferring the electrons to the apoplastic side to reduce O_2 to $O_2^{\cdot-}$ (see Chapter 8; Murphy & Auh, 1996). Several different $O_2^{\cdot-}$-producing enzyme activities are observed to be associated with plant plasma membranes: enzymes utilising NADH, NADPH, or both, or requiring quinones for activity (Murphy & Auh, 1996; van Gestelen et al., 1998).

$O_2^{\cdot-}$ production by wall-bound peroxidases is inhibited by KCN and NaN_3, and most membrane-bound NAD(P)H oxidases are inhibited by diphenyleneiodonium (DPI) and imidazole. To use these inhibitors to determine what kind of $O_2^{\cdot-}$-producing system is involved in a specific situation is, however, questionable. In tobacco plasma membrane vesicles, two types of $O_2^{\cdot-}$-producing enzymes are active: one that can be inhibited by DPI and imidazole, but also one that is insensitive to these inhibitors but sensitive to KCN and NaN_3 (Papadakis & Roubelakis-Angelakis, 1999). Using DPI to 'prove' the involvement of a membrane-bound NAD(P)H oxidase is dubious since $O_2^{\cdot-}$ production by horseradish peroxidase has been shown to be inhibited by DPI (Frahry & Schopfer, 1998). The use of DPI as inhibitor of plasma membrane-bound NAD(P)H oxidases was inspired by the inhibitory action of DPI in a mammalian system. Phagocytes produce a burst of $O_2^{\cdot-}$ upon pathogen invasion, using a membrane-bound NAD(P)H oxidase. The inhibition of both animal- and plant-membrane-bound NAD(P)H oxidases indicates a certain level of homology between the two systems. This is discussed in more detail in Chapter 8.

9.4.4   Role of $O_2^{\cdot-}$ in plant cell walls

Evidence suggests strongly that $O_2^{\cdot-}$ radicals do occur in plant cell walls, but what might be their role? $O_2^{\cdot-}$ can be converted to H_2O_2 and O_2 both non-enzymically (Elstner, 1987) and catalysed by SOD. The maximum non-enzymic rate occurs at pH 4.8, a typical apoplastic acidity, which is the pK_a of $O_2^{\cdot-}$ – i.e. the pH at which equimolar $O_2^{\cdot-}$ and $HO_2^{\cdot}$ exist, giving the maximum opportunity for these two species to interact:

$$O_2^{\cdot-} + HO_2^{\cdot} \rightarrow HO_2^{-} + O_2 \quad \text{(rate constant } \sim 1.5 \times 10^7 \text{ M}^{-1}\text{ s}^{-1}\text{)}$$

(9.5)

$$HO_2^{-} + H^+ \rightarrow H_2O_2$$

(9.6)

The H_2O_2 thus generated is a substrate for peroxidases (see earlier). For example, $O_2^{\cdot-}$ production in spinach vascular bundles is likely to be a source of H_2O_2 used for lignin biosynthesis, since the production of $O_2^{\cdot-}$ was closely related with the distribution of Zn-SOD, H_2O_2 and lignification (Ogawa et al., 1997). Besides a role in H_2O_2 production, $O_2^{\cdot-}$ is also a precursor for hydroxyl radicals (see later).

$O_2^{\cdot-}$ can thus be a precursor for both H_2O_2 and $\cdot OH$, which have two opposite proposed roles in the cell wall. H_2O_2 is associated with lignification and cell wall cross-linking, making the wall less extensible (see earlier). $\cdot OH$ on the other hand is linked with polymer scission, leading to cell wall loosening (see later). No direct effect of $O_2^{\cdot-}$ on cell wall polymers is known yet, suggesting that the downstream fate of $O_2^{\cdot-}$ determines the role of its production in the apoplast.

9.4.5 Regulation of $O_2^{\cdot-}$ production in plant cell walls

The production of $O_2^{\cdot-}$ in the plant cell wall is regulated by the developmental programme and by internal and external stimuli. An example of $O_2^{\cdot-}$ production regulated by the developmental programme is the lignification of vascular tissue in spinach hypocotyls. Generation of $O_2^{\cdot-}$ is associated with the sites where lignification of xylem vessels occurs, and with the sites of CuZn-SOD distribution (Ogawa et al., 1997). Besides the close correlation between lignification and $O_2^{\cdot-}$ production, $O_2^{\cdot-}$ is also produced in non-lignifying germinating seeds (Schopfer et al., 2001), where the $O_2^{\cdot-}$ is suggested to protect emerging seedlings against pathogen attack.

Hormonal regulation of $O_2^{\cdot-}$ production has been found in maize coleoptiles (Schopfer et al., 2002). Addition of exogenous auxin resulted in enhanced $O_2^{\cdot-}$ production by coleoptile segments, compared with non-auxin-supplied controls. The auxin-stimulated $O_2^{\cdot-}$ producing enzyme was believed to be a plasma-membrane-bound NAD(P)H oxidase. From this viewpoint, it is interesting that an auxin-stimulated NADH oxidase has been found in soybean plasma membranes (Morré et al., 1995). Auxin-induced $O_2^{\cdot-}$ production, however, is not a universal response, since 4-day-old suspension cultured rose cells do not show an auxin-induced $O_2^{\cdot-}$ production (Murphy & Auh, 1996).

Apoplastic $O_2^{\cdot-}$ production is also evoked by various external stimuli, such as mechanical stress in potato tubers (Johnson et al., 2003), or the Zn-nutrition of a plant (Pinton et al., 1994). Of all the external stimuli known to induce enhanced $O_2^{\cdot-}$ production, pathogens and pathogen-related stimuli are most widely studied. In this chapter, the regulatory role of these stimuli will be described only briefly. Fungal elicitor molecules from Mycosphaerella pinodes have been shown to stimulate wall-catalysed $O_2^{\cdot-}$ production in pea and cowpea (Kiba et al., 1997), suggesting that wall-bound peroxidases can respond to fungal elicitors, without a necessary role for the cytoplasm. On the other hand, fungal elicitors can also enhance membrane-catalysed $O_2^{\cdot-}$ production. Treatment of potato tubers with Phytophthora infestans, or with elicitors from this oomycete fungus, resulted in enhanced $O_2^{\cdot-}$ production in membrane fractions subsequently isolated from these tubers (Doke, 1985). The elicitors also induced $O_2^{\cdot-}$ production in membrane fractions isolated from untreated potato tubers (Doke & Miura, 1995). Fungal elicitors are thus capable of enhancing $O_2^{\cdot-}$ production by wall-bound peroxidases and membrane-bound NAD(P)H oxidases without de novo synthesis of regulatory components or enzymes. Besides a direct activation of enzyme (complexes) that are already present, fungal elicitors also induce translocation of components of membrane-bound NAD(P)H-oxidases from the cytosol to the plasma membrane (Xing et al., 1997).

Production of $O_2^{\cdot-}$ by membrane-bound enzymes of bean roots showed a pH optimum around 7.5, and the rate of $O_2^{\cdot-}$ production was much reduced at pH values less than 6 (Pinton et al., 1994). The regulation of $O_2^{\cdot-}$ production by pH might be of importance in regulating its activity in relation to other cell wall related processes, since many wall-modifying enzymes also show pH-regulated activity.

9.4.6   Lifetime and fate of $O_2^{\cdot-}$ in plant cell walls

The lifetime of $O_2^{\cdot-}$ in the cell wall is believed to be very short. The major scavenger of $O_2^{\cdot-}$ in the apoplast is SOD, which catalyses the dismutation of $O_2^{\cdot-}$ to yield O_2 and H_2O_2. Most SOD found in the cell wall is of the CuZn-SOD type. Four isoforms of CuZn-SOD were found in the apoplast of Scots pine needles (Streller & Wingsle, 1993), and CuZn-SOD has been localised in close proximity of the plasma membrane, as well as deep within the secondary cell wall (Ogawa et al., 1996). In spinach leaves and hypocotyls, the presence of SOD correlates with lignin biosynthesis, and it is proposed to play a role in the production of H_2O_2 from $O_2^{\cdot-}$. SOD has also been proposed to be situated near the membrane-bound NAD(P)H oxidases, to dismutate the $O_2^{\cdot-}$ produced by these enzymes (Ogawa et al., 1997).

$O_2^{\cdot-}$ can also react with phenolics that are present in the cell wall. Scavenging of $O_2^{\cdot-}$ by caffeate, p-coumarate and ferulate has been observed and the reported rate constants range from 10^4 to 10^5 M^{-1} s^{-1}, which is at least four orders of magnitude lower than the SOD activity measured in the same study (Taubert et al., 2003). The rate constants were determined for the individual phenolic acids; information about the $O_2^{\cdot-}$ scavenging ability of feruloyl-polysaccharide esters remains to be elucidated. Superoxide also reacts with ascorbate and glutathione, with rate constants of 10^5–10^6 M^{-1} s^{-1} (compare ~10^{10} M^{-1} s^{-1} for reaction of these compounds with $\cdot OH$).

9.5 $\cdot OH$ in plant cell walls

9.5.1   Properties of $\cdot OH$

$\cdot OH$ is an exceedingly short-lived species: its typical half-life in a biological milieu is estimated at approximately 1 ns. It is neutral at all physiologically relevant pH values, and is a very strongly oxidising species. It reacts with most organic compounds, but especially aromatics, carbohydrates and lipids; it can thus cause mutation (by reacting with the aromatic bases of DNA and cleaving deoxyribose residues), membrane damage, enzyme inactivation and damage to structural carbohydrates e.g. the muco-polysaccharides of mammalian connective tissues. $\cdot OH$ is, therefore, a 'biological hazard' if produced in the wrong place and at the wrong time. For this reason, anti-oxidant systems are in place in many compartments of the cell, as discussed elsewhere in this book.

However, since $\cdot OH$ is so short-lived, it is unable to diffuse significant distances from its site of production. In the nanosecond for which an $\cdot OH$ radical exists in a cell (or wall), it would be unlikely to diffuse further than about 1 nm. For comparison, a single glucose residue in a cellulose molecule is about 0.5 nm long. Thus, $\cdot OH$'s site of production will be very close to its site of action. Primary cell walls are thin (often only ~100 nm thick), but an $\cdot OH$ radical generated near the centre of such a wall would be very unlikely to reach the plasma membrane where it could potentially damage the protoplast. There is rapidly growing interest in the possible positive roles which ˙OH may serve within the plant cell wall.

9.5.2 Proposed mechanisms of apoplastic ˙OH formation

Non-enzymic production of ˙OH from H_2O_2. Although ˙OH is produced by the action of γ-rays on water, and may thus be formed artefactually during the irradiation of food products, the best established mechanism for biological production of ˙OH is by Fenton-type reactions, in which H_2O_2 reacts with a reduced transition metal ion. These reactions can be exemplified by

$$Cu^+ + H_2O_2 \rightarrow \text{˙OH} + OH^- + Cu^{2+} \tag{9.7}$$

and

$$Fe^{2+} + H_2O_2 \rightarrow \text{˙OH} + OH^- + Fe^{3+} \tag{9.8}$$

The rate constant for Cu^+ is approximately 60 times greater than that for Fe^{2+} (Halliwell & Gutteridge, 1999). Thus, although Cu is less abundant biologically than Fe, the former may, nevertheless, play a predominant role in ˙OH production. The plant cell wall contains very strong Cu^{2+}-binding sites (Graham et al., 1981), which could help to position this metal appropriately for controlled ˙OH production. It is likely that histidine-rich glycoproteins, known to be present in the cell wall (Sommer-Knudsen et al., 1997; J. Sommer-Knudsen, M. Verkuijlen, A. Bacic & A.E. Clarke, personal communication), or the negatively charged uronic acids of the pectic fraction (van Cutsem & Gillet, 1982), act to fix the location of the Cu^{2+}.

The requisites for the Fenton reaction are that the transition metal ion should become reduced (e.g. $Cu^{2+} \rightarrow Cu^+$) and that H_2O_2 should also be present. There is plenty of evidence for H_2O_2 in the apoplast (see earlier), and it is well established *in vitro* that Cu^{2+} (and Fe^{3+}) can be reduced by certain important apoplastic solutes such as $O_2^{\cdot-}$ or ascorbate (Halliwell & Gutteridge, 1999).

Evidence for a role of $O_2^{\cdot-}$ in this respect was adduced on the basis that exogenous SOD inhibited ˙OH production in rice cell cultures (Kuchitsu et al., 1995). Catalase and DTPA (a metal-chelator) were also inhibitory, as expected if the Fenton reaction is responsible for ˙OH production. Kuchitsu et al. (1995) suggested a mechanism in which $O_2^{\cdot-}$ reduces Fe^{3+} to Fe^{2+}, which then undergoes a Fenton reaction with H_2O_2 to form ˙OH.

The possible role of ascorbate is of particular interest because it can both reduce Cu^{2+} to Cu^+ and generate H_2O_2 from O_2 non-enzymically at physiological pH:

$$AH_2 + 2Cu^{2+} \rightarrow A + 2Cu^+ + 2H^+ \tag{9.9}$$

$$AH_2 + O_2 \rightarrow A + H_2O_2 \tag{9.10}$$

[where AH_2 = ascorbate and A = dehydroascorbate]. The two products, Cu^+ and H_2O_2, can then undergo a non-enzymic Fenton reaction to form ˙OH. Very low concentrations of Cu^{2+} are adequate for this process because the metal is constantly recycled and thus effectively acts as a 'catalyst' of the net stoichiometry:

$$3AH_2 + 2O_2 \xrightarrow{[Cu^{2+}]} 3A + 2\text{·OH} + 2H_2O \qquad (9.11)$$

It is interesting that ascorbate, a well-known anti-oxidant, can, by reacting with a mild oxidant (O_2), generate ROS (H_2O_2 and ultimately ·OH) which are much more reactive than O_2 itself. This 'pro-oxidant' function of ascorbate is not exhibited during the reaction catalysed by ascorbate oxidase (AAO), which generates only water:

$$2AH_2 + O_2 \rightarrow 2A + 2H_2O$$

Ascorbate is well established as an apoplastic solute. For example, the total ascorbate plus dehydroascorbate concentration in the spinach leaf apoplast was estimated to be 0.15–0.6 mM (Takahama & Oniki, 1992). Ascorbate is metabolically unstable in the apoplast, itself undergoing oxidative degradation to a range of breakdown products, including ultimately oxalate. Oxalate, and possibly some intermediates thereto, can lead to H_2O_2 production (Green & Fry, 2005).

The production of ·OH by ascorbate-driven processes is blocked by peroxidase (M.A. Green and S.C. Fry, unpublished). Thus, if ascorbate-generated ·OH acts as a wall-loosening agent, peroxidase could, in principle, oppose this process – in agreement with the common observation that apoplastic peroxidase activity correlates negatively with the rate of cell expansion. Peroxidase is generally assumed to restrict growth by oxidative cross-linking of wall phenolics; but suppression of ·OH production would push in the same direction.

Non-enzymic production of ·OH from $O_2^{·-}$. Chen and Schopfer (1999) have developed the ideas of Kuchitsu *et al.* (1995) and suggested that peroxidases play a role in the production of ·OH as a wall-loosening agent, in contradiction to the general assumption that peroxidases act as wall-tightening agents. They report that some peroxidases can catalyse three distinct types of reaction – peroxidatic, oxidatic and hydroxylic:

$$\text{peroxidatic:} \quad H_2O_2 + 2NADH \rightarrow 2NAD\text{·} + 2H_2O \qquad (9.12)$$

$$\text{oxidatic:} \quad NAD\text{·} + O_2 \rightarrow NAD^+ + \quad O_2^{·-} \quad \text{and} \qquad (9.13)$$

$$NADH + H^+ + O_2^{·-} \rightarrow NAD\text{·} + H_2O_2 \qquad (9.14)$$

$$\text{hydroxylic:} \quad H_2O_2 + O_2^{·-} \rightarrow \text{·OH} + OH^- + O_2 \qquad (9.15)$$

The hydroxylic (also known as Haber–Weiss) reaction can be catalysed by some peroxidases at 10–100 times the rate achieved by equimolar inorganic Fe (Chen & Schopfer, 1999; Liszkay *et al.*, 2003). To this list of reactions may be added the SOD-catalysed and/or non-enzymic dismutation of $O_2^{·-}$,

$$2O_2^{·-} + 2H^+ \rightarrow O_2 + H_2O_2 \qquad (9.16)$$

It is then possible to deploy Equations 9.12–9.16 in a ratio such that the overall stoichiometry is

$$7NADH + 7H^+ + 4O_2 \xrightarrow{\ [peroxidase]\ } 7NAD^+ + 6H_2O + 2^\cdot OH \qquad (9.17)$$

i.e. to achieve the peroxidase-catalysed net production of $^\cdot$OH from raw materials of NADH, H^+ and O_2. [The equation given by Liszkay *et al.* (2003), proposing that the NADH : O_2 stoichiometry is 1 : 1 rather than 7 : 4, is not correctly balanced.] Exogenous H_2O_2 was beneficial but not strictly necessary for $^\cdot$OH production (from NADH and O_2 by peroxidase *in vitro*) since the peroxidase can function as an NADH oxidase, generating H_2O_2.

$$NADH + H^+ + O_2 \rightarrow NAD^+ + H_2O_2 \qquad (9.18)$$

The ability of peroxidase to generate $^\cdot$OH was demonstrated *in vitro*, with deoxyribose or benzoate used as molecular probes for detecting $^\cdot$OH (Chen & Schopfer, 1999). However, it remains uncertain whether NADH is routinely available to apoplastic peroxidases *in vivo*.

Peroxidase inhibitors (cyanide, azide, sulphide, hydroxylamine, 1,10-phenanthroline or 2,2′-bipyridyl) were shown to block IAA-induced growth in maize coleoptile segments (Schopfer *et al.*, 2002) and to block $^\cdot$OH production *in vivo* (Liszkay *et al.*, 2003). This evidence was used to suggest that peroxidase is responsible for $^\cdot$OH generation which leads to wall loosening *in vivo*. However, the inhibitors used are not completely specific for peroxidase: they may also chelate Fenton-active metal ions and they are $^\cdot$OH scavengers (Figure 9.4). Thus, although these data are compatible with a role for $^\cdot$OH in wall loosening, they do not appear to prove that peroxidase is responsible for $^\cdot$OH production.

9.5.3 Evidence for presence of $^\cdot$OH in the wall/apoplast of the living plant cell

Electron paramagnetic resonance. $^\cdot$OH is the most reactive known molecule. Since $^\cdot$OH is exceptionally short-lived under biological conditions, it is not feasible to purify this ROS prior to its assay. However, it is possible to attempt detection of an ongoing turnover of $^\cdot$OH (i.e. simultaneous production and consumption) that leads to a low steady-state concentration, such as may occur in the culture medium (essentially a continuum of the cell wall) of a cell culture. The most clear-cut evidence for the presence of $^\cdot$OH in solution would be its characteristic electron paramagnetic resonance (EPR) spectrum. However, the steady-state concentration of apoplastic $^\cdot$OH is likely to be well below the limit of detection by this method.

The difficulty of the low steady-state concentration of $^\cdot$OH can be circumvented by use of 'spin trapping'. In this technique, reporter substances are added which react with $^\cdot$OH to form a much more long-lived radical that can gradually accumulate to concentrations detectable by EPR. For example, suspension-cultured rice cells were incubated in a medium containing 0.78% (w/v) ethanol plus 10 mM

α-(4-pyridyl 1-oxide)-N-*tert*-butylnitrone (4-POBN) (Kuchitsu *et al.*, 1995). (Few plant cell cultures will grow in the presence of 0.78% ethanol, and thus toxicity symptoms should be watched for.) The ˙OH reacts with the excess ethanol

$$CH_3–CH_2OH + ˙OH \rightarrow ˙CH_2–CH_2OH + H_2O \qquad (9.19)$$

and the carbon-centred radical (˙CH$_2$–CH$_2$OH) then rapidly reacts with 4-POBN to form a stable nitroxide spin-adduct, detectable by EPR. A difficulty with this approach is that the ethanol and possibly the 4-POBN may penetrate the plasma membrane and thus report total ˙OH rather than just apoplastic ˙OH. The approach thus does not immediately indicate whether the ˙OH detected was in the apoplast, where it could potentially act on wall polysaccharides, or in the symplast. However, the finding (Kuchitsu *et al.*, 1995) that exogenous enzymes (catalase or SOD) blocked the observed production of ˙OH supports the notion that the processes occurred apoplastically, since enzymes would not readily penetrate the plasma membrane. In similar experiments with maize coleoptiles, Schopfer *et al.* (2002) used 50 mM 4-POBN in a medium containing 3.9% (w/v) ethanol, which may be a high enough concentration to permeabilise membranes.

Membrane-permeant chromogenic and fluorogenic probes. An alternative means of detecting ˙OH in biological samples is the use of chromogenic or fluorogenic indicators of this radical. Two potentially useful fluorogenic probes have recently been devised (Setsukinai *et al.*, 2003). The non-fluorescent compound 2-(6-(4′-hydroxy)phenoxy-3H-xanthen-3-on-9-yl)benzoic acid (HPF) is reported to be relatively specific for ˙OH, generating a highly fluorescent product, fluorescein, in its presence. Peroxynitrite (ONOO$^-$) also induced the formation of fluorescein from HPF, which might be caused by the generation of ˙OH by ONOO$^-$. The closely similar but aminated compound, 2-(6-(4′-amino)phenoxy-3H-xanthen-3-on-9-yl)benzoic acid (APF), yields fluorescein not only with ˙OH and ONOO$^-$ but also with hypochlorite (OCl$^-$) and is therefore less specific. However, both HPF and APF (like many phenols and aromatic amines) are substrates for peroxidase in the presence of H$_2$O$_2$, and would thus be expected to generate fluorescent products in the plant apoplast whether or not ˙OH is generated there. Furthermore, both HPF and APF can readily permeate the plasma membrane, and will, therefore, report intraprotoplasmic ˙OH as well as apoplastic.

Benzoate has also been used as a fluorogenic probe for ˙OH, but it too is highly membrane-permeable. However, even if a probe (e.g. benzoate) is not excluded by the plasma membrane, the data can be consolidated by the demonstration that scavengers selected for their membrane-impermeance can diminish apparent ˙OH production. For example, Schopfer *et al.* (2001) used 100 mM mannitol in this way and observed a 92% inhibition of benzoate hydroxylation.

Membrane-impermeant chromogenic and fluorogenic probes. A strategy to test specifically for *apoplastic* ˙OH radicals is based on their ability to react with exogenous, membrane-impermeant probes. Aromatic compounds could be suitable 'radical traps' for this purpose because they undergo exceedingly rapid addition reactions with ˙OH. For example, in the case of benzene, the reaction is reported to be as shown in Equation 9.20 (see later).

Ideally, such a 'trap' should meet the following criteria (Fry *et al.*, 2002):

(i) it reacts with ˙OH to form an easily recognisable, stable product;
(ii) it does not react with other oxygen species such as $O_2^{˙-}$ or H_2O_2;
(iii) it is unable to cross the plasma membrane;
(iv) it is not subject to enzymic modification in the apoplast.

Aromatic 'traps' react with ˙OH to form phenols (e.g. phenylalanine forms *o*-, *m*- and *p*-tyrosines), which give evidence for ˙OH (Grootveld & Halliwell, 1986; Ghiselli, 1998) [and/or peroxynitrite (Halliwell & Kaur, 1997)] in animals. Unfortunately, in the plant apoplast, phenolic products may be rapidly oxidised by peroxidases and are thus difficult to quantify.

Schopfer *et al.* (2001) used 2′,7′-dichlorofluorescin as a relatively membrane-impermeant probe, which reacts with ROS to generate a highly fluorescent product, 2′,7′-dichlorofluorescein. According to this test, the embryos and seed coats of radish seeds were found to generate apoplastic ROS during germination. This process was proposed to contribute to protection against pathogen attack, but could also contribute to loosening of the embryo and seed-coat cell walls to facilitate the tremendous cell expansion required for germination. Part of the ROS population detected by 2′,7′-dichlorofluorescin was deduced to be ˙OH on the basis that it also caused degradation of deoxyribose, which may also be a membrane-impermeant ˙OH probe (Schopfer *et al.*, 2001).

Probes for detecting ˙OH that have been applied to animal tissues include D-phenylalanine [which reacts with ˙OH to generate *o*-, *m*- and *p*-tyrosine (Halliwell & Kaur, 1997), the first two of which are not usually found *in vivo*], benzoate (which generates hydroxybenzoates, including salicylate) and salicylate itself [which generates dihydroxybenzoates (Ghiselli, 1998)]. However, the major detectable products formed from these particular probes are phenolic and would be unstable in plant tissues owing to the abundance of peroxidases; in addition, salicy-late is not ideal either as a probe or as a reaction-product because of its potent physiological effects on plant cells. Also, phenylalanine, salicylate and benzoate may cross the plasma membrane and, therefore, also react with any ˙OH produced intra-protoplasmically.

Membrane-impermeant or wall-impermeant radioactive probes. Membrane-impermeant ˙OH-traps carrying a radioactive marker could have the advantage of great sensitivity because of the ability of the scintillation counter to detect very small quantities of isotopes, especially ^3H. A commercially available compound that we (Miller & Fry, 2004) have tested as a sensitive new probe for apoplastic ˙OH is N-[4-^{3}H]benzoylglycylglycylglycine (BzG$_3$; also known as hippuryl glycyl glycine), which would be expected to react with ˙OH to release an easily measurable, stable end-product: tritiated water (^3H_2O). 'OH-generating solutions of aerated ascorbate, especially in the presence of H_2O_2 and traces of Cu^{2+}, caused the oxidation of [^3H]BzG$_3$, yielding hydroxylated by-products and generating ^3H_2O (Miller & Fry, 2004). The ^3H_2O can be quantitatively separated from organic products and remaining [^3H]BzG$_3$ on a small column of Dowex-1 (OH$^-$ form). Production of ^3H_2O from [^3H]BzG$_3$ was much greater with 'OH-generating mixtures than with other ROS such as H_2O_2 or O_2^{-}. The slight production of ^3H_2O from [^3H]BzG$_3$ that did occur in the presence of H_2O_2 or O_2^{-} was efficiently blocked by 'OH-scavengers, e.g. 500 mM glycerol, mannitol, butan-1-ol or dimethylsulphoxide, which are not good scavengers of H_2O_2 or O_2^{-}. This indicates that ^3H_2O production from [^3H]BzG$_3$ was caused specifically by 'OH and that other ROS only generated any ^3H_2O because of their ability to form traces of 'OH. [^3H]BzG$_3$ has an excellent sensitivity for 'OH at biologically relevant levels (i.e. sufficient to cause polysaccharides scission *in vitro*), and is thus a simple, specific, sensitive and robust method for detecting 'OH production *in vitro* (Miller & Fry, 2004); it may, therefore, also be applicable for the *in-vivo* detection of 'OH in the apoplast.

The reliable application of [^3H]BzG$_3$ as an apoplastic probe to any particular tissue requires a demonstration that the cells do not take up the intact [^3H]BzG$_3$ or hydrolyse its amide bond (which would yield freely membrane-permeable [^3H]benzoic acid). To overcome this potential problem, we have also investigated two alternative traps: *N*-[^3H]benzoyl-pentalysine methyl ester (BzK$_5$Me) and *N*-[^3H]benzoyl-polyallylamine (Bz-PA; Figure 9.2), synthesised in our laboratory. BzK$_5$Me and Bz-PA are polycationic probes and thus become anchored to acidic polysaccharides, facilitating detection of apoplastic 'OH. Furthermore, Bz-PA appeared to be completely resistant to apoplastic benzoyl amidase action, and thus unable to release potentially membrane-permeant [^3H]benzoic acid.

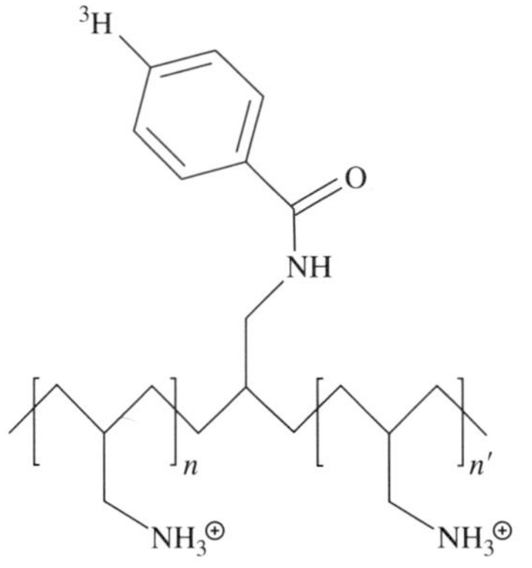

Figure 9.2   *N*-[^3H]Benzoyl-polyallylamine: a potential 'trap' for reporting apoplastic hydroxyl radicals.

In preliminary studies, 1-cm segments of various organs (maize coleoptiles and mesocotyls; stems, petioles or styles of various dicots) were incubated in MES buffer, pH 6.1, containing [³H]Bz-PA, with or without auxin. After incubation for 0.2–5.5 h, ³H₂O was detected in all samples, suggesting apoplastic ˙OH (Fry *et al.*, 2002). There was no evidence for enhanced ³H₂O production accompanying auxin-promoted growth. In parallel studies, [³H]Bz-PA was supplied to tomato fruits, which were also shown to generate ³H₂O, indicating the presence of ˙OH (J. Dumville, unpublished).

A probe for extracellular (as distinct from wall-localised) ˙OH is [¹⁴C]phenylethyl polyacrylate (Cohen *et al.*, 2002). This polymer is proposed to be too large (~10 kDa) to permeate the wall of a fungal hypha, and, therefore, only able to detect truly extracellular ˙OH radicals. After putative ˙OH attack, Cohen *et al.* (2002) hydrolysed the probe molecule to release the ester-linked [¹⁴C]phenylethanol moieties, which were analysed by HPLC. Among the products were 2-, 3- and 4-hydroxy-[¹⁴C]phenylethanols, indicative of ˙OH attack.

Fingerprinting of wall polysaccharides. Another approach for the detection of (recent past) apoplastic ˙OH is to look for a chemical 'fingerprint' characteristic of ˙OH-attacked wall polysaccharides. Such attack, unlike hydrolysis, cleaves polysaccharide chains primarily by the introduction of mid-chain oxo groups, which can be detected by treatment with NaB³H₄, yielding unusual ³H-aldose (as opposed to ³H-alditol) residues (Miller & Fry, 2001; Figure 9.3). When authentic pectin, xyloglucan or mixed-linkage β-glucan were treated with ˙OH (supplied as ascorbate plus H₂O₂) *in vitro*, chain scission occurred; also, as 'collateral damage', relatively stable oxo groups were introduced which were detectable by staining

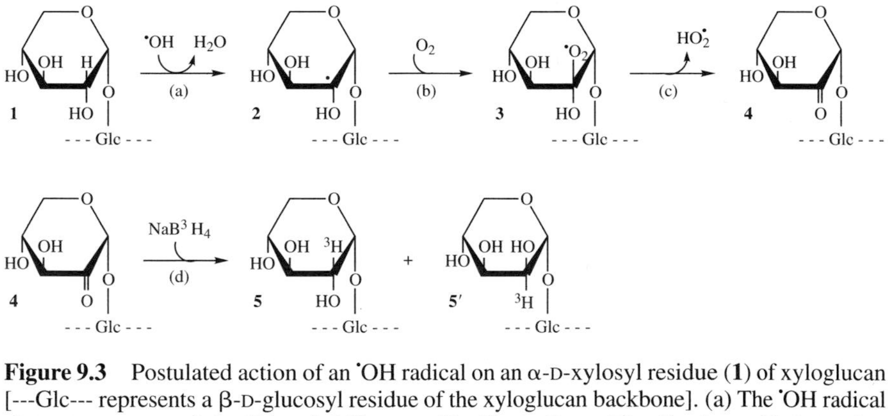

Figure 9.3 Postulated action of an ˙OH radical on an α-D-xylosyl residue (**1**) of xyloglucan [---Glc--- represents a β-D-glucosyl residue of the xyloglucan backbone]. (a) The ˙OH radical abstracts any one of the C-bonded H-atoms (the illustration arbitrarily shows H-abstraction from C-2) to form an α-hydroxyalkyl radical (**2**); (b) reaction with atmospheric O₂; (c) elimination of a hydroperoxyl (or superoxide) radical to form a xylos-2-ulosyl residue (**4**). If (**4**) is then treated with NaB³H₄ (d), a mixture of [2-³H]xylosyl (**5**) and [2-³H]lyxosyl residues (**5′**) would be expected, which would be released as free ³H-monosaccharides upon acid hydrolysis.

with aniline hydrogen-phthalate and by reaction with NaB^3H_4 (Fry et al., 2001, 2002). The NaB^3H_4-products included ^3H-sugar residues, including epimers (e.g. ribose, mannose and lyxose) not present in the original polysaccharides (Miller & Fry, 2001). The products (released by hot acid) were ^3H-sugars, not ^3H-alditols, and were thus distinguished from products of reducing termini. The chromatographic pattern of the ^3H-products forms a 'fingerprint', diagnostic of a polysaccharide that has previously been damaged by hydroxyl radicals.

Using a similar method, we have obtained preliminary evidence that pear fruit cell wall polysaccharides are subjected to ˙OH attack during ripening (Fry et al., 2001). Work is currently in progress in our laboratory to develop fluorescent rather than radiochemical labelling to obtain these fingerprints.

9.5.4   Factors that trigger formation of apoplastic ˙OH

Microbial elicitors. EPR spectral evidence for the presence of apoplastic ˙OH in cultured rice cells has been reported (Kuchitsu et al., 1995). The ˙OH was elicited in these cells by administration of an oligosaccharin of the N-acetylchito-oligosaccharide class. Thus, ˙OH production may be an integral part of the well-known oxidative burst which often accompanies a plant's defence response to pathogen attack. In agreement with this, von Tiedemann (1997) showed by use of deoxyribose as a probe that bean leaf discs generated ˙OH in response to infection by *Botrytis*. The more aggressive the race of *Botrytis*, the more effectively did the fungus suppress peroxidase activity in the plant, suggesting that peroxidase is a negative, rather than positive, factor in ˙OH production.

Developmental programme. Effusion of ROS by embryos and seed coats of radish seeds is observed to coincide approximately with germination (emergence of the radicle). This ROS production was enhanced by darkness or gibberellin, which promoted germination, and inhibited by germination inhibitor, abscisic acid (Schopfer et al., 2001). Application of ˙OH-scavengers (benzoate, adenine, mannitol, formate and thiourea but not urea) suggested that the $2',7'$-dichlorofluorescin method used was detecting not only H_2O_2 but also ˙OH.

The *temporal* programme of root cell development can be observed at any given time as a *spatial* pattern by examination of segments cut at various distances from the root tip. Production of apoplastic H_2O_2 in different zones of the onion root correlated positively with the rate of cell expansion (Córdoba-Pedregosa et al., 2003); the authors suggested that this observation is compatible with a role for ˙OH (formed from H_2O_2) in wall loosening. Furthermore, in the vicinity of the elongation zone, apoplastic ascorbate was largely (87%) in its oxidised form (dehydroascorbate), compatible with the proposal that it participates in oxidising reactions, which may generate ROS.

There is preliminary evidence for developmentally programmed ˙OH production in the apoplast of the pear fruit during ripening (Fry et al., 2001). This observation is also compatible with the observation that tomato fruit cells begin to 'leak' ascorbate into the apoplast early during the ripening programme (Dumville & Fry, 2003). The apoplastic ascorbate thereby accumulated would be able to generate 'OH non-enzymatically (see earlier).

9.5.5   Proposed roles for apoplastic 'OH in vivo

Proposed evidence for a role of 'OH in cell expansion. Schopfer (2001) presented several lines of evidence that apoplastic 'OH could loosen the cell wall and promote cell expansion. He showed that an 'OH-generating mixture (ascorbate + H_2O_2) was able to loosen 'isolated walls' (defined as frozen/thawed, abraded organ segments; preferably pre-loaded with Fe^{3+} or Cu^{2+}) of dicots, gramineous monocots and a gymnosperm, facilitating irreversible creep when the 'isolated walls' were stretched in an extensiometer. This supports the view that 'OH can loosen plant cell walls non-enzymatically, at least *in vitro*. Similarly, growth was promoted by ascorbate plus H_2O_2 in living organ segments. Furthermore, auxin promoted the production of O_2^- [a possible precursor of 'OH, but equally also a *product* of the reaction of 'OH with wall polysaccharides (von Sonntag, 1987)] in maize coleoptile epidermal cells (Schopfer, 2001; Schopfer *et al.*, 2002). Finally, 0.1–10 mM adenine*, histidine or salicylate blocked the growth-promoting action of auxin: this effect was attributed (Schopfer, 2001) to the ability of these compounds to scavenge 'OH; however, at such low concentrations, their effect seems more likely to have been due to the compounds' strong ability to chelate metal ions such as Cu^{2+} (Dawson *et al.*, 1986). Published rate-constants (Figure 9.4) indicate that 1% (w/v) sucrose would be a more effective scavenger of 'OH than 10 mM of adenine (Buxton *et al.*, 1988), yet 1% sucrose certainly does not block auxin-induced growth. In a related paper (Schopfer *et al.*, 2002), it was shown that 10 mM benzoate completely blocked IAA-induced growth, whereas 14 mM DMSO, a slightly more effective 'OH-scavenger (Figure 9.4; Buxton *et al.*, 1988) had no effect. Schopfer's (2001) data show that artificially generated 'OH can loosen the wall and promote cell expansion, and also that apoplastic ROS, probably including 'OH, are generated *in vivo* during auxin-induced growth. However, we consider that the experiments with 'scavengers' (possibly acting instead as chelators) provide evidence more strongly supporting a role for transition metals in apoplastic ROS generation and cell expansion than in deciding whether or not 'OH serves a role in wall loosening.

NADH (200 μM) was able to mimic auxin (20 μM IAA) in promoting apoplastic O_2^- production (Schopfer *et al.*, 2002) and cell elongation (Liszkay *et al.*, 2003) in living maize coleoptiles. NADH has often been reported as a substrate for H_2O_2 production, both *in vivo* and *in vitro*, so the H_2O_2 thus generated may indeed lead to the formation of wall-loosening amounts of 'OH. NAD<u>P</u>H, however, inhibited IAA-induced O_2^- production, indicating considerable (enzyme-like) specificity. Desferrioxamine (complexed with manganese) also blocked apoplastic

---

* Figure 10 of Schopfer (2001) states 0.1–10 μM, but this should probably read 0.1–10 mM; compare Table 3 of Schopfer *et al.* (2002).

O_2^{-} accumulation, possibly because of its ability to cause the dismutation of O_2^{-} (Schopfer et al., 2002), although the specificity of Mn-desferrioxamine as a 'SOD mimic' is unclear, and an ability to mop up other (Fenton active) metals such as Cu and Fe cannot be ruled out as the reason for prevention of ROS production by desferrioxamine.

In-vivo production of $^{\cdot}OH$ in maize coleoptiles was demonstrated by use of 850 mM ethanol in the presence of 4-POBN (Schopfer et al., 2002). The $^{\cdot}OH$ signal was almost abolished by treatment of the segments with 1 mM thiourea. This effect was interpreted as due to the ability of thiourea to scavenge $^{\cdot}OH$. However, a consideration of rate constants (Buxton et al., 1988) indicates that 1 mM thiourea would cause only an approximately 0.3% inhibition of the ability of 850 mM ethanol to react with, and thus report the presence of, $^{\cdot}OH$. It again seems more tenable that the thiourea blocked $^{\cdot}OH$ *production* by chelating the Fenton-active metal ions present in the apoplast.

The 'fingerprinting' of wall polysaccharides in maize coleoptile segments did not provide any evidence that there was more extensive $^{\cdot}OH$ attack of these poly-saccharides after auxin treatment (Fry et al., 2002). Thus, we consider that it remains to be proven whether the non-enzymic scission of wall polysaccharides is responsible for auxin-induced cell expansion.

High expression of AAO often correlates with rapid cell expansion (Lin & Varner, 1991; Kato & Esaka, 1996; Al-Madhoun et al., 2003), an observation compatible with a pro-oxidant role for ascorbate's oxidation-products e.g. semide-hydroascorbate (Hidalgo et al., 1991) or dehydroascorbate (Lin & Varner, 1991) in wall loosening. One attractive possibility is that dehydroascorbate generates some wall-loosening factor, possibly H_2O_2 and hence (via Fenton reactions) $^{\cdot}OH$. Overexpression of AAO in tobacco cell cultures had little effect on growth of the (walled) cells, but protoplasts isolated from these cells exhibited increased swelling in the high-AAO line (although, surprisingly, the initial protoplast diameter was reported to be 3 μm, which is smaller than most chloroplasts!) (Kato & Esaka, 2000). The swelling-response of the isolated 'protoplasts' initially suggested that the effect of AAO on growth is not mediated via an effect on the cell wall; however, in the protoplast suspensions, only a minority of the AAO was released into the culture medium, and, furthermore, the protoplasts were cultured under conditions favouring wall regeneration, so the possibility remains that the AAO was located in a newly forming cell wall, and that a pro-oxidant effect of dehydroascorbate on this nascent wall was responsible for growth regulation. It seems unlikely that the widely observed correlation between AAO activity and cell expansion is due to an increase in the osmotic pressure of the cell sap (the only factor likely to promote the expansion of truly wall-less protoplasts).

The release of ROS by maize leaves (into an agar-solidified medium) was observed by their ability to convert 2',7'-dichlorofluorescin (in the agar) to the highly fluorescent product, 2',7'-dichlorofluorescein. This ROS effusion correlated positively with the local elemental growth rate along the length of the leaf (Rodríguez et al., 2002). DPI (200 μM) inhibited leaf segment growth (Rodríguez *et al.*, 2002). Since DPI inhibits both membrane-bound NADPH oxidase and also the wall-peroxidase-catalysed NADH-dependent production of H_2O_2 (Frahry & Schopfer, 1998), the observations support a role for a ROS cascade (possibly culminating in ˙OH production) in leaf cell elongation. Added 50 μM H_2O_2 partially restored DPI-blocked leaf growth (Rodríguez *et al.*, 2002), again supporting a role for apoplastic ROS in cell wall loosening.

In addition, when leaf elongation was partially inhibited by salinity (NaCl), it could be partially restored by ROS-generation (Rodríguez *et al.*, 2004). NaCl-treated maize leaf segments were promoted two-fold in cell expansion by an ˙OH-generating mixture of ascorbate, Cu^{2+} and H_2O_2 (Rodríguez *et al.*, 2004), again supporting a role for ROS, especially ˙OH, in cell expansion.

Other evidence that exogenous ROS, including ˙OH, may promote cell expansion has been provided by studies on root hairs but interpreted as being due to an ability of apoplastic ˙OH to promote Ca^{2+} uptake (Foreman *et al.*, 2003). This would appear to require ˙OH at the plasma membrane rather than in the cell wall.

Further evidence possibly supporting a role of ˙OH in wall loosening includes the following observations:

• Exogenous ascorbate (especially in the presence of equimolar dehydroascorbate, a mixture described as [containing] semidehydroascorbate), which can generate ˙OH, promotes cell expansion in onion roots (Hidalgo *et al.*, 1991).
• In soya hypocotyls, dithiothreitol inhibits both NADH oxidase (which forms apoplastic $O_2^{˙-}$, a potential precursor of ˙OH) and auxin-stimulated cell extension (Morré *et al.*, 1995).
• In maize coleoptiles, ferricyanide promotes elongation, possibly by activating a plasmalemmar NADH oxidase (Carrasco-Luna *et al.*, 1995).

These observations are all compatible with roles for ˙OH in wall-loosening, although direct evidence for ˙OH-attacked wall polysaccharides in growing walls currently remains lacking.

Proposed evidence for a role of ˙OH in fruit ripening. It was suggested some time ago (Brennan & Frenkel, 1977) that endogenous H_2O_2 formation contributes to pear fruit softening (Fry *et al.*, 2001), although the mechanism for this was not elucidated. A role for H_2O_2 in auxin breakdown, leading indirectly to fruit 'senescence' (ripening), was suggested. Brennan and Frenkel's work was largely ignored for many years.

However, the wall polysaccharides of the pear fruit, in contrast to those of auxin-treated maize coleoptiles, did exhibit an increase in 'fingerprint' features suggestive of ˙OH attack during ripening-associated softening (Fry *et al.*, 2001). Further work is required to test the hypothesis that apoplastic ˙OH contributes to fruit softening. It is intriguing, however, that certain fruits possessing little [e.g. banana (Pathak *et al.*, 2000), strawberry (Nogata *et al.*, 1993) and transgenic tomatoes (Smith *et al.*, 1990; Redgwell & Fischer, 2002)] or no [e.g. persimmon (Cutillas-Iturralde *et al.*, 1993)] detectable pectin-cleaving enzymes somehow manage to degrade the pectic polysaccharides in their walls (and/or middle lamellae) during ripening.

Hydroxyl radicals are widely regarded as biologically hazardous. However, in 'throw-away' tissues such as fruit pericarp (whose biological raison d'être is to be eaten by an animal), some damage to DNA, proteins and lipids may be tolerated. Indeed, some oxidative damage and loss of membrane integrity is a normal feature of fruit ripening (Brady, 1987; Jiménez *et al.*, 2002). These considerations invite a reassessment of the prevailing view that proteins are the sole agents of wall loosening in ripening fruits *in vivo*.

Proposed evidence for a role of ·OH in the oxidative burst. N-Acetylchito-oligosaccharide treatment of cultured rice cells elicited ·OH production (Kuchitsu *et al.*, 1995), supporting a role for ·OH in the oxidative burst often elicited by potential pathogens. Furthermore, based on the inhibitory effect of ·OH-scavengers, it has been suggested that ·OH production, part of an oxidative burst, is a necessary step in the abiotic elicitation of phytoalexins (Epperlein *et al.*, 1986).

Role of ·OH in plant cell wall decomposition by saprophytic fungi. Brown rot fungi, e.g. *Gloeophyllum trabeum* and *Postia placenta*, have been shown to generate extracellular ·OH, which serves to attack the crystalline cellulose on which these fungi live. The extracellular (as distinct from wall-localised) ·OH was demonstrated by use of the wall-impermeant probe, [^{14}C]phenylethyl polyacrylate (see earlier) (Cohen *et al.*, 2002).

9.5.6 Chemical effects and fate of ·OH in the apoplast

Non-enzymic reactions with wall components. ·OH, the most reactive known molecule, reacts exceedingly rapidly with most organic compounds, the two main types of reaction being (Buxton *et al.*, 1988):

(i)   Addition to aromatics and alkenes, forming hydroxylated products. For many such reactions, the rate-constant is so high ($k \approx 10^{10}\,\text{M}^{-1}\,\text{s}^{-1}$) that the reaction is essentially diffusion-limited. In the case of benzene, the reaction is reported to be

$$C_6H_6 + \dot{O}H \rightarrow \dot{C}_6H_6OH \tag{9.20}$$
(benzene)   (hydroxycyclohexadienyl radical)

then either

$$\dot{C}_6H_6OH + \dot{C}_6H_6OH \rightarrow (C_6H_5)_2 + 2\,H_2O \tag{9.21}$$
(biphenyl)

or

$$\dot{C}_6H_6OH + \dot{C}_6H_6OH \rightarrow C_6H_6 + C_6H_5OH + H_2O \tag{9.22}$$
(benzene) (phenol)

(ii)   Abstraction of an H atom from a carbon atom in any of a wide range of
        organic molecules, including polysaccharides. Again, rate-constants are typ-
        ically very high ($k \approx$ 1–4 $\times$ $10^9 M^{-1} s^{-1}$ for monosaccharides). As a result of
        this reaction, ˙OH (and thus also ascorbate) is capable of causing polysac-
        charide scission; ˙OH can even be generated simply by freeze-drying (e.g. of a
        sample of the animal polysaccharide, hyaluronic acid) – calling for extreme
        care against this type of artefact in such investigations (Tokita *et al.*, 1997).
            When H is abstracted from a polysaccharide, further reactions quickly
        lead to chain scission and a significant by-product is O_2^- (von Sonntag,
        1987). We showed (Fry, 1998) that, in the presence of an ˙OH-generating
        solution (O_2, ascorbate and Cu^{2+}), all polysaccharides tested (xyloglucan,
        pectin, alginate, carboxymethylcellulose, etc.) underwent non-enzymic
        scission at 20°C and pH 3–7.
            In aerobic environments, ˙OH is thought (Schuchmann & von Sonntag,
        1978; von Sonntag, 1987; Zegota, 2002) to attack sugar residues as follows:

(a) $>CH–OH + ˙OH \rightarrow >C˙–OH + H_2O$ [= abstraction   of   any
                                                    C-bonded H atom; Equation
                                                    9.23]
(b) $>C˙–OH + O_2 \rightarrow >C(OO˙)–OH$ [= reaction   with   atmos-
                                                    pheric O_2; Equation 9.24]
(c) $>C(OO˙)–OH \rightarrow >C=O + H^+ + O_2^-$ [= elimination of a superox-
                                                    ide radical; Equation 9.25]

Depending on *which* H atom was abstracted in step (a), the oxo group
($>C=O$) formed in step (c) may be unstable (leading to polysaccharide
chain scission) or relatively stable (in the form of a glycosulose residue). For
the animal polysaccharide hyaluronic acid it has been proposed that hydrogen abstraction from C-1 (the carbon atom contributing to the glycosidic
bond) leads to the addition of a hydroxyl group to the carbon-centred radical.
The resulting product could than lead to strand scission (Deeble *et al.*, 1990).
When pectin, mixed-linkage β-glucan or xyloglucan were treated with ˙OH
in vitro, chain scission occurred, and relatively stable oxo groups were
simultaneously introduced which were detectable by staining with aniline
hydrogen-phthalate and by reaction with NaB^3H_4 (Miller & Fry, 2001).

Protection against apoplastic ˙OH by scavengers. In principle, evidence for ˙OH
action can be sought by use of scavengers. This approach has been successfully
applied *in vitro*, where high concentrations can be used (Fry, 1998). However,
in vivo, particularly high concentrations would be required because cellular
compartments contain high concentrations of organic substances (e.g. polysaccharides in the wall; sugars and proteins in the symplast) and the necessary
concentrations of a scavenger would often be toxic, or at least inhibitory to
growth for osmotic reasons. The effectiveness of an exogenous scavenger can be
predicted from its concentration and its rate constant for reaction with ˙OH

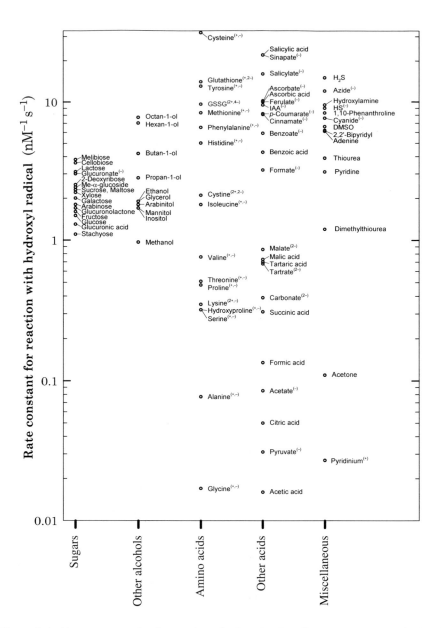

Figure 9.4 Rate constants for the reaction of selected cell wall components, and of related substances, with ˙OH. The values are from the much more extensive lists collated by Buxton *et al.* (1988). Additional relevant values (off the scale) include H_3BO_3, <0.00005; $HPO_4^{(2-)}$, 0.00015; urea, 0.00079; oxalate$^{(2-)}$, 0.0077; and bicarbonate$^{(-)}$, 0.0085 nM^{-1} s^{-1}. Note that the values for sugars and other alcohols are tightly clustered, whereas amino acids and other compounds cover a much wider range of values. Note also that, in many cases, a carboxylic acid has a rate constant different from that of its conjugate base (e.g. acetic acid versus acetate), and this fact must be taken into account in calculations of a buffer's scavenging ability. Abbreviations: IAA$^{(-)}$, indole-3-acetate; GSSG, oxidised glutathione; DMSO, dimethylsulphoxide.

(Figure 9.4), this information being taken in comparison with similar data on the naturally occurring endogenous substances that react with ˙OH (e.g. wall polysaccharides or apoplastic sucrose).

The use of ˙OH-scavengers is further hampered by the fact that ˙OH may react with the actual organic ligand holding the transition metal ion that was responsible for ˙OH formation. The effective concentration of this ligand at the site of ˙OH production will be exceedingly high and, therefore, the ligand will be a very vulnerable 'target', not appreciably out-competed by exogenous scavengers (Halliwell & Gutteridge, 1990).

9.6 Concluding remarks

It will be clear that there is now substantial evidence for ROS production in the plant cell wall, and for the hypothesis that this ROS production causes wall loosening. The only ROS that we believe can directly cause wall loosening (by causing the scission of structural polysaccharides) is ˙OH. However, the other ROS described in this chapter ($O_2^{˙-}$ and H_2O_2) are both precursors and products of ˙OH (see Figure 9.5). Thus, the detection of any ROS may indicate that ˙OH is, or was, present. In addition, ozone readily generates ˙OH in aqueous solution.

ROS (especially H_2O_2) can also tighten the cell wall, e.g. by enabling the oxidative coupling of polysaccharide-bound phenolics or by driving lignification. Thus, H_2O_2 is a two-edged sword: it seems possible that in the presence of high peroxidase (e.g. in salinated rice roots; Lin & Kao, 2001), H_2O_2 couples phenolics as well as preventing ascorbate-driven ˙OH production, and leads to wall tightening, whereas in tissues with lower peroxidase levels, the H_2O_2 can lead to ˙OH production (via ascorbate and the Fenton reaction), resulting in wall loosening, e.g. in maize leaves (Rodríguez et al., 2002).

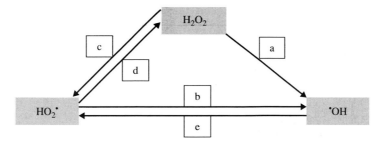

Figure 9.5 Potential routes for the interconversion of apoplastic ROS.
Note: the interconversions given for $HO_2^˙$ also apply to $O_2^{˙-}$. The reactions are:

   (a)   Fenton reaction and/or Haber–Weiss reaction;
   (b)   Haber–Weiss reaction;
   (c)   action of peroxidase + NADH;
   (d)   SOD-catalysed or non-enzymic dismutation;
   (e)   reaction of ˙OH with polysaccharide or similar substance in presence of O_2.

˙OH is often regarded as detrimental to life and its proposed physiological role might, therefore, be doubted. However, in a biological compartment, ˙OH can diffuse no more than approximately 1 nm before reacting with an organic molecule (Griffiths & Lunec, 1996). This span is small compared with the thickness of a primary wall. ˙OH is thus a *site-specific* oxidant: if produced at a given site in the wall matrix or middle lamella, it would be very likely to attack a nearby molecule, e.g. a polysaccharide chain, and thus to promote wall loosening, before reaching the plasma membrane.

The site-specificity of ˙OH production, e.g. at wall-bound Cu atom, could be controlled by the synthesis and secretion of His-rich glycoproteins (Sommer-Knudsen *et al.*, 1997) and other Cu-binding ligands (Graham, 1981; Fry *et al.*, 2002). The initiation and termination of ˙OH production could be controlled by the activity of proteins, e.g. the synthesis, secretion and enzymic oxidation of ascorbate and the re-absorption of dehydroascorbate (Smirnoff & Wheeler, 2000); the enzymic production of $O_2^{\cdot-}$ and H_2O_2 (Morré *et al.*, 1995; Bolwell *et al.*, 2002); the presence of peroxidase; and the removal of H_2O_2. Therefore, non-enzymic polysaccharide scission could be just as tightly controlled by proteins, and thus genes, as any other postulated wall-loosening mechanism such as the action of hydrolases, transglycosylases and expansins.

Acknowledgements

We thank the BBSRC for a grant in support of the previously unpublished work.

References

Al-Madhoun, A.S., Sanmartin, M. and Kanellis, A.K. (2003) 'Expression of ascorbate oxidase isoenzymes in cucurbits and during development and ripening of melon fruit', *Postharvest Biology Technology* **27**, 137–146.

Angelini, R., Manes, F. and Federico, R. (1990) 'Spatial and functional correlation between diamine-oxidase and peroxidase activities and their dependence upon de-etiolation and wounding in chickpea stems', *Planta* **182**, 89–96.

Asthir, B., Duffus, C.M., Smith, R.C. and Spoor, W. (2002) 'Diamine oxidase is involved in H_2O_2 production in the chalazal cells during barley grain filling', *Journal of Experimental Botany* **53**, 677–682.

Berridge, M.V. and Tan, A.S. (1998) 'Trans-plasma membrane electron transport: a cellular assay for NADH-and NADPH-oxidase based on extracellular; superoxide-mediated reduction of the sulfonated tetrazolium salt WST-1', *Protoplasma* **205**, 74–82.

Bestwick, C.S., Brown, I.R. and Mansfield, J.W. (1998) 'Localized changes in peroxidase activity accompany hydrogen peroxide generation during the development of a nonhost hypersensitive reaction in lettuce', *Plant Physiology* **118**, 1067–1078.

Bindschedler, L.V., Minibayeva, F., Gardner, S.L., Gerrish, C., Davies, D.R. and Bolwell, G.P. (2001) 'Early signalling events in the apoplastic oxidative burst in suspension cultured French bean cells involve cAMP and Ca^{2+}', *New Phytologist* **151**, 185–194.

Bolwell, G.P., Bindschedler, L.V., Blee, K.A. *et al.* (2002) 'The apoplastic oxidative burst in response to biotic stress in plants: a three-component system', *Journal of Experimental Botany* **53**, 1367–1376.

Bolwell, G.P., Page, A., Pislewska, M. and Wojtaszek, P. (2001) 'Pathogenic infection and the oxidative defences in plant apoplast', *Protoplasma* **217**, 20–32.

Boudet, A.M. (2003) Towards an understanding of the supramolecular organization of the ligni-fied wall, in *The Plant Cell Wall* (ed. J.K.C. Rose), Blackwell, Oxford, pp. 155–182.

Brady, C.J. (1987) 'Fruit ripening', *Annual Review of Plant Physiology* **38**, 155–178.

Brady, J.D., Sadler, I.H. and Fry, S.C. (1997) 'Pulcherosine, an oxidatively coupled trimer of tyrosine in plant cell walls: its role in cross-link formation', *Phytochemistry* **47**, 349–353.

Brennan, T. and Frenkel, C. (1977) 'Involvement of hydrogen peroxide in the regulation of senescence in pear', *Plant Physiology* **59**, 411–416.

Buxton, G.V., Greenstock, C.L., Helman, W.P. and Ross, A.B. (1988) 'Critical review of rate constants for reactions of hydrated electrons, hydrogen atoms and hydroxyl radicals ($^{\bullet}OH/^{\bullet}O^{-}$) in aqueous solution', *Journal of Physical and Chemical Reference Data* **17**, 513–886.

Caliskan, M., Ozcan, B., Turan, C. and Cuming, A.C. (2004) 'Localization of germin genes and their products in developing wheat coleoptiles', *Journal of Biochemistry and Molecular Biology* **37**, 339–342.

Carrasco-Luna, J., Calatayud, A., González-Darós, F. and del Valle-Tascón, S. (1995) 'Hexacyanoferrate (III) stimulation of elongation in coleoptile segments from *Zea mays* L', *Protoplasma* **184**, 63–71.

Chen, S.X. and Schopfer, P. (1999) 'Hydroxyl-radical production in physiological reactions. A novel function of peroxidase', *European Journal of Biochemistry* **260**, 726–735.

Cohen, R., Jensen, K.A., Houtman, C.J. and Hammerl, K.E. (2002) 'Significant levels of extracellular reactive oxygen species produced by brown rot basidiomycetes on cellulose', *FEBS Letters* **531**, 483–488.

Cona, A., Cenci, F., Cervelli, M. *et al.* (2003) 'Polyamine oxidase, a hydrogen peroxide-producing enzyme, is up-regulated by light and down-regulated by auxin in the outer tissues of the maize mesocotyl', *Plant Physiology* **131**, 803–813.

Córdoba-Pedregosa, M.D., Cordoba, F., Villalba, J.M. and González-Reyes, J.A. (2003) 'Differential distribution of ascorbic acid, peroxidase activity, and hydrogen peroxide along the root axis in *Allium cepa* L. and its possible relationship with cell growth and differentiation', *Protoplasma* **221**, 57–65.

Córdoba-Pedregosa, M.C., González-Reyes, J.A., Serrano, A., Villalba, J.M., Navas, P. and Córdoba, F. (1998) 'Plasmalemma-associated malate dehydrogenase activity in onion root cells', *Protoplasma* **205**, 29–36.

Cutillas-Iturralde, A., Zarra, I. and Lorences, E.P. (1993) 'Metabolism of cell wall polysaccharides from persimmon fruit. Pectin solubilization during fruit ripening occurs in the apparent absence of polygalacturonase activity', *Physiologia Plantarum* **89**, 369–375.

Dawson, R.M.C., Elliott, D.C., Elliott, W.H. and Jones, K.M. (1986) *Data for Biochemical Research*, Clarendon, Oxford, UK.

De Agazio, M. and Zacchini, M. (2001) 'Dimethylthiourea, a hydrogen peroxide trap, partially prevents stress effects and ascorbate peroxidase increase in spermidine-treated maize roots', *Plant Cell and Environment* **24**, 237–244.

Deeble, D.J., Bothe, E., Schuchmann, H.P., Parsons, B.J., Phillips, G.O. and von Sonntag, C. (1990) 'The kinetics of hydroxyl-radical-induced strand breakage of hyaluronic acid. A pulse radioly-sis study using conductometry and laser-light-scattering', *Zeitschrift für Naturforschung* **45c**, 1031–1043.

Doke, N. (1985) 'NADPH-dependent O_2^- generation in membrane fractions isolated from wounded potato tubers inoculated with *Phytophtora infestans*', *Physiological Plant Pathology* **27**, 311–322.

Doke, N. and Miura, Y. (1995) '*In vitro* activation of NADPH-dependent O_2^- generating system in a plasma membrane-rich fraction of potato tuber tissues by treatment with an elicitor from *Phytophthora infestans* or with digitonin', *Physiological and Molecular Plant Pathology* **46**, 17–28.

Donaldson, P.A., Anderson, T., Lane, B.G., Davidson, A.L. and Simmonds, D.H. (2001) 'Soybean plants expressing an active oligomeric oxalate oxidase from the wheat *gf-2.8* (germin) gene are resistant to the oxalate-secreting pathogen *Sclerotina sclerotiorum*', *Physiological and Molecular Plant Pathology* **59**, 297–307.

Dumville, J.C. and Fry, S.C. (2003). 'Solubilisation of tomato fruit pectins by ascorbate: a possible non-enzymic mechanism of fruit softening', *Planta* **217**, 951–961.

Elstner, E.F. (1987) Metabolism of activated oxygen species, in *The Biochemistry of Plants, A Comprehensive Treatise*, vol. 11 (ed. D.D. Davies), Academic Press, San Diego, pp. 253–315.

Epperlein, M.M., Noronhadutra, A.A. and Strange, R.N. (1986) 'Involvement of the hydroxyl radical in the abiotic elicitation of phytoalexins in legumes', *Physiological and Molecular Plant Pathology* **28**, 67–77.

Epstein, L. and Lamport, D.T.A. (1984) 'An intramolecular linkage involving isodityrosine in extensin', *Phytochemistry* **23**, 1241–1246.

Federico, R. and Angelini, R. (1988) 'Distribution of polyamines and their related catabolic enzymes in etiolated and light-grown leguminosae seedlings', *Planta* **173**, 317–321.

Foreman, J., Demidchik, V., Bothwell, J.H.F. *et al.* (2003) 'Reactive oxygen species produced by NADPH oxidase regulate plant cell growth', *Nature* **422**, 442–446.

Frahry, G. and Schopfer, P. (1998) 'Inhibition of O_2-reducing activity of horseradish peroxidase by diphenyleneiodonium', *Phytochemistry* **48**, 223–227.

Frahry, G. and Schopfer, P. (2001) 'NADH-stimulated, cyanide-resistant superoxide production in maize coleoptiles analyzed with a tetrazolium-based assay', *Planta* **212**, 175–183.

Fry, S.C. (1998) 'Oxidative scission of plant cell wall polysaccharides by ascorbate-induced hydroxyl radicals', *Biochemical Journal* **332**, 507–515.

Fry, S.C., Dumville, J.C. and Miller, J.G. (2001) 'Fingerprinting of polysaccharides attacked by hydroxyl radicals *in vitro* and in the cell walls of ripening pear fruit', *Biochemical Journal* **357**, 729–737.

Fry, S.C., Miller, J.G. and Dumville, J.C. (2002) 'A proposed role for copper ions in cell wall loosening', *Plant and Soil* **247**, 57–67.

Fry, S.C., Willis, S.C. and Paterson, A.E.J. (2000) 'Introprotoplastic and wall-localised formation of arabinoxylan-bound diferulates and larger ferulate coupling-products in maize cell-suspension cultures', *Planta* **211**, 679–692.

Gabriel, R. and Kesselmeier, J. (1999) 'Apoplastic solute concentrations of organic acids and mineral nutrients in the leaves of several fagaceae', *Plant Cell Physiology* **40**, 604–612.

Ghiselli, A. (1998) Aromatic hydroxylation: salicylic acid as a probe for measuring hydroxyl radical production, in *Free Radical and Antioxidant Protocols* (ed. D. Armstrong), Humana Press, Totowa, New Jersey, pp. 89–100.

Gómez, L.D., Casano, L.M. and Trippi, V.S. (1995) 'Effect of hydrogen peroxide on degreadation of cell wall associated proteins in growing bean hypocotyls', *Plant Cell Physiology* **36**, 1259–1264.

Graham, R.D. (1981) *Copper in Soils and Plants* (ed. J.F. Loneragan, A.D. Robson and R.D. Graham), Academic Press, Sydney, pp. 141–163.

Grant, J.J. and Loake, G.J. (2000) 'Role of reactive oxygen intermediates and cognate redox signaling in disease resistance', *Plant Physiology* **124**, 21–29.

Grant, J.J., Yun, B.W. and Loake, G.J. (2000) 'Oxidative burst and cognate redox signalling reported by luciferase imaging: identification of a signal network that functions independently of ethylene, SA and Me-JA but is dependent on MAPKK activity', *The Plant Journal* **24**, 569–582.

Green, M.A. and Fry, S.C. (2005) 'Vitamin C degradation in plant cells via enzymatic hydrolysis of 4-*O*-oxalyl-L-threonate', *Nature* **433**, 83–87.

Griffiths, H.R. and Lunec, J. (1996) Investigating the effects of oxygen free radicals on carbohydrates in biological systems, in *Free Radicals: A Practical Approach* (ed. N.A. Punchard and F.J. Kelly), IRL Press, Oxford, pp. 185–200.

Grootveld, M. and Halliwell, B. (1986) 'Aromatic hydroxylation as a potential measure of hydroxyl-radical formation *in vivo*', *Biochemical Journal* **237**, 499–504.

Gross, G.G. (1977) 'Cell wall-bound malate dehydrogenase from horseradish', *Phytochemistry* **16**, 319–321.

Gross, G.G., Janse, C. and Elstner, E.F. (1977) 'Involvement of malate, monophenols, and the superoxide radical in hydrogen peroxide formation by isolated cell walls from horseradish (*Armoracia lapthifolia* Gilib.)', *Planta* **136**, 271–276.

Halliwell, B. (1978) 'Lignin synthesis: the generation of hydrogen peroxide and superoxide by horseradish peroxidase and its stimulation by manganese (II) and phenols', *Planta* **140**, 81–88.

Halliwell, B. and Gutteridge, J.M.C. (1990) 'Role of free radicals and catalytic metal ions in human disease: an overview', *Methods in Enzymology* **189**, 1–85.

Halliwell, B. and Gutteridge, J.M.C. (1999) *Free Radicals in Biology and Medicine*, 3rd edn., Oxford Science Publications, Oxford, UK.

Halliwell, B. and Kaur, H. (1997) 'Hydroxylation of salicylate and phenylalanine as assays for hydroxyl radicals: a cautionary note visited for the third time', *Free Radical Research* **27**, 239.

Hatfield, R.D. and Ralph, J. (1999) 'Modelling the feasibility of intramolecular dehydrodiferulate formation in grass walls', *Journal of the Science of Food and Agriculture* **79**, 425–427.

Hidalgo, A., Garcia-Herdugo, G., Gonzalez-Reyes, J.A., Morré, D.J. and Navas, P. (1991) 'Ascorbate free radical stimulates onion root growth by increasing cell elongation', *Botanical Gazette* **152**, 282–288.

Ishida, A., Ookubo, K. and Ono, K. (1987) 'Formation of hydrogen peroxide by NAD(P)H oxidation with isolated cell wall-associated peroxidase from cultured liverwort cells, *Marchantia polymorpha* L', *Plant Cell Physiology* **28**, 723–726.

Jiménez, A., Creissen, G., Kular, B. *et al.* (2002) 'Changes in oxidative processes and components of the antioxidant system during tomato fruit ripening', *Planta* **214**, 751–758.

Johnson, S.M., Dorhery, S.J. and Croy, R.R.D. (2003) 'Biphasic superoxide generation in potato tubers. A self-amplifying response to stress', *Plant Physiology* **131**, 1440–1449.

Kato, N. and Esaka, M. (1996) 'cDNA cloning and gene expression of ascorbate oxidase in tobaccco', *Plant Molecular Biology* **30**, 833–837.

Kerr, E.M. and Fry, S.C. (2003) 'Pre-formed xyloglucans increase in molecular weight in three distinct compartments of a maize cell-suspension culture', *Planta* **217**, 327–339.

Kerr, E.M. and Fry, S.C. (2004) 'Extracellular cross-linking of xylan and xyloglucan in maize cell-suspension cultures: the role of oxidative phenolic coupling', *Planta* **219**, 73–83.

Kiba, A., Miyake, C., Toyoda, K., Ichinose, Y., Yamada, T. and Shiraishi, T. (1997) 'Superoxide generation in extracts from isolated plant cell walls is regulated by fungal signal molecules', *Phytopathology* **87**, 846–852.

Kim, H.J., Pesacreta, T.C. and Triplett, B.A. (2004) 'Cotton-fiber germin-like protein. II: Immunolocalization, purification, and functional analysis', *Planta* **218**, 525–535.

Kuchitsu, K., Kosaka, H., Shiga, T. and Shibuya, N. (1995) 'EPR evidence for generation of hydroxyl radical triggered by *N*-acetylchitooligosaccharide elicitor and a protein phosphatase inhibitor in suspension-cultured rice cells', *Protoplasma* **188**, 138–142.

Lane, B.G., Dunwell, J.M., Ray, J.A., Schmitt, M.R. and Cuming, A.C. (1993) 'Germin, a protein marker of early plant development, is an oxalate oxidase', *Journal of Biological Chemistry* **268**, 12239–12242.

Le Deunff, E., Davoine, C., Le Dantec, C., Billard, J.P. and Huault, C. (2004) 'Oxidative burst and expression of *germin/oxo* genes during wounding of ryegrass leaf blades: comparison with senescence of leaf sheaths', *The Plant Journal* **38**, 421–431.

Li, Y., Zhu, H., Kuppusamy, P., Roubaud, V., Zweier, J.L. and Trush M.A. (1998) 'Validation of lucigenin (bis-*N*-methylacridinium) as a chemilumigenic probe for detecting superoxide anion radical production by enzymatic and cellular systems', *Journal of Biological Chemistry* **273**, 2015–2023.

Lin, W. (1982) 'Response of corn root protoplasts to exogenous reduced nicotinamide adenine dinucleotide: oxygen consumption, ion uptake, and membrane potential', *Proceedings of the National Academy of Sciences USA* **79**, 3773–3776.

Lin, C.C. and Kao, C.H. (2001) 'Cell wall peroxidase activity, hydrogen peroxide level and NaCl-inhibited root growth of rice seedlings', *Plant and Soil* **230**, 135–143.

Lin, L.S. and Varner, J.E. (1991) 'Expression of ascorbic acid oxidase in zucchini squash (*Cucurbita pepo* L.)', *Plant Physiology* **96**, 159–165.

Liskay, A., Kenk, B. and Schopfer, P. (2003) 'Evidence for the involvement of cell wall peroxidase in the generation of hydroxyl radicals mediating extension growth', *Planta* **217**, 658–667.

Liu, L., Eriksson, K.E.L. and Dean, J.F.D. (1995) 'Localization of hydrogen peroxide production in *Pisum sativum* L. using epi-polarization microscopy to follow cerium perhydroxide deposition', *Plant Physiology* **107**, 501–506.

Miller, J.G. and Fry, S.C. (2001) 'Characteristics of xyloglucan after attack by hydroxyl radicals', *Carbohydrate Research* **332**, 389–403.

Miller, J.G. and Fry, S.C. (2004) '*N*-[^{3}H]Benzoylglycylglycylglycine as a probe for hydroxyl radicals', *Analytical Biochemistry* **335**, 126–134.

Møller, S.G. and McPherson, M.J. (1998) 'Developmental expression and biochemical analysis of the *Arabidopsis* atao1 gene encoding an H_2O_2-generating diamine oxidase', *The Plant Journal* **13**, 781–791.

Morré, D.J., Brightman, A.O., Hidalgo, A. and Navas, P. (1995) 'Selective inhibition of auxin-stimulated NADH oxidase activity and elongation growth of soybean hypocotyls by thiol reagents', *Plant Physiology* **107**, 1285–1291.

Murphy, T.M. and Auh, C.K. (1996) 'The superoxide synthase of plasma membrane preparations for cultured rose cells', *Plant Physiology* **110**, 621–629.

Murphy, T.M., Vu, H. and Nguyen, T. (1998) 'The superoxide synthase of rose cells. Comparison of assays', *Plant Physiology* **117**, 1301–1305.

Nogata, Y., Ohta, H. and Voragen, A.G.J. (1993) 'Polygalacturonase in strawberry fruit', *Phytochemistry* **34**, 617–620.

Ogawa, K., Kanematsu, S. and Asada, K. (1996) 'Intra- and extra-cellular localization of "cytosolic" CuZn-superoxide dismutase in spinach leaf and hypocotyl', *Plant Cell Physiology* **37**, 790–799.

Ogawa, K., Kanematsu, S. and Asada, K. (1997) 'Generation of superoxide anion and localization of CuZn-superoxide dismutase in the vascular tissue of spinach hypocotyls: their association with lignification', *Plant Cell Physiology* **38**, 1118–1126.

Olson, P.D. and Varner, J.E. (1993) 'Hydrogen peroxide and lignification', *The Plant Journal* **4**, 887–892.

Otter, T. and Pollę, A. (1997) 'Characterisation of acidic and basic apoplastic peroxidases from needles of Norway spruce (*Picea abies* L., Karsten) wih respect to lignifying substrates', *Plant Cell Physiology* **38**, 595–602.

Papadakis, A.K. and Roubelakis-Angelakis, K.A. (1999) 'The generation of active oxygen species differs in tobacco and grapevine mesophyll protoplasts', *Plant Physiology* **121**, 197–205.

Pathak, N., Mishra, S. and Sanwal, G.G. (2000) 'Purification and characterization of polygalac-turonase from banana fruit', *Phytochemistry* **54**, 147–152.

Pinton, R., Cakmack, I. and Marschner, H. (1994) 'Zinc deficiency enhanced NAD(P)H-dependent superoxide radical production in plama membrane vesicles isolated from roots of bean plants', *Journal of Experimental Botany* **45**, 45–50.

Rea, G., de Pinto, M.C., Tavazza, R. *et al.* (2004) 'Ectopic expression of maize polyamine oxidase and pea copper amine oxidase in the cell wall of tobacco plants', *Plant Physiology* **134**, 1414–1426.

Redgwell, R.J. and Fischer, M. (2002) Fruit texture, cell wall metabolism and consumer percep-tions, in *Fruit Quality and its Biological Basis* (ed. M. Knee), Sheffield Academic Press, Sheffield, UK, pp. 46–88.

Rodríguez, A.A., Cordoba, A.R., Ortega, L. and Taleisnik, E. (2004) 'Decreased reactive oxygen species concentration in the elongation zone contributes to the reduction in maize leaf growth under salinity', *Journal of Experimental Botany* **55**, 1383–1390.

Rodríguez, A.A., Grunberg, K.A. and Taleisnik, E.L. (2002) 'Reactive oxygen species in the elongation zone of maize leaves are necessary for leaf extension', *Plant Physiology* **129**, 1627–1632.

Rodríguez-Lopez, M., Baroja-Fernandez, E., Zandueta-Criado, A. *et al.* (2001) 'Two isoforms of a nucleotide-sugar pyrophosphatase/phosphodiesterase from barley leaves (*Hordeum vulgare* L.) are distinct oligomers of HvGLP1, a germin-like protein', *FEBS Letters* **490**, 44–48.

Rouau, X., Cheynier, V., Surget, A. *et al.* (2003) 'A dehydrotrimer of ferulic acid from maize bran', *Phytochemistry* **63**, 899–903.

Schopfer, P. (2001) 'Hydroxyl radical-induced cell-wall loosening *in vitro* and *in vivo*: implica-tions for the control of elongation growth', *The Plant Journal* **28**, 679–688.

Schopfer, P., Liszkay, A., Bechtold, M., Frahry, G. and Wagner, A. (2002) 'Evidence that hydroxyl radicals mediate auxin-induced extension growth', *Planta* **214**, 821–828.

Schopfer, P., Plachy, C. and Frahry, G. (2001) 'Release of reactive oxygen intermediates (super-oxide radicals, hydrogen peroxide and hydroxyl radicals) and peroxidase in germinating radish seeds controlled by light, gibberellin and abscisic acid', *Plant Physiology* **125**, 1591–1602.

Schuchmann, M.N. and von Sonntag, C. (1978) 'The effect of oxygen on the OH-radical-induced scission of the glycosidic linkage of cellobiose', *International Journal of Radiation Biology* **34**, 397–400.

Setsukinai, K.I., Urano, Y., Kakinuma, K., Majima, H.J. and Nagno, T. (2003) 'Development of novel fluorescence probes that can reliably detect reactive oxygen species and distinguish specific species', *Journal of Biological Chemistry* **278**, 3170–3175.

Shinkle, J.R., Swoap, S.J., Simon, P. and Jones, R.L. (1992) 'Cell wall free space of *Cucumis* hypocotyls contains NAD and a blue light-regulated peroxidase activity', *Plant Physiology* **98**, 1336–1341.

Smirnoff, N. and Wheeler, G.L. (2000) 'Ascorbic acid in plants: biosynthesis and function', *Critical Review of Biochemical and Molecular Biology* **35**, 291–314.

Smith, C.J.S., Watson, C.F., Morris, P.C. *et al.* (1990) 'Inheritance and effect on ripening of antisense polygalacturonase genes in transgenic tomatoes', *Plant Molecular Biology* **14**, 369–379.

Sommer-Knudsen, J., Bacic, A. and Clarke, A.E. (1997) 'A metal-binding, cell-wall glycoprotein from *Nicotiana alata*', *Journal of Experimental Botany* **48** (Suppl), 102.

Streller, S. and Wingsle, G. (1994) '*Pinus sylvestris* L. needles contain extracellular CuZn superoxide dismutase', *Planta* **192**, 195–201.

Sutherland, M.W. and Learmonth, B.A. (1997) 'The tetrazolium dyes MTS and XTT provide new quantitative assays for superoxide and superoxide dismutase', *Free Radical Research* **27**, 283–289.

Takahama, U. and Oniki, T. (1992) 'Regulation of peroxidase-dependent oxidation of phenolics in the apoplast of spinach leaves by ascorbate', *Plant Cell Physiology* **33**, 379–387.

Tamás, L., Simonovicova, M., Huttova, J. and Mistrik, I. (2004) 'Aluminium stimulated hydrogen peroxide production of germinating barley seeds', *Environmental and Experimental Botany* **51**, 281–288.

Taubert, D., Breitnback, T., Lazar, A. *et al.* (2003) 'Reaction rate constants of superoxide scavenging by plant antioxidants', *Free Radical Biology and Medicine* **35**, 1599–1607.

Tokita, Y., Ohshima, K. and Okamoto, A. (1997) 'Degradation of hyaluronic acid during freeze drying', *Polymer Degradation Stability* **55**, 159–164.

van Cutsem, P. and Gillet, C. (1982) 'Activity coefficients and selectivity values of Cu^{++}, Zn^{++} and Ca^{++} ions adsorbed in the *Nitella flexilis* L. cell wall during triangular ion exchanges', *Journal of Experimental Botany* **33**, 847–853.

van Gestelen, P., Asard, H., Horeman, N. and Caubergs, R.J. (1998) 'Superoxide-producing NAD(P)H oxidases in plasma membrane vesicles from elicitor responsive bean plants', *Physiologia Plantarum* **104**, 653–660.

von Sonntag, C. (1987) *The Chemical Basis of Radiation Biology*, Taylor and Francis, London.

von Tiedemann, A. (1997) 'Evidence for a primary role of active oxygen species in induction of host cell death during infection of bean leaves with *Botrytis cinerea*', *Physiological and Molecular Plant Pathology* **50**, 151–166.

Wisniewski, J.P., Rathburn, E.A., Knox, J.P. and Brewin, N.J. (2000) 'Involvement of diamine oxidase and peroxidase in insolubilization of the extracellular matrix: implications for pea nodule initiation by *Rhizobium leguminosarum*', *Molecular Plant–Microbe Interactions* **13**, 413–420.

Wojtaszek, P. (1997) 'Oxidative burst: an early plant response to pathogen infection', *Biochemical Journal* **322**, 681–692.

Woo, E.J., Dunwell, J.M., Goodenough, P.W., Marvier, A.C. and Pickersgill, R.W. (2000) 'Germin is a manganese containing homohexamer with oxalate oxidase and superoxide dismutase activities', *Nature Structural Biology* **7**, 1036–1040.

Xing, T., Higgins, V.J. and Blumwald, E. (1997) 'Race-specific elicitors of *Cladosporium fulvum* promote translocation of cytosolic components of NADPH oxidase to the plasma membrane of tomato cells', *The Plant Cell* **9**, 249–259.

Yoda, H., Yamaguchi, Y. and Sano, H. (2003) 'Induction of hypersensitive cell death by hydrogen peroxide produced through polyamine degradation in tobacco plants', *Plant Physiology* **132**, 1973–1981.

Zegota, H. (2002) 'Some quantitative aspects of hydroxyl radical induced reactions in γ-irradiated aqueous solutions of pectins', *Food Hydrocolloids* **16**, 353–361.

10 Reactive oxygen species and photosynthesis

Barry A. Logan

10.1 Introduction

The photosynthetic pathway enables green plants to use sunlight to generate the chemical energy and reductant necessary to undertake the thermodynamically unfavorable reactions that are involved in synthesizing sugars from CO_2 via the Calvin cycle. Water oxidation provides the electrons consumed in sugar synthesis, yielding molecular diatomic oxygen (O_2) as a by-product. Much of this O_2 diffuses away from the leaf passively through the stomata. During illumination, however, photosynthetic cells can possess very high O_2 concentrations. The proximity of high O_2 concentrations and the energy and electron transfer events of light harvesting and photosynthetic electron transport render chloroplasts exceptionally susceptible to oxidative damage. In all but the deepest shade, the photosynthetic pathway is likely to act as a primary source of reactive oxygen species (ROS) in disease-free leaves.

Plants are not at the mercy of the macromolecular damage that can be inflicted by ROS. Acclimatory adjustments in leaf morphology and the contents of constituents that make up the photosynthetic pathway preserve the balance between light absorption and photosynthetic light use and thus limit the absorption of problematic 'excess light'. In addition, biochemical mechanisms that fall under the general heading 'photoprotection' minimize ROS production and detoxify those that form. Photoprotective mechanisms include thermal energy dissipation, which safely eliminates excess light by converting it to heat, and antioxidation via a complex array of enzymes and redox molecules, which convert superoxide to water. Both thermal energy dissipation and antioxidation are woven into the regulatory regimes of the chloroplast. This chapter reviews the photogeneration of ROS and insights into the molecular mechanisms of thermal energy dissipation and antioxidation, as well as their acclimation to prevailing environmental conditions.

10.2 Light absorption and allocation

10.2.1 Triplet-chlorophyll mediated singlet O_2 formation

The photobiophysics of chlorophyll render it well suited for its role in supporting the energy needs of photosynthesis. Chlorophyll absorbs wavelengths which are among the most intense that arrive at the earth's surface from the sun (Taiz & Zeiger, 2002). Unlike some pigments, such as carotenoids, chlorophyll does not readily undergo thermal de-excitation. Therefore, chlorophyll can be maintained in the excited state long enough for electron transfer and charge separation, which take place on far slower time-scales, to occur. However, these same features of chlorophyll create problems when light absorption exceeds photosynthetic light utilization. The absorption of excess light is a regular occurrence for most plants in the field, given that the biochemical reactions that make up the Calvin cycle saturate at light intensities well below full sunlight in most plants. When excitation energy is not consumed by photosynthesis, the lifetime of singlet-excited state chlorophyll in the light harvesting antennae increases. This increases the probability that chlorophyll will undergo an intersystem crossing in which the spin of the excited electron 'flips', forming triplet chlorophyll (Foote, 1976). Molecular diatomic oxygen possesses the unusual feature of being triplet in the ground state, which enables it to accept energy from triplet chlorophyll. Singlet O_2 results from such an energy transfer event (Asada, 1996; Niyogi, 1999). Singlet O_2 is a highly unstable ROS that damages macromolecules, primarily via addition reactions that yield endoperoxides (Halliwell & Gutteridge, 1999). Plants reduce singlet-O_2-mediated cellular damage via an exquisitely responsive and elegant protective mechanism known as thermal energy dissipation (reviewed in Demmig-Adams & Adams, 1992, 1996; Niyogi, 1999).

10.2.2 Thermal energy dissipation

Thermal energy dissipation, or 'feedback de-excitation', protects plants from the dangers of excess light absorption by safely converting excitation energy to heat before it can sensitize singlet O_2 formation. Thus, thermal energy dissipation can be thought of as a proactive protective mechanism, intervening to prevent the formation of ROS rather than detoxifying existing ROS, as in the case of antioxidation. Over the last two decades, the research tools of ecophysiology and plant genetics have revealed much about the mechanism of thermal energy dissipation, although significant gaps in our understanding remain (Holt et al., 2004).

Thermal energy dissipation requires the presence of: (i) a low thylakoid lumen pH; (ii) de-epoxidized carotenoids of the xanthophyll cycle [zeaxanthin (Z) or antheraxanthin (A)] and (iii) PsbS, a minor light harvesting complex protein (Gilmore & Yamamoto, 1993a,b; Li et al., 2000). A low thylakoid lumen pH presumably develops when proton pumping via photosynthetic electron transport outpaces Calvin cycle activity under conditions of excess light. Z and A are formed from the di-epoxidized member of the xanthophyll cycle, violaxanthin (V), by the activity of violaxanthin de-epoxidase (VDE) (Figure 10.1). VDE is localized to the thylakoid lumen and utilizes ascorbic acid (as opposed to the ascorbate anion) as a reductant (Bratt et al., 1995). This requirement for the protonated form of ascorbic acid imposes a low pH optimum on VDE and can be interpreted as one means by which plants regulate levels of thermal energy dissipation to meet their needs, but not exceed them. Zeaxanthin epoxidase, a stromal enzyme, catalyzes the formation of V from Z and A.

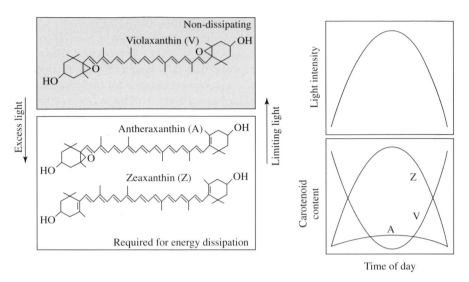

Figure 10.1 Molecular structures of the carotenoids of the xanthophyll cycle and a schematic depiction of diurnal changes in xanthophyll cycle composition in a full-sun exposed horizontally oriented leaf.

The mechanism by which Z and A potentiate thermal energy dissipation remains in question. Some evidence suggests that these carotenoids play an indirect role by facilitating chlorophyll: chlorophyll self-quenching via internal conversion (see Horton *et al.*, 1996, 2000). However, models that favor a direct role for de-epoxidized carotenoids are gaining support. Z (and perhaps A) possesses an S_1 excited state that is below that of chlorophyll (Frank *et al.*, 1994; Owens, 1997; Ma *et al.*, 2003). Thus, it is thermodynamically possible for singlet-excited state chlorophyll to donate its energy to Z. Z, like all carotenoids, then readily undergoes thermal de-excitation. Alternatively, Holt *et al.* (2005) observed the formation of a carotenoid radical cation after excitation of chlorophyll and suggest that thermal energy dissipation occurs through energy transfer to a chlorophyll-zeaxanthin heterodimer which then undergoes charge separation.

Using a forward genetics approach that selected *Arabidopsis* mutants with aberrant chlorophyll fluorescence quenching, Li *et al.* (2000) determined that the minor light harvesting complex protein PsbS [which is also referred to as CP22 (Funk *et al.*, 1994)] is required for thermal energy dissipation. Transgenic overexpression of PsbS was subsequently shown to lead to enhanced rates of thermal energy dissipation (Li *et al.*, 2002). PsbS contains four transmembane helices and is a member of the Light Harvesting Complex protein superfamily (Li *et al.*, 2000). The two lumenal loops that connect helices each contain glutamate residues that are critical for thermal energy dissipation. Conversion of these glutamates to glutamines via site-directed mutagenesis leads to a loss of thermal energy dissipation; conversion of one glutamate leads to partial loss (Li *et al.*, 2004). Taken together, these observations strongly suggest that a low lumenal pH is sensed via protonation of these glutamates. Protonation, presumably, then brings about a conformational change in PsbS, and possibly the light harvesting complex overall, that puts it in an energy dissipating state. This model is consistent with the requirement for a low lumenal pH not only for VDE activity, but also for energy dissipation itself. Although it remains to be demonstrated, it is possible that acidification leads to Z binding and/or brings chlorophyll and Z into positions that permit resonance energy or electron transfer.

10.2.3   Acclimation of the light harvesting complex to the environment

Plants can adjust levels of thermal energy dissipation over very short time-scales (seconds) and their capacities for thermal energy dissipation over somewhat longer time-scales (days–weeks). Short-term adjustments in thermal energy dissipation are mediated primarily through conversions among the pigments of the xanthophyll cycle. Over the course of a day, plants accumulate Z at the expense of V until peak irradiance near midday and then undergo reconversion of Z to V as the sun sets (Adams *et al.*, 1992) (Figure 10.1). The extent of this conversion correlates with the intensity of the growth light environment (Figure 10.2). Thus, regulation of thermal energy dissipation is elegant in its speed and metabolic efficiency. Plants adjust their capacities for thermal energy dissipation in response to their environment by altering the size of their xanthophyll cycle pool and also their expression of PsbS. Full sun-acclimated plants possess xanthophyll cycle pools [typically expressed as millimoles $V + A + Z$ (mole chl $a + b)^{-1}$] that are several-fold larger than those of shade-acclimated plants (Grace & Logan, 1996; Logan *et al.*, 1996) (Figure 10.2). Furthermore, since photosynthetic light utilization influences the level of excess light absorption, full-sun acclimated plants with lower photosynthetic capacities possess larger xanthophyll cycle pools than plants in the same environment possessing higher photosynthetic capacities (Logan *et al.*, 1998a). Environmental stresses such as chilling and nutrient deficiency reduce photosynthetic light utilization, consequently increase excess light absorption, and tend to bring about an increase in xanthophyll cycle pool size and midday conversion to the dissipating forms Z and A (Adams & Demmig-Adams, 1994; Verhoeven *et al.*, 1996, 1997; Logan *et al.*, 1998c; Burkle & Logan, 2003) (Figure 10.2). The level of PsbS expression also acclimates to the environment. Seasonally colder temperatures and high growth light intensities lead to increased PsbS contents (Ottander *et al.*, 1995; Ebbert *et al.* (2005); B. Logan & K. Niyogi, unpublished data).

During winter, many long-lived evergreen plants greatly reduce photosynthetic capacity and maintain high levels of thermal energy dissipation (Adams & Demmig-Adams, 1994; Logan *et al.*, 1998c; Öquist & Huner, 2003). Thermal energy dissipation in winter does not exhibit patterns of diurnal responsiveness to incident sunlight that one observes during the growing season; rather, it remains engaged through the low-light and dark periods and is associated with the nocturnal retention of Z and A. Interestingly, although sustained thermal energy dissipation in

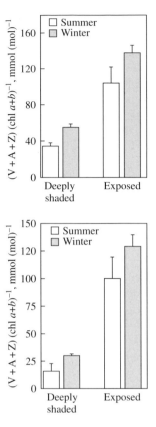

Figure 10.2 Xanthophyll cycle pool size (upper panel) and the conversion state of the xanthophyll cycle at midday (lower panel), both expressed per total chlorophyll, for deeply shaded and exposed populations of *Mahonia repens* in summer (open bars) and winter (closed bars). These plants were found growing in Skunk Canyon in the eastern foothills of the Colorado Rocky Mountains. Error bars represent standard errors of the mean; n = 3–6. V + A + Z = violaxanthin + antheraxanthin + zeaxanthin. Modified from Logan *et al.* (1998c).

overwintering evergreens involves large increases in xanthophyll cycle pool size (per unit chlorophyll) (Adams & Demmig-Adams, 1994) and up-regulation of PsbS content (Ottander *et al.*, 1995), it is not associated with the preservation of a low lumenal pH. Therefore, energy dissipation in winter must involve a mechanism that does not require conformational changes to PsbS driven by protonation during lumen acidification. Gilmore and Ball (2000) demonstrated the existence of features in the 77-K fluorescence spectrum of *Eucalyptus pauciflora* that are unique to the winter-acclimated state. They named the most prominent feature the 'cold-hard band' (Gilmore & Ball, 2000). Thus, it appears that *E. pauciflora* and probably other long-lived evergreens (although this remains to be demonstrated) massively reconfigure their light harvesting complexes such that they can be 'stored' in a continuously dissipating state over winter.

10.3   Singlet O_2 generation at photosystem II

In addition to the antenna processes described above, photosystem II (PSII) reaction centers, themselves, can be a source of potentially harmful ROS. The PSII reaction center core proteins (D1 and D2) bind electron carriers that mediate electron transfer from the reaction center chlorophyll known as P680, through pheophytin (Phe), to a stably bound quinone accepter known as Q_A, and ultimately to a reversibly bound quinone acceptor known as Q_B (Andersson & Barber, 1996; Melis, 1999). Once reduced, Q_B diffuses within the membrane to the cytochrome b/f complex, which bridges electron transfer events at PSII and photosystem I (PSI). Conditions of environmental stress that limit photosynthetic utilization of reducing equivalents can lead to an over-reduction of the Q_A pool (Huner et al., 1998; Melis, 1999). If electron transfer is driven by further light absorption while Q_A is reduced, Phe^- can undergo charge recombination with $P680^+$, which results in the formation of triplet P680 (Andersson & Barber, 1996; Asada, 1996; Melis, 1999). Triplet P680 is capable of sensitizing oxidative damage directly, but also readily reacts with ground-state O_2 to form singlet O_2. Under conditions of stress, the kinetics of D1 protein degradation and singlet O_2 production correlate closely, supporting a leading role for singlet O_2 in oxidative damage (Hideg et al., 1994).

Recent findings suggest that, under some circumstances, the effects of singlet O_2 are best attributed to its ability to trigger stress-response signal transduction cascades and not its direct toxic potential. op den Camp et al. (2003) showed that Arabidopsis flu mutants accumulate protochlorophyllide – a potent photosensitizer of singlet O_2 production. When exposed to light, flu mutants generated measurable singlet O_2, growth was inhibited and necrotic lesions appeared on the leaves. Interestingly, specific membrane breakdown products appeared (i.e. stereospecific hydroxyoctadecatienoic acid), suggesting that membrane peroxidation was primarily via enzymatic oxidation of linolenic acid, and not direct, non-specific reactions between membrane lipids and singlet O_2. Furthermore, a suite of early stress-response genes was selectively expressed.

The results of the study by op den Camp et al. (2003) and those by Hideg et al. (1994) and others that report a direct role for singlet O_2 can be reconciled by the possibility that the rate and location of singlet O_2 production differs in flu mutants versus wild-type plants. Flu mutants generate singlet O_2 at appreciably greater rates than wild-type plants (op den Camp et al., 2003). In addition, in wild-type plants, there is evidence to suggest that singlet O_2 generated at PSII rarely escapes the protein matrix of the reaction center before rendering damage (Andersson & Barber, 1996). Thus, the effect of singlet O_2 production on signal transduction in flu mutants may be exaggerated relative to that of wild-type plants by its magnitude and access to the aqueous phase of the cell.

10.4   Electron transport and O_2 photoreduction

10.4.1   The water–water cycle

PSI is capable of utilizing O_2 as an electron acceptor, resulting in the formation of superoxide, the one-electron reduction product. In plants experiencing environmental conditions to which they are acclimated, much of the superoxide that is generated via O_2 photoreduction at PSI is safely reduced further to form water. This detoxification is accomplished by a pathway referred to as the 'water–water cycle' (Asada, 1999) (Figure 10.3). It earns its name from the fact that water is the source of electrons (via water oxidation at the Oxygen Evolving Complex of PSII) as well as the final product of the pathway. Hence, it can be thought of as a quasi-futile cycle that serves only to consume reducing equivalents. The water–water cycle is comprised of enzymatic and non-enzymatic reactions and some degree of redundancy. This has made it very difficult to model fluxes through the pathway (see Polle, 2001). To further complicate matters, it may be that different constituents in the pathway are emphasized under different environmental conditions (i.e. non-enzymatic reactions may play a greater role at lower temperatures; see later).

Photogenerated superoxide can disproportionate into hydrogen peroxide (H_2O_2) and O_2 non-enzymatically; however, in the chloroplast, this reaction is catalyzed by superoxide dismutase (SOD) (McCord & Fridovich, 1969). The chloroplast possesses stromal and thylakoid-associated isoforms of SOD which employ various metal co-factors, including a thylakoid-associated CuZn-SOD and a stromal Fe-SOD (Kurepa et al., 1997; Asada, 1999).

H_2O_2 is converted to water via the activity ascorbate peroxidase (APX), also yielding the one-electron oxidation product of ascorbate, monodehydroascorbate (MDA) (Jablonski & Anderson, 1982). Although H_2O_2 is less reactive than superoxide,

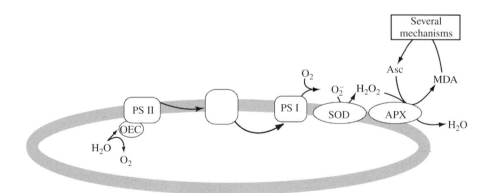

Figure 10.3 Electron flow through the water–water cycle. APX = ascorbate peroxidase, Asc = ascorbate, MDA = monodehydroascorbate radical, OEC = oxygen evolving complex, SOD = superoxide dismutase.

its removal from the chloroplast is critical, as H_2O_2 can deactivate thiol-activated *bis*phosphatases of the Calvin cycle (Charles & Halliwell, 1981). In addition, in the presence of superoxide to maintain transition-metal cations in the reduced state, H_2O_2 can decompose into the hydroxyl radical via the Fenton reaction (Halliwell & Gutteridge, 1999). The hydroxyl radical is among the most powerful oxidizing agents ever detected in biological systems (Buettner, 1993) and is thought to damage cellular constituents at 'diffusion-controlled' rates.

Immunolabeling experiments suggest that thylakoid isoforms of APX (and even stromal forms to some degree) co-localize with PSI reaction centers. In addition, CuZn-SOD and APX have been found in equimolar concentrations with PSI. This evidence led Asada to suggest the existence of a thylakoid-associated PSI/CuZn-SOD/APX super-enzyme complex (Asada, 1996). Such a complex could further reduce ROS-mediated cell damage by organizing the detoxification pathway and limiting the release of ROS.

The chloroplast possesses several mechanisms that recycle MDA back to reduced ascorbate, including direct photoreduction at cytochrome b/f or PSI (Miyake & Asada, 1992; Grace *et al.*, 1995). The enzyme MDA reductase can also recycle ascorbate, primarily using NADH as a reductant (Hossain *et al.*, 1984). Alternatively, two MDA molecules can disproportionate to form reduced ascorbate and dehydroascorbate (DHA), the two-electron oxidation product of ascorbate. DHA, in turn, can undergo a double reduction catalyzed by DHA reductase, which uses reduced glutathione as a reductant. Finally, the resultant oxidized glutathione can be reduced via glutathione reductase, which utilizes NADPH as a reductant (see Chapter 1).

In addition to the reactions described above, a detoxification/antioxidant recycling pathway that is non-enzymatic with the exception of glutathione reduction (via GR) is thermodynamically possible. Ascorbate can react with superoxide non-enzymatically (Halliwell & Gutteridge, 1999), and glutathione can reduce the resultant MDA non-enzymatically under the slightly alkaline conditions that prevail in the stroma during illumination (Foyer & Halliwell, 1976; Winkler *et al.*, 1994). This series of reactions would then depend only on GR to maintain the reduction state of the glutathione pool. The role of such non-enzymatic reactions might be greatest during exposure to low temperatures that would inhibit enzyme activities. If this were so, one may expect significant up-regulation of GR activity during cold acclimation. Indeed, high foliar GR activities often correlate with winter tolerance (Collén & Davison, 1999) and transgenic cotton that overproduce chloroplastic GR exhibit enhanced resistance to photoinhibition during short-termed exposures to chilling temperatures (Kornyeyev *et al.*, 2001, 2003b).

Not only does O_2 photoreduction act as a sink for photogenerated reductant, so too does MDA recycling, with MDA photoreduction consuming electrons directly and DHA recovery consuming NADPH, and consequently generating $NADP^+$, the preferred electron acceptor in photosynthetic electron transport. If the capacity for ROS detoxification is sufficient to keep pace with the rate of O_2 photoreduction, then the water–water cycle could be thought of as photoprotective in nature because it safely consumes electrons.

10.4.2   The extent of O_2 photoreduction

The water–water cycle can play a significant role in photoprotection only if there is an appreciable rate of O_2 photoreduction. However, determining the electron flux to O_2 presents challenges, in large part because direct measurement is complicated by the fact that linear electron transport also results in O_2 evolution and that Rubisco is capable of oxygenation. As a consequence, the extent of O_2 photoreduction remains an issue of debate, with different methodological approaches yielding different estimates.

A recent attempt to resolve this question employed transgenic tobacco with suppressed Rubisco activities achieved via the expression of an antisense construct for the Rubisco small subunit (reviewed in Badger *et al.*, 2000). These antisense plants exhibit an inhibition of CO_2 assimilation to differing degrees, with no attendant inhibition of photosynthetic electron transport capacity (Ruuska *et al.*, 2000a). Since Rubisco oxygenation is suppressed along with carboxylation, these plants enable one to separate direct O_2 photoreduction from Rubisco oxygenation. Using chlorophyll fluorescence to estimate the rate of electron transport, Ruuska *et al.* (2000a) reported that the relationship between electron transport and gas exchange was linear over a broad range of O_2 and CO_2 concentrations and that the slope did not differ significantly from that of wild-type tobacco. Calculated rates of electron transport necessary to support Calvin cycle activity corresponded closely with actual estimated rates of electron transport, suggesting that O_2 photoreduction was not a significant sink for reducing equivalents at steady state (Ruuska *et al.*, 2000a).

Historically, the rate of O_2 photoreduction has been measured using isotopically labeled $^{18}O_2$ at CO_2 and O_2 concentrations that suppress Rubisco oxygenation. Using $^{18}O_2$ enables one to distinguish O_2 produced via water-splitting from O_2 consumed via direct photoreduction. In the transgenic tobacco described earlier, this approach also suggested low rates of direct O_2 photoreduction (Ruuska *et al.*, 2000a). However, others have reached different conclusions. Using this method, electron flow to O_2 representing 10–30% of total electron flow has been reported (e.g. Canvin *et al.*, 1980; Furbank *et al.*, 1982).

Comparing rates of electron transport measured using fluorescence to those measured as the rate of O_2 evolution provides yet another means of estimating the rate of O_2 photoreduction. This is because O_2 photoreduction silences a fraction of O_2 evolution and, therefore, leads to an underestimation of electron transport. O_2 photoreduction does not affect rates of electron transport estimated via fluorescence in a similar fashion. Thus, the slope of the relationship between rates of electron transport measured via these two methods under non-photorespiratory conditions yields an estimate of electron flux to O_2. Lovelock and Winter (1996) applied this approach to tropical trees and concluded that as much as 30% of overall electron transport leads to O_2 photoreduction in tropical trees at light saturation.

Some of the approaches described earlier require the imposition of physiologically unrealistic gas concentrations upon leaves. Pleiotropic effects associated with antisense plants may complicate the interpretation of experiments involving their use. Finally, estimating whole-leaf rates of photosynthetic electron transport from chlorophyll fluorescence emission may over-emphasize the influence of the upper layer of green cells. Hence, none of these methodologies is free of complicating considerations. Whether direct O_2 photoreduction is a substantial sink for reducing equivalents at steady state remains to be resolved. However, multiple methods do agree that significant electron flow to O_2 occurs during photosynthetic induction (Neubauer & Yamamoto, 1992; Ruuska et al., 2000b). Electron flow to O_2 during induction may play an essential role in poising the electron transport chain and augmenting the trans-thylakoid membrane pH gradient that is required to induce thermal energy dissipation. Even if O_2 photoreduction ultimately proves to be a relatively minor sink at steady state, the balance between photosynthetic ROS production and scavenging could still have profound effects on the redox status of key chloroplast constituents, including regulatory elements.

10.4.3 O_2 metabolism and the regulation of PSII excitation pressure

Antioxidants, such as the constituents of the water–water cycle, may protect against oxidative stress not just directly by scavenging ROS, but also indirectly by facilitating the formation of a downstream electron sink that eases the so-called excitation pressure on PSII (Huner et al., 1998). PSII excitation pressure refers to the reduction state of the final quinone acceptor of PSII, Q_A. PSII reaction centers with Q_A in the reduced state are more vulnerable to photoinhibitory inactivation because such centers are more likely to experience charge recombination events that yield triplet chlorophyll and ultimately singlet O_2 (Melis, 1999; see Section 10.3). The reduction state of Q_A is influenced by the levels of light absorption, thermal energy dissipation and the rate at which reducing equivalents are consumed downstream of PSII (reviewed in Ort & Baker, 2002). Under environmental conditions such as chilling that tend to increase PSII excitation pressure, transgenic overproduction of antioxidant enzymes of the water–water cycle (SOD, APX or GR) led to higher rates of electron transport through PSII, lower Q_A reduction states and, consequently, lower levels of photoinhibition (Kornyeyev et al., 2001, 2003a,b). The enhancement in chilling tolerance was abolished when the leaves of transgenic plants were infiltrated with DCMU, an inhibitor of electron transport from PSII, which leads to a massive over-reduction of Q_A (Kornyeyev et al., 2001). This suggests that the effect of the transgenic manipulations on rates of electron transport, and not their direct effect on ROS scavenging, partially explains their protective role. Hence, ROS production and scavenging appear to have been recruited into the regulatory regimes of the photosynthetic pathway and it would seem that, ironically, protecting PSII against photoinactivation depends, in part, on the production of ROS (and subsequent detoxification) at PSI.

10.4.4 Acclimation of the water–water cycle to the environment

Leaves adjust their capacities for antioxidant enzyme activities and their contents of low-molecular weight antioxidants in response to prevailing environmental conditions. Such responses can be observed over a time-scale of days to weeks

(Logan *et al.*, 1998b, 2003) and presumably bring ROS scavenging capacity into balance with ROS generation. Acclimation to a broad range of environmental stresses can be understood in terms of their overall effects on excess light absorption. Clearly, the intensity of the growth light environment influences the absorption of excess light and studies of the scavenging systems of numerous plant species report a strong correlation between growth light intensity and the activities/contents of various foliar antioxidants (Grace & Logan, 1996; Logan *et al.*, 1998a,c) (Figure 10.4). Chilling temperatures limit Calvin cycle activity, reducing photosynthetic light utilization and consequently increasing excess light absorption. Many evergreen plants, particularly conifers, lower excess light absorption in winter by decreasing chlorophyll content (Adams & Demmig-Adams, 1994). However, massive up-regulation of antioxidant scavenging capacity is also a typical response to the onset of winter (Anderson *et al.*, 1992; Logan *et al.*, 1998b; Collén & Davison, 2001) or experimentally imposed cold temperatures (de Kok & Oosterhuis, 1983; Schöner & Krause, 1990; Mishra *et al.*, 1993; Xin & Browse, 2000; Logan *et al.*, 2003). A portion of chilling induction of antioxidant enzyme activity can be interpreted as a means of compensating for the inhibitory effect of low temperatures on antioxidant enzymes, themselves. Nutrient-deficient substrates often reduce rates of whole-plant growth. This reduces the requirement for photosynthate and often brings about a strong down-regulation of photosynthetic capacity (Logan *et al.*, 1999). When the primary limiting nutrient is nitrogen, a drastic reduction in chlorophyll content is also commonly observed (Verhoeven *et al.*, 1997; Logan *et al.*, 1999). Decreasing light absorption in this manner appears to re-establish the balance between light absorption and light use such that up-regulation of ROS scavenging is not observed in N-limited spinach (Logan *et al.*, 1999). In contrast, it should be noted that deficiencies in other minerals such as magnesium (Cakmak & Marschner, 1991) and manganese (Polle *et al.*, 1992) are reported to result in decreased foliar chlorophyll content as well as up-regulation of ROS scavenging enzymes and low-molecular weight compounds.

10.5   Linkages between photosynthesis and extrachloroplastic oxidative metabolism

Chloroplastic oxidative metabolism is influenced by anatomical and morphological leaf features, primarily through their effect on light absorption. Some plants possess opaque surface waxes (Barker *et al.*, 1997), reflective pubescence (Ehleringer & Björkman, 1978) or heavily pigmented epidermal layers (Grace *et al.*, 1998), all of which serve to reduce light absorption by the chloroplasts below. Many plants that develop in full sunlight exposure orient their leaves at a steep angle (relative to horizontal orientation) in order to minimize excess light absorption during peak midday irradiance (Mooney *et al.*, 1977). This is particularly common among plants adapted to stressful environments. Some plants, including many legumes, possess pulvinar cells that enable them to adjust the angle of their leaflets or whole leaves

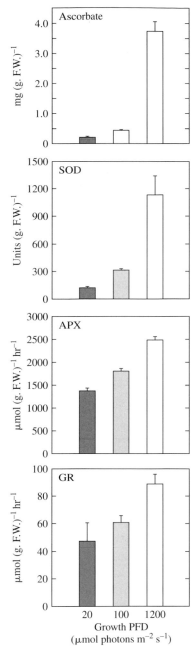

Figure 10.4 Ascorbate content and SOD, APX and GR activities of leaves of *Vinca major* grown under three different photon flux densities in controlled-environment chambers (direct full sunlight has a photon flux density of ~2000 μmolphotons m^{-2} s^{-1}). Error bars represent standard errors of the mean; $n = 3$. APX = ascorbate peroxidase, FW = fresh weight, GR = gluathione reductase, SOD = superoxide dismutase. Modified from Grace and Logan (1996).

over short time-scales. After rainfall, some desert plants practice diaheliotropism, or sun-tracking, in order to maximize light interception at a time when readily available water allows for high rates of photosynthetic CO_2 assimilation (Ehleringer & Forseth, 1980). At the other end of the spectrum, *Macroptilium atropurpureum* minimizes light interception by keeping its leaflets parallel to the sun's rays, a response known as paraheliotropism (Ludlow & Björkman, 1984). Nitrogen-limited soybean has been shown to exhibit midday paraheliotropism along with morning and afternoon diaheliotropism in order to position leaves such that light absorption is maintained very near the intensity where electron transport and the Calvin cycle co-limit photosynthesis and their nitrogen investment in the photosynthetic apparatus is maximized (Kao & Forseth, 1992). *Oxalis oregana*, an understory herb in redwood forests of the northwestern United States, folds its leaflets downward within approximately 5 minutes of exposure to bright sunflecks (Powles & Björkman, 1989), presumably to protect against excess light-mediated damage. Leaflets that were experimentally restrained in the horizontal position suffered almost twice as much sunfleck-induced photoinhibition (Powles & Björkman, 1989).

Chloroplast position within the cell is dynamic and under the control of the actin cytoskeleton. Over relatively short time-scales (i.e. minutes) chloroplast position can adjust to prevailing light conditions. In shade, chloroplasts are generally found on the lower and upper planes of leaf mesophyll cells, where light interception would be greatest, whereas under intense light chloroplasts take advantage of self-shading by organizing along the lateral walls of the cell (Haupt & Scheuerlein, 1990; Brugnoli & Björkman, 1992; Park *et al.*, 1996; Kasahara *et al.*, 2002; Williams *et al.*, 2003).

Superoxide does not readily cross biological membranes and is sufficiently unstable so as to undergo reaction relatively near to its site of generation. In contrast, H_2O_2 is less reactive and capable of diffusing passively across membranes. Thus, extrachloroplastic scavenging systems may be involved in controlling the damage caused by photogenerated ROS. In fact, cytosolic isoforms of APX are up-regulated during high-light stress (Karpinski *et al.*, 1997; Yoshimura *et al.*, 2000). In most plant cells, the vacuole is the largest cellular compartment. Vacuoles possess guaiacol peroxidase, for whom phenolic compounds serve as efficient reductants (Yamasaki *et al.*, 1997; Yamasaki & Grace, 1998; Chapter 6). Furthermore, many plant species respond to environmental stress by greatly up-regulating leaf phenolic content (Grace & Logan, 2000). Thus, vacuolar H_2O_2 detoxification employing phenolic reductants may be a significant pathway (see Chapter 6).

10.6   Concluding remarks

The last two decades of research have revealed that photoprotection is as complex and intricately regulated as the photosynthetic pathway itself. Indeed, photosynthesis and photoprotection are tightly interwoven and evolution has created a role for ROS generation and scavenging in regulating photosynthetic electron flow. It remains to be seen whether this new knowledge of photoprotection will lead to the creation of crop varieties with enhanced stress tolerance in an agricultural setting. Antioxidation does not readily lend itself to 'improvement' via single-gene transgenic up-regulation, due to its complexity and its role in the regulation of chloroplast metabolism. At present, evidence for crop improvement via transgenic manipulation of antioxidant enzymes derives from controlled laboratory-based studies often employing unrealistically intense or abrupt environmental stresses (discussed in Logan *et al.*, 2003). Even if we never learn to harness photoprotection for our own gain, we can appreciate the balancing act performed by plants as they adjust to an environment that changes over sub-diurnal, diurnal and seasonal time-scales in order to exploit sunlight and simultaneously protect against its potentially hazardous effects.

References

Adams, W.W. and Demmig-Adams, B. (1994) 'Carotenoid composition and down regulation of photosystem II in three conifer species during the winter', *Physiologia Plantarum* **92**, 451–458.

Adams, W.W., Volk, M., Hoehn, A. and Demmig-Adams, B. (1992) 'Leaf orientation and the response of the xanthophyll cycle to incident light', *Oecologia* **90**, 404–410.

Anderson, J.V., Chevone, B.I. and Hess, J.L. (1992) 'Seasonal variation in the antioxidant system of eastern white pine needles', *Plant Physiology* **98**, 501–508.

Andersson, B. and Barber, J. (1996) Mechanisms of photodamage and protein degradation during photoinhibition of photosystem II, in *Photosynthesis and the Environment* (ed. N.R. Baker), Kluwer Academic Publishers, Dordrecht, pp. 101–121.

Asada, K. (1996) Radical production and scavenging in chloroplasts, in *Photosynthesis and the Environment* (ed. N.R. Baker), Kluwer Academic Publishers, Dordrecht, pp. 123–150.

Asada, K. (1999) 'The water–water cycle in chloroplasts: scavenging of active oxygens and dissipation of excess photons', *Annual Review of Plant Physiology and Plant Molecular Biology* **50**, 601–639.

Badger, M.R., von Caemmerer, S., Ruuska, S. and Nakano, H. (2000) 'Electron flow to oxygen in higher plants and algae: rates and control of direct photoreduction (Mehler reaction) and rubisco oxygenase', *Philosophical Transactions of the Royal Society of London, Series B* **355**, 1433–1446.

Barker, D.H., Seaton, G.G.R. and Robinson, S.A. (1997) 'Internal and external photoprotection in developing leaves of the CAM plant *Cotyledon orbiculata*', *Plant, Cell and Environment* **20**, 617–624.

Bratt, C.E., Arvidsson, P.-O., Carlsson, M. and Åkerlund, H.-E. (1995) 'Regulation of violaxanthin de-epoxidase activity by pH and ascorbate concentration', *Photosynthesis Research* **45**, 169–175.

Brugnoli, E. and Björkman, O. (1992) 'Chloroplast movements in leaves: influence on chlorophyll fluorescence and measurements of light-induced changes related to ΔpH and zeaxanthin formation', *Photosynthesis Research* **32**, 23–35.

Buettner, G.R. (1993) 'The pecking order of free radicals and antioxidants: lipid peroxidation, α-tocopherol, and ascorbate', *Archives of Biochemistry and Biophysics* **300**, 535–543.

Burkle, L.A. and Logan, B.A. (2003) 'Seasonal acclimation of photosynthesis in eastern hemlock and partridgeberry growing in different light environments', *Northeastern Naturalist* **10**, 1–16.

Cakmak, I. and Marschner, H. (1992) 'Magnesium deficiency and high light intensity enhance activities of superoxide dismutase, ascorbate peroxidase, and glutathione reductase in bean leaves', *Plant Physiology* **98**, 1222–1227.

Canvin, D.T., Berry, J.A., Badger, M.R., Fock, H. and Osmond, C.B. (1980) 'Oxygen exchange in leaves in the light', *Plant Physiology* **66**, 302–307.

Charles, S.A. and Halliwell, B. (1981) 'Light activation of fructose bisphosphatase in isolated spinach chloroplasts and deactivation by hydrogen peroxide', *Planta* **151**, 242–246.

Collén, J. and Davison, I.R. (1999) 'Stress tolerance and reactive oxygen metabolism in the intertidal red seaweeds *Mastocarpus stellatus* and *Chondrus crispus*', *Plant, Cell and Environment* **22**, 1143–1151.

Collén, J. and Davison, I.R. (2001) 'Seasonality and thermal acclimation of reactive oxygen metabolism in *Fucus vesiculosus* (Phaeophyceae)', *Journal of Phycology* **37**, 474–481.

de Kok, L.J. and Oosterhuis, F.A. (1983) 'Effects of frost-hardening and salinity on glutathione and sulfhydryl levels and on glutathione reductase activity in spinach leaves', *Physiologia Plantarum* **58**, 47–51.

Demmig-Adams, B. and Adams, W.W. (1992) 'Photoprotection and other responses of plants to high light stress', *Annual Review of Plant Physiology and Plant Molecular Biology* **43**, 599–626.

Demmig-Adams, B. and Adams, W.W. (1996) 'The role of xanthophyll cycle carotenoids in the protection of photosynthesis', *Trends in Plant Science* **1**, 21–26.

Ebbert, V., Adams, W.W., Mattoo, A.K., Sokolenko, A. and Demmig-Adams, B. (2005) 'Up-regulation of a photosystem II core protein phosphatase inhibitor and sustained D1 phosphorylation in zeaxanthin-retaining, photoinhibited needles of overwintering Douglas fir', **28**, 232–240.

Ehleringer, J. and Björkman, O. (1978) 'Pubescence and leaf spectral characteristics in a desert shrub, *Encelia farinosa*', *Oecologia* **36**, 151–162.

Ehleringer, J. and Forseth, I. (1980) 'Solar tracking by plants', *Science* **210**, 1094–1098.

Foote, C.S. (1976) Photosensitized oxidation and singlet oxygen: consequences in biological systems, in *Free Radicals in Biology*, Vol. 2 (ed. W.A. Pryor), Academic Press, New York, pp. 85–124.

Foyer, C.H. and Halliwell, B. (1976) 'The presence of glutathione and glutathione reductase in chloroplasts: a proposed role in ascorbic acid metabolism', *Planta* **133**, 21–25.

Frank, H.A., Cua, A., Chynwat, V., Young, A., Gosztola, D. and Wasielewski, M.R. (1994) 'Photophysics of carotenoids associated with the xanthophyll cycle in photosynthesis', *Photosynthesis Research* **41**, 389–395.

Funk, C., Schröder, W.P., Green, B.R., Renger, G. and Andersson, B. (1994) 'The intrinsic 22 kD is a chlorophyll-binding subunit of photosystem II', *FEBS Letters* **342**, 261–266.

Furbank, R.T., Badger, M.R. and Osmond, C.B. (1982) 'Photosynthetic oxygen exchange in isolated cells and chloroplasts of C_3 plants', *Plant Physiology* **70**, 927–931.

Gilmore, A.M. and Ball, M.C. (2000) 'Protection and storage of chlorophyll in overwintering evergreens', *Proceedings of the National Academy of Sciences USA* **97**, 11098–11101.

Gilmore, A.M. and Yamamoto, H.Y. (1993a) 'Linear models relating xanthophylls and lumen acidity to non-photochemical fluorescence quenching. Evidence that antheraxanthin explains zeaxanthin-independent quenching', *Photosynthesis Research* **35**, 67–78.

Gilmore, A.M. and Yamamoto, H.Y. (1993b) Biochemistry of xanthophyll-dependent nonradiative energy dissipation, in *Photosynthetic Responses to the Environment, Vol. 8: Current Topics in Plant Physiology* (ed. H.Y. Yamamoto and C.M. Smith), American Society of Plant Physiologists, Maryland, pp. 160–165.

Grace, S.C. and Logan, B.A. (1996) 'Acclimation of foliar antioxidant systems to growth irradiance in three broad-leaved evergreen species', *Plant Physiology* **112**, 1631–1640.

Grace, S.C. and Logan, B.A. (2000) 'Energy dissipation and radical scavenging by the plant phenylpropanoid pathway', *Philosophical Transactions of the Royal Society of London, Series B* **355**, 1499–1510.

Grace, S.C., Logan, B.A. and Adams, W.W. (1998) 'Seasonal differences in foliar content of chlorogenic acid, a phenylpropanoid antioxidant, in *Mahonia repens*', *Plant, Cell and Environment* **21**, 513–521.

Grace, S., Pace, R. and Wydrzynski, T. (1995) 'Formation and decay of monodehydroascorbate radicals in illuminated thylakoids as determined by EPR spectroscopy', *Biochimica et Biophysica Acta* **1229**, 155–165.

Halliwell, B. and Gutteridge, J.M.C. (1999) *Free Radicals in Biology and Medicine*, 3rd edn., Oxford University Press, Oxford, UK.

Haupt, W. and Scheuerlein, R. (1990) 'Chloroplast movement', *Plant, Cell and Environment* **13**, 595–614.

Hideg, E., Spetea C. and Vass, I. (1994) 'Singlet oxygen production in thylakoid membranes during photoinhibition as detected by EPR spectroscopy', *Photosynthesis Research* **39**, 191–199.

Holt, N.E., Fleming, G.R. and Niyogi, K.K. (2004) 'Toward an understanding of the mechanism of nonphotochemical quenching in green plants', *Biochemistry* **43**, 8281–8289.

Holt, N.E., Zigmantas, D., Valkunas, L., Li, X.P., Niyogi, K.K. and Fleming, G.R. (2005) 'Carotenoid cation formation and the regulation of photosynthetic light harvesting', *Science* **307**, 433–436.

Horton, P., Ruban, A.V. and Walters, R.G. (1996) 'Regulation of light harvesting in green plants', *Annual Review of Plant Physiology and Plant Molecular Biology* **47**, 655–684.

Horton, P., Ruban, A. and Wentworth, M. (2000) 'Allosteric regulation of the light harvesting system of PSII', *Philosophical Transactions of the Royal Society of London, Series B* **355**, 1361–1370.

Hossain, H.A., Nakano, Y. and Asada, K. (1984) 'Monodehydroascorbate reductase in spinach chloroplasts and its participation in regeneration of ascorbate for scavenging hyrogen peroxide', *Plant and Cell Physiology* **25**, 385–395.

Huner, N.P.A., Öquist, G. and Sarhan, F. (1998) 'Energy balance and acclimation to light and cold', *Trends in Plant Science* **3**, 224–230.

Jablonski, P.P. and Anderson, J.W. (1982) 'Light-dependent reduction of hydrogen peroxide by ruptured pea chloroplasts', *Plant Physiology* **69**, 1407–1413.

Kao, W.-Y. and Forseth, I.N. (1992) 'Diurnal leaf movement, chlorophyll fluorescence and carbon assimilation in soybean grown under different nitrogen and water availabilities', *Plant, Cell and Environment* **15**, 703–710.

Karpinski, S., Escobar, C., Karpinska, B., Creissen, G. and Mullineaux, P.M. (1997) 'Photosynthetic electron transport regulates the expression of cytosolic ascorbate peroxidase genes in *Arabidopsis* during light stress', *The Plant Cell* **9**, 627–640.

Kasahara, M., Kagawa, T., Oikawa, K., Suetsugu, N., Miyao, M. and Wada, M. (2002) 'Chloroplast avoidance movement reduces photodamage in plants', *Nature* **420**, 829–832.

Kornyeyev, D., Logan, B.A., Allen, R.D. and Holaday, A. (2003a) 'Effect of chloroplastic over-production of ascorbate peroxidase on photosynthesis and photoprotection in cotton leaves subjected to low temperature photoinhibition', *Plant Science* **165**, 1033–1041.

Kornyeyev, D., Logan, B.A., Payton, P., Allen, R.D. and Holaday, A.S. (2001) 'Enhanced photochemical light utilization and decreased chilling-induced photoinhibition of photosystem II in cotton overexpressing genes encoding chloroplast-targeted antioxidant enzymes', *Physiologia Plantarum* **113**, 323–331.

Kornyeyev, D., Logan, B.A., Payton, P.R., Allen, R.D. and Holaday, A.S. (2003b) 'Elevated chloroplastic glutathione reductase activities decrease chilling-induced photoinhibition by increasing rates of photochemistry, but not thermal energy dissipation, in transgenic cotton', *Functional Plant Biology* **30**, 101–110.

Kurepa, J., Hérouart, D., Van Montagu, M. and Inzé, D. (1997) 'Differential expression of CuZn- and Fe-superoxide dismutase genes of tobacco during development, oxidative stress and hormonal treatments', *Plant and Cell Physiology* **38**, 463–470.

Li, X.-P., Björkman, O., Shih, C. *et al.* (2000) 'A pigment-binding protein essential for regulation of photosynthetic light harvesting', *Nature* **403**, 391–395.

Li, X.-P., Gilmore, A.M., Caffarri, S. *et al.* (2004) 'Regulation of photosynthetic light harvesting involves intrathylakoid lumen pH sensing by the PsbS protein', *Journal of Biological Chemistry* **279**, 22866–22874.

Li, X.-P., Müller-Moulé, P., Gilmore, A.M. and Niyogi, K.K. (2002) 'PsbS-dependent enhancement of feedback de-excitation protects photosystem II from photoinhibition', *Proceedings of the National Academy of Sciences USA* **99**, 15222–15227.

Logan, B.A., Barker, D.H., Demmig-Adams, B. and Adams, W.W. (1996) 'Acclimation of leaf carotenoid composition and ascorbate levels to gradients in the light environment within an Australian rainforest', *Plant, Cell and Environment* **19**, 1083–1090.

Logan, B.A., Demmig-Adams, B., Adams, W.W. and Grace, S.C. (1998a) 'Antioxidation and xanthophyll cycle-dependent energy dissipation in *Cucurbita pepo* and *Vinca major* acclimated to four growth irradiances in the field', *Journal of Experimental Botany* **49**, 1869–1879.

Logan, B.A., Demmig-Adams, B. and Adams, W.W. (1998b) 'Antioxidation and xanthophyll cycle dependent energy dissipation in *Cucurbita pepo* and *Vinca major* during a transfer from low to high irradiance in the field', *Journal of Experimental Botany* **49**, 1881–1888.

Logan, B.A., Grace, S.C., Adams, W.W. and Demmig-Adams, B. (1998c) 'Seasonal differences in xanthophyll cycle characteristics and antioxidants in *Mahonia repens* growing in different light environments', *Oecologia* **116**, 9–17.

Logan, B.A., Demmig-Adams, B., Adams, W.W. and Rosenstiel, T.N. (1999) 'Effect of nitrogen limitation on foliar antioxidants in relationship to other metabolic characteristics', *Planta* **209**, 213–220.

Logan, B.A., Monteiro, G., Kornyeyev, D., Payton, P., Allen, R. and Holaday, A. (2003) 'Transgenic overproduction of glutathione reductase does not protect cotton, *Gossypium hirsutum* (Malvaceae), from photoinhibtion during growth under chilling conditions', *American Journal of Botany* **90**, 1400–1403.

Lovelock, C.E. and Winter, K. (1996) 'Oxygen-dependent electron transport and protection from photoinhibition in leaves of tropical tree species', *Planta* **198**, 580–587.

Ludlow, M.M. and Björkman, O. (1984) 'Paraheliotropic leaf movement in *Siratro* as a protective mechanism against drought-induced damage to primary photosynthetic reactions: damage by excessive light and heat', *Planta* **161**, 505–518.

Ma, Y.-Z., Holt, N.E., Li, X.-P., Niyogi, K.K. and Flemming, G.R. (2003) 'Evidence for direct carotenoid involvement in the regulation of photosynthetic light harvesting', *Proceedings of the National Academy of Sciences USA* **100**, 4377–4382.

McCord, J.M. and Fridovich, I. (1969) 'Superoxide dismutase. An enzymic function for erthyrocuprein (hemocuprein)', *Journal of Biological Chemistry* **244**, 6049–6055.

Melis, A. (1999) 'Photosystem-II damage and repair cycle in chloroplasts: what modulates the rate of photodamage *in vivo*?', *Trends in Plant Science* **4**, 130–135.

Mishra, N.P., Mishra, R.K. and Singhal, G.S. (1993) 'Changes in the activities of antioxidant enzymes during exposure of intact wheat leaves to strong visible light at different temperatures in the presence of different protein synthesis inhibitors', *Plant Physiology* **102**, 867–880.

Miyake, C. and Asada, K. (1992) 'Thylakoid-bound ascorbate peroxidase in spinach chloroplasts and photoreduction of its primary oxidation product monodehydroascorbate radicals in thylakoids', *Plant and Cell Physiology* **33**, 541–553.

Mooney, H.A., Ehleringer, J. and Björkman, O. (1977) 'The leaf energy balance of leaves of the evergreen desert shrub *Atriplex hymenelytra*', *Oecologia* **29**, 301–310.

Neubauer, C. and Yamamoto, H.Y. (1992) 'Mehler-peroxidase reaction mediates zeaxanthin formation and zeaxanthin-related fluorescence quenching in intact chloroplasts', *Plant Physiology* **99**, 1354–1361.

Niyogi, K.K. (1999) 'Photoprotection revisited: genetic and molecular approaches', *Annual Review of Plant Physiology and Plant Molecular Biology* **50**, 333–359.

op den Camp, R.G.L., Przybyla, D., Ochsenbein, C. *et al.* (2003) 'Rapid induction of distinct stress responses after the release of singlet oxygen in *Arabidopsis*', *The Plant Cell* **15**, 2320–2332.

Öquist, G. and Huner, N.P.A. (2003) 'Photosynthesis in overwintering evergreen plants', *Annual Review of Plant Biology* **54**, 329–355.

Ort, D.R. and Baker, N.R. (2002) 'A photoprotective role for O_2 as an alternative electron sink in photosynthesis?', *Current Opinion in Plant Biology* **5**, 193–198.

Ottander, C., Campbell, D. and Öquist, G. (1995) 'Seasonal-changes in photosystem-II organization and pigment composition in *Pinus sylvestris*', *Planta* **197**, 176–183.

Owens, T.G. (1997) Processing of excitation energy by antenna pigments, in *Photosynthesis and the Environment* (ed. N.R. Baker), Kluwer Academic Publishers, Dordrecht, pp. 1–23.

Park, Y.I., Chow, W.S. and Anderson, J.M. (1996) 'Chloroplast movement in the shade plant *Tradescantia albiflora* helps protect photosystem II against light stress', *Plant Physiology* **111**, 867–875.

Polle, A. (2001) 'Dissecting the superoxide dismutase-ascorbate-glutathione-pathway in chloroplasts by metabolic modeling. Computer simulations as a step towards flux analysis', *Plant Physiology* **126**, 445–462.

Polle, A., Chakrabarti, K., Chakrabarti, S., Seifert, F., Schramel, P. and Renneberg, H. (1992) 'Antioxidants and manganese deficiency in needles of norway spruce (*Picea abies* L.) trees', *Plant Physiology* **99**, 1084–1089.

Powles, S.B. and Björkman, O. (1989) 'Leaf movement in the shade species *Oxalis oregana*. II. Role in protection against injury by intense light', *Carnegie Institution of Washington Yearbook* **80**, 63–66.

Ruuska, S.A., Badger, M.R., Andrews, T.J. and von Caemmerer, S. (2000a) 'Photosynthetic electron sinks in transgenic tobacco with reduced amounts of rubisco: little evidence for significant Mehler reaction', *Journal of Experimental Botany* **51**, 357–368.

Ruuska, S.A., von Caemmerer, S., Badger, M.R., Andrews, T.J., Price, G.D. and Robinson, S.A. (2000b) 'Xanthophyll cycle, light energy dissipation and electron transport in transgenic tobacco with reduced carbon assimilation capacity', *Australian Journal of Plant Physiology* **27**, 289–300.

Schöner, S. and Krause, G.H. (1990) 'Protective systems against active oxygen species in spinach: response to cold acclimation in excess light', *Planta* **180**, 383–389.

Taiz, L. and Zeiger, E. (2002) *Plant Physiology*, 3rd edn., Sinauer Associates, Inc., Sunderland.

Verhoeven, A.S., Adams, W.W. and Demmig-Adams, B. (1996) 'Close relationship between the state of the xanthophyll cycle pigments and photosystem II efficiency during recovery from winter stress', *Physiologia Plantarum* **96**, 567–576.

Verhoeven, A.S., Demmig-Adams, B. and Adams, W.W. (1997) 'Enhanced employment of the xanthophyll cycle and thermal energy dissipation in spinach exposed to high light and nitrogen stress', *Plant Physiology* **113**, 817–824.

Williams, W.E., Gorton, H.L. and Witiak, S.M. (2003) 'Chloroplast movements in the field', *Plant, Cell and Environment* **26**, 2005–2014.

Winkler, B.S., Orselli, S.M. and Rex, T.S. (1994) 'The redox couple between glutathione and ascorbic acid: a chemical and physiological perspective', *Free Radicals in Biology and Medicine* **17**, 333–349.

Xin, Z. and Browse, J. (2000) 'Cold comfort farm: the acclimation of plants to freezing temperatures', *Plant, Cell and Environment* **23**, 893–902.

Yamasaki, H. and Grace, S.C. (1998) 'EPR detection of phytophenoxyl radicals stabilized by zinc ions: evidence for the redox-coupling of plant phenolics with ascorbate in the H_2O_2-peroxidase system', *FEBS Letters* **422**, 377–380.

Yamasaki, H., Sakihama, Y. and Ikehara, N. (1997) 'Flavonoid-peroxidase reaction as a detoxification mechanism of plant cells against H_2O_2', *Plant Physiology* **115**, 1405–1412.

Yoshimura, K., Yabuta, Y., Ishikawa, T. and Shigeoka, S. (2000) 'Expression of spinach ascorbate peroxidase isoenzymes in response to oxidative stresses', *Plant Physiology* **123**, 223–233.

11 Plant responses to ozone

Pinja Jaspers, Hannes Kollist, Christian Langebartels and
Jaakko Kangasjärvi

11.1 Introduction

Practically all adverse environmental conditions (heat, cold, high light and UV-B
radiation, ozone, water deficit, high salinity and heavy metals) and several biotic
challenges (attack by different pathogens) lead to the production of excess reactive
oxygen species (ROS). The site of ROS formation and the oxygen species produced
depend on the nature of the stress. For example, in chilling temperatures, the chloroplast electron transport chain is especially vulnerable (Salinas, 2002), whereas
during pathogen attack and exposure to ozone, ROS concentrate in the apoplastic
space (Lamb & Dixon, 1997). Excess light creates singlet oxygen (Fryer et al.,
2002; Hideg et al., 2002) and peroxisomal photorespiration produces hydrogen
peroxide (H_2O_2) (Igamberdiev & Lea, 2002).

Despite their potentially harmful nature, ROS accumulation is not always
unwanted or accidental. It has become evident that ROS also have a vital signaling
function (Vranová et al., 2002; Overmyer et al., 2003; Chapter 7). They are used
as cellular second messengers in plant cells during, e.g. hypersensitive response,
ABA signal transduction (reviewed by Himmelbach et al., 2003), and senescence
(del Rio et al., 2003). The subcellular site of ROS production differs in these cases
and both the nature and the location of ROS production determine the outcome
(Laloi et al., 2004).

Ozone (O_3) is the triatomic form of oxygen. It is an important protective component against UV radiation in the stratosphere, but, in the troposphere, it is one of
the most notorious air pollutants. In plants, long-term ozone exposure of relatively
low concentrations causes reduction in photosynthesis and growth, as well as premature senescence in sensitive species and cultivars. These, in turn, lead to lowered
growth rates and crop yields, as well as decreased pathogen tolerance and possibly
other ecological alterations (Heath & Taylor, 1997; Pell et al., 1997; Langebartels &
Kangasjärvi, 2004; Timonen et al., 2004). In contrast to the fairly subtle effects of
this so-called chronic ozone exposure, a short, high-level exposure causes immediately visible cell death in plants, demonstrated as lesions on leaves. The appearance
of the lesions resembles hypersensitive cell death and recent research has shown
that the resemblance is not only superficial: as reviewed in Kangasjärvi et al. (1994)
and Rao et al. (2000a), these phenomena share many physiological and molecular
features.

In this chapter, the mechanisms involved in plant sensitivity to acute, high-level O_3 are discussed. Five individual processes can be separated when the lesion formation is dissected to its components: (i) the stomatal control of O_3 entry to the leaf intercellular air spaces, (ii) reactions of O_3 with the apoplastic components, formation of ROS from O_3 degradation and the detoxification of the ROS by apoplastic antioxidants, (iii) the control of initial cell death – necrotic trauma or regulation of programmed cell death (PCD), (iv) propagation of the cell death and (v) containment of the lesion formation. The mechanisms and processes involved in these are discussed separately later. We concentrate especially on the experiments performed in controlled environments addressing and identifying the mechanisms involved in O_3 lesion formation. Thus, most of the research reviewed here has been performed using model plants and especially mutants of *Arabidopsis thaliana*, which has greatly expanded our knowledge about the basis of O_3 sensitivity.

11.1.1 Regulation of O_3 flux to leaves

O_3 is formed in light-driven reactions of atmospheric oxygen with air impurities, such as NO_x and hydrocarbons (Stockwell *et al.*, 1997). The concentration of O_3 in the lower troposphere varies according to many factors, such as season, weather and anthropogenic activities. O_3 entry through the leaf cuticle is negligible (Kerstiens & Lendzian, 1989) and, therefore, stomata play a vital role in determining the flux of O_3 into the apoplastic space. As the stomatal guard cells are the first structure to regulate O_3 entry, their sensitivity to external stimuli is an important factor in the overall plant responses and sensitivity/tolerance to O_3. If the stomatal aperture is wider, and consequently the flux of gaseous substances is higher in one plant when compared to another, the dose of O_3 delivered to the intercellular spaces is also larger (Heath & Taylor, 1997; Sandermann & Matyssek, 2004).

Stomatal closure is the result of active release of solutes from the guard cells that surround the stomatal pore. When talking about stomatal regulation, one has to bring up abscisic acid (ABA), the β-carotene-derived plant hormone responsible for the control of stomatal opening and closure. The signal transduction chain starting from ABA and leading to this response has been studied extensively and several signaling components have been identified (Finkelstein *et al.*, 2002; Himmelbach *et al.*, 2003). The ABA signal is transmitted through several (interacting) signal transduction pathways, most of which utilize cytosolic calcium as a second messenger. Together with calcium, at least phospholipases C and D, cyclic nucleotide derivatives, inositol phosphates and protein kinases and phosphatases have been demonstrated as mediators of ABA signaling (Himmelbach *et al.*, 2003).

Although stomatal closure in response to O_3 was documented several decades ago (Hill & Littlefield, 1969), the question of whether it is caused by the direct effect of O_3 on guard cells or by a signal triggered from mesophyll cells affected by O_3 remained open until recently. O_3 can affect the stomatal function through changes in photosynthesis, ABA signaling, by generating an 'artificial' ROS burst directly in the guard cells, or by inducing ethylene emission, which will then lead to stomatal closure. Moldau *et al.* (1990) showed that O_3 induced stomatal closure within the first 12 min of O_3 exposure and concluded that this was caused by a direct effect of O_3 on guard cells rather than by dysfunction of mesophyll photosynthesis since mesophyll conductance to CO_2 remained unchanged during the 4-h exposure period. However, the regulation of stomata by ABA in response to O_3 is also involved. When the wild-type Col-0 *Arabidopsis* was compared to mutants with different degrees of ABA insensitivity, the O_3-induced stomatal closure was significantly faster in the wild-type Col-0 than in the mutants. However, O_3 caused the closure of stomata also in the partially ABA insensitive *rcd1* and in the strongly ABA insensitive *abi2* mutant (Ahlfors *et al.*, 2004a). Furthermore, O_3-induced ethylene emission has also been shown to cause stomatal closure (Gunderson & Taylor, 1991). These results show that O_3 affects the stomata with several mechanisms that are both dependent and independent of ABA.

How can O_3 be indirectly involved in the regulation of stomata? It was discovered recently that ABA signaling utilizes ROS also (Pei *et al.*, 2000; Murata *et al.*, 2001; Zhang *et al.*, 2001). Therefore, O_3 also – or the ROS formed from the degradation of O_3 in the apoplast – could directly have relevance in the regulation of stomatal function. *Arabidopsis* NADPH oxidases AtrbohD and AtrbohF (*Arabidopsis thaliana respiratory burst oxidase homolog*), which generate ROS during the hypersensitive response, are required for elevation in cytosolic calcium concentration in stomatal guard cells. In addition to this, knocking out both genes made plants insensitive to inhibition of root elongation and seed germination by ABA suggesting that ROS production is a general feature of ABA signaling in plants (Kwak *et al.*, 2003). ROS have also a link to two protein phosphatases, ABI1 and ABI2, negative regulators of ABA signaling. In biochemical studies, H_2O_2 inhibited the function of these proteins (Meinhard & Grill, 2001; Meinhard *et al.*, 2002). This could be one mechanism of positive regulation of plasma membrane Ca^{2+} influx by ROS. It has also been shown that O_3 can affect the aperture of stomata by directly affecting the K^+ fluxes in the guard cells (Torsethaugen *et al.*, 1999). However, only a few reports have addressed the effect of O_3 on ion fluxes across the plasma membrane in general (Castillo & Heath, 1990; McAinsh *et al.*, 1996; Clayton *et al.*, 1999; Torsethaugen *et al.*, 1999). Thus, the mechanisms by which O_3 affects the stomatal function still requires further study.

11.1.2 O_3 degradation to ROS and removal by antioxidants in the apoplast

After the entry of O_3 through the stomata to the apoplastic space, the reactions in the cell wall and the close proximity of the plasma membrane are the next factors that affect plant responses to O_3. The antioxidative capacity of the apoplast determines the fate of the leaf cells exposed to radicals. O_3 concentration inside the leaf during O_3 exposure is close to zero (Laisk *et al.*, 1989), which means that O_3 is rapidly degraded in the apoplast (see Figure 11.1). It is well established that

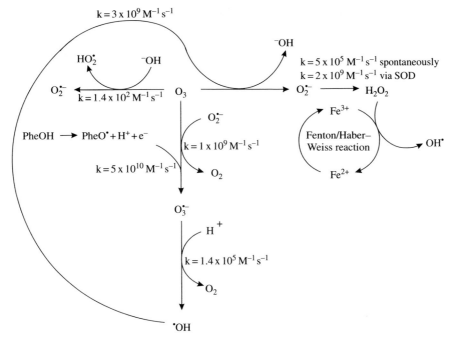

Figure 11.1 Possible degradation reactions of O_3 in the apoplast. Decomposition of O_3 in the reaction with the hydroxyl ion is slow due to the low rate constant over the common range of cell wall pH 5.0–6.5. Instead, the reaction between O_3 and $O_2^{\cdot-}$ is very fast and gives rise to an ozonide radical, $O_3^{\cdot-}$, which can acquire a proton and decompose to $^{\cdot}OH$ and O_2. The $^{\cdot}OH$ radicals formed can restart the cycle by reaction with O_3 and formation of $O_2^{\cdot-}$. Alternatively, O_3 may be converted to $O_3^{\cdot-}$ by trapping an electron from phenolic compounds present in the apoplast since a considerable rise in $^{\cdot}OH$ production was detected when caffeic or ferulic acid was added to ozonated water solution (Grimes *et al.*, 1983). $^{\cdot}OH$ can also be produced in the Fenton/Haber–Weiss reaction in the presence of metal ions and H_2O_2. For further details and references for rate constants, see Moldau (1998).

ascorbic acid (AA) provides important protection from oxidative injury by removing the harmful ROS generated from O_3 (Noctor & Foyer 1998; Chapter 3). It was shown 40 years ago that spraying of plants or feeding of detached leaves via petiole with AA-containing solutions diminished leaf cell death in O_3 exposure experiments (Menser, 1964). The protective role of AA is also supported by the enhanced O_3 sensitivity of mutants deficient in ascorbate biosynthesis (Conklin *et al.*, 1996) and by transgenic ascorbate peroxidase antisense tobacco plants, which were also more sensitive to O_3 (Örvar & Ellis, 1997). As proposed by Chameides (1989), apoplastic AA could form an effective defense barrier against O_3. Recently, clear evidence for that was presented by Sanmartin *et al.* (2003) who showed that removal of the reduced form of AA from the apoplast by overexpressing apoplastic ascorbate oxidase made plants substantially more sensitive to O_3. On the other hand, it has been shown that, at least in cereals like wheat and barley, the removal of O_3 by direct reaction with apoplastic AA is minor since mesophyll cell walls are thin and thus the effective path length for the reaction is short (Kollist *et al.*, 2000). Similarly, Ranieri *et al.* (1999) concluded that the different O_3 sensitivity of two poplar clones was not due to their different ability to remove O_3 by apoplastic AA, but that other processes were more important in determining their O_3 sensitivity. Also in an attempt to examine the *in vivo* reactivity of ascorbate with O_3, AA was not capable of inhibiting the O_3-mediated oxidation of fluorescent dyes, which were pre-infiltrated into the apoplast (Jakob & Heber, 1998). Nevertheless, Moldau (1998) has pointed out that if O_3 is converted into other ROS like hydroxyl radical, singlet oxygen or ozonide (Figure 11.1), the scavenging activity of AA is much higher since reaction constants between AA and these radicals are faster than with O_3 (Figure 11.2).

The apoplastic oxidation/reduction reactions of AA are shown in Figure 11.2. The first step of AA oxidation, either directly by ROS or catalyzed by ascorbate oxidase and/or ascorbate peroxidase, forms monodehydroascorbate (MDHA), which can be reduced to AA by specific transmembrane cytochrome b_{561} in the

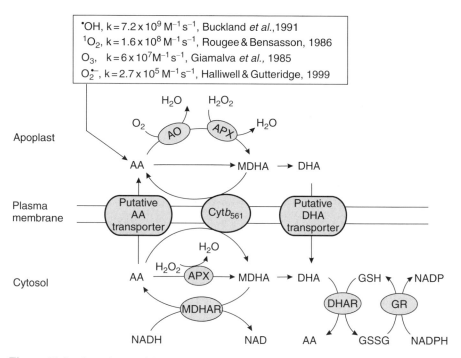

Figure 11.2   Ascorbate oxidation and reduction in the apoplast and cytosol. Ascorbate is synthesized in the cytosol and mitochondria and is transported via a putative transporter to the apoplast where it can react with most of the ROS directly or enzymatically. Reaction constants between AA and ROS are given in the text box above. AO, ascorbate oxidase; APX, ascorbate peroxidase; MDHAR, monodehydroascorbate reductase; DHAR, dehydroascorbate reductase; GR, glutathione reductase; Cyt b_{561}, transmembrane b-type cytochrome.

apoplastic space (Horemans *et al.*, 1994). Further oxidation of MDHA produces dehydroascorbate (DHA). Although there is some evidence for the DHA reduction by DHA reductase in the apoplast, it still remains a matter of debate since the activities measured are very low (Pignocci & Foyer, 2003). It is more likely that DHA is transported back to the cytosol where it is recycled to AA by the well-known ascorbate–glutathione cycle (Foyer & Halliwell, 1976; Figure 11.2; Chapters 1 and 3). Consequently, to maintain the antioxidative capacity of AA in the apoplast, this transmembrane transport must be faster than oxidation of AA by O_3. Comparison of O_3 flux into the leaf with ascorbate flux through the plasma membrane in silver birch and spring barley indicated that, at present, ambient concentrations of O_3 (40 ppb) ascorbate flux through the plasma membrane was more than three times higher than flux of O_3 reaching plasma membrane (Kollist, 2001). However, if O_3 concentration of 100 ppb was applied to spring barley, the flux of O_3 exceeded the flux of ascorbate across the plasma membrane. This indicates that, at moderate concentrations of O_3, its detoxification by apoplastic AA is not limited by the ascorbate flux across the plasma membrane but, at elevated concentrations, flux of O_3 readily exceeds ascorbate flux and accumulation of O_3 or its degradation products may occur. Consequently, other processes are involved in the responses to the increased oxidant load.

Taken together, when the formation of ROS from O_3 degradation at high O_3 concentrations exceeds the antioxidant capacity of the apoplast, the redox balance of the cellular environment will be perturbed. This 'redox shift' activates signal transduction, which results in processes similar to the hypersensitive response and/or systemically acquired resistance. The latter two are tightly controlled by plant hormones, as discussed later.

11.1.3   Induction of the active oxidative burst and sensing of ROS

When the production of ROS from the breakdown of O_3 in the apoplast exceeds the apoplastic antioxidative capacity, an endogenous, active, self-propagating ROS generation that continues after the end of the O_3 exposure is induced. This oxidative burst is similar to the one induced by the perception of an incompatible pathogen (Figure 11.3; Schraudner *et al.*, 1998; Rao & Davis, 1999; Overmyer *et al.*, 2000) and is an integral factor in the cell death by the process of PCD. The involvement of PCD in O_3-lesion formation is widely accepted (Beers & McDowell, 2001; Rao & Davis, 2001; Berger, 2002), although the morphological and biochemical hallmarks of PCD have been demonstrated directly only recently in tobacco and *Arabidopsis* (Pasqualini *et al.*, 2003; Overmyer *et al.*, 2005). In tobacco, the O_3-sensitive cultivar Bel W3 exhibited a similar biphasic oxidative burst in the response to O_3 as during HR, but the tolerant cultivar Bel B had only a modest rise in endogenous ROS production (Schraudner *et al.*, 1998). Similar induction of active ROS production has been seen in O_3-exposed sensitive *Arabidopsis*, birch, tomato, *Malva* and *Rumex* cultivars and species, while the tolerant counterparts only exhibited ROS formation during the exposure period (Wohlgemuth *et al.*, 2002). Consequently, O_3

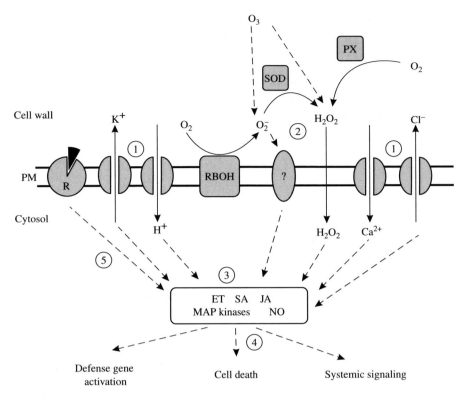

Figure 11.3   Reactive oxygen species (ROS) in O_3 exposure and hypersensitive cell death. Both pathogen attack and O_3 exposure induce ion fluxes through the plasma membrane (1) and an endogenous oxidative burst, the production of ROS, such as superoxide and hydrogen peroxide, in the apoplast (2). These ROS are sensed through yet unidentified mechanisms and they initiate further responses, such as activation of MAP kinases and accumulation of plant hormones (3). Ultimately, this signaling leads to activation of defense genes, systemic signaling and cell death (4). Pathogen perception by a receptor (R) also utilizes ROS-independent signaling (5). PM, plasma membrane; PX, peroxidase; R, receptor for elicitor; RBOH, NADPH oxidase (Respiratory Burst Oxidase Homolog); SOD, superoxide dismutase.

has been recognized as an abiotic elicitor of plant defense responses (Kangasjärvi *et al.*, 1994; Sharma *et al.*, 1996; Sandermann *et al.*, 1998; Rao *et al.*, 2000a) and is a good tool to study signaling cascades that involve extracellular ROS formation in the regulation of gene expression and cell death without wounding the plant.

The current view is that the ROS eliciting the downstream responses during the HR and O_3-induced oxidative burst are superoxide (O_2^-), H_2O_2 or both. For example, results with parsley cell culture indicated O_2^- as the signal for defense induction (Jabs *et al.*, 1997). In the *Arabidopsis* mutants *lsd1* and *rcd1*, O_2^- was both necessary and sufficient to drive hypersensitive-like cell death (Jabs *et al.*, 1997; Overmyer *et al.*, 2000). In contrast, in tobacco, tomato, birch and soybean, H_2O_2 seemed to be the molecule mediating defense gene induction and/or cell death (Levine *et al.*, 1994; Schraudner *et al.*, 1998; Moeder *et al.*, 2002; Pellinen *et al.*, 2002).

An O_2^--generating NADPH oxidase (Figure 11.3) has long been suspected to be a prominent source of ROS during the oxidative burst in the HR. Indeed, it was recently shown that AtrbohD and AtrbohF were required for the oxidative burst in incompatible interaction between *Arabidopsis* and *Pseudomonas syringae* (Torres *et al.*, 2002) and in elicitor-stimulated tobacco (Simon-Plas *et al.*, 2002). Apoplastic O_2^- accumulation has also been suggested to be the driving force in the O_3 lesion formation in *Arabidopsis*; DPI, an inhibitor of flavin-containing oxidases reduced the extent of O_3 lesion formation (Rao & Davis, 1999; Overmyer *et al.*, 2000). Even though DPI is regarded as inhibitor of the NADPH oxidase, it also affects the O_3-induced H_2O_2 production in birch and tobacco, which is not surprising since O_2^- will be dismutated to H_2O_2. Furthermore, in tobacco, O_3 activated genes encoding subunits of the NADPH oxidase, and it has been shown that H_2O_2 accumulation in tobacco apoplast after O_3 induction is a result of dismutation of O_2^- generated by the NADPH oxidase complex (Langebartels *et al.*, 2002). All these results support the importance of the NADPH oxidase in the O_3-induced oxidative burst.

In addition to NADPH oxidases, cell wall peroxidases and oxalate oxidases have been proposed as possible sources of ROS (Lamb & Dixon, 1997; Bolwell *et al.*, 2002). The subcellular location of H_2O_2 accumulation in birch was, in addition to the surface of plasma membrane, also in the cell wall. This suggests that two distinct O_3-induced sources for the H_2O_2 generation were operational and, accordingly, DPI did not completely block the H_2O_2 accumulation (Pellinen *et al.*, 1999).

Several redox sensors are suggested to monitor apoplastic ROS/redox state, but the specific mechanisms involved and the molecular identity of these sensors is, however, less well understood (Laloi *et al.*, 2004). Intracellular ROS are not likely to be sensed through a direct receptor–ligand interaction but, instead, their accumulation alters the redox balance of the cell (Lamb & Dixon, 1997). This can directly affect the activity of transcription factors, second messengers or enzymes involved in biochemical pathways. Alternatively, the cellular redox status can be sensed by the abundant redox-sensitive molecules thioredoxin and glutathione, which, in turn, transmit the signal forward (Vranová *et al.*, 2002). The ultimate targets in all these cases are either the thiol groups of cysteine residues or the iron–sulfur clusters in the catalytic centers of enzymes. See Chapters 2 and 7 for a further discussion of signaling and the role of thiols.

Well characterized examples of redox-regulated signaling components exist mainly in yeast and bacteria, but, recently, the function of NPR1, an essential regulator of systemic acquired resistance (SAR) in *Arabidopsis*, was shown to be regulated through changes in the redox-state of specific cysteine residues (Mou *et al.*, 2003). In the uninduced (oxidized) state, the protein forms homo-oligomers through cysteine bridging and is localized in the cytosol. After an initial oxidative burst, the induction of SAR leads to the accumulation of antioxidants and a more reductive state of the cell. This reduces the cysteine residues in NPR1 and results in monomerization of the protein followed by nuclear localization and activation of defense genes. As discussed later, the O_3-induced responses, changes in gene expression and, ultimately, the cell death visible as O_3-lesions also use this NPR1-dependent signaling. Accordingly, several recent reports have shown that plant O_3 sensitivity or tolerance are intimately linked to, and determined by, several hormones.

11.2   Hormonal control of plant O_3 responses

Plant hormones play key roles in O_3 lesion development, although all the signal transduction pathways involved and the ultimate executioners of cell death itself are unknown. ABA, salicylic acid (SA), ethylene and jasmonic acid (JA) are important in determining the degree of plant ozone sensitivity and O_3 lesion initiation, propagation and containment (reviewed by Overmyer et al., 2003). The role of ABA lies mostly in the regulation of stomatal conductance and O_3 influx, although there are indirect indications that ABA may be involved in other processes as well. Ethylene and SA are needed for the development of O_3-induced cell death and JA acts to counteract the process. This is illustrated by the fact that the mutants and accessions of *Arabidopsis* and also other species, first described as O_3-sensitive, have turned out to be either partially JA insensitive (*Arabidopsis* mutants *oji1*, *rcd1*, ecotype Cvi-0, poplar clone NE-388), or ethylene overproducers (*Arabidopsis* mutant *rcd1* and ecotypes Ws and Kas-1) (Koch et al., 2000; Overmyer et al., 2000; Rao et al., 2000b; Kanna et al., 2003; Tamaoki et al., 2003). Recently, an O_3 sensitive *Arabidopsis* mutant *rcd1* (Overmyer et al., 2000; Fujibe et al., 2004) was also shown to be partially ABA insensitive (Ahlfors et al., 2004a), which suggests that ABA has roles other than stomatal regulation in the mechanisms involved in tolerance to O_3.

The three processes that determine O_3 lesion formation after stomatal control and apoplastic antioxidants, initiation, propagation and containment of the O_3 lesion, can be depicted as a self-amplifying loop termed the oxidative cell death cycle (Van Camp et al., 1998; Overmyer et al., 2003; Figure 11.4). In this cycle, the endogenous ROS production triggered by O_3 is ethylene-dependent (Overmyer et al., 2000; Moeder et al., 2002; Wohlgemuth et al., 2002; Kanna et al., 2003). The ROS production drives lesion propagation and SA-dependent (Örvar et al., 1997; Rao & Davis, 1999) cell death, which continues until the third component, antagonistic to lesion propagation, contains the enlargement of the lesion.

11.2.1   Lesion initiation

Experimental evidence suggests that a certain number of lesion initiation sites exist in leaves. O_3 has been shown to induce death of a small number of cells, visible at the microscopic level also in resistant plants (Overmyer et al., 2000), and when tobacco plants were exposed to O_3 in consecutive days, the number of lesions was determined during the first day and, in the subsequent exposures, the lesions just expanded in perimeter without any new lesion initiations (Wohlgemuth et al., 2002). Whether

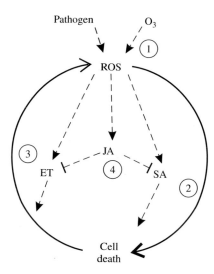

Figure 11.4 The oxidative cell death cycle. Reactive Oxygen Species (ROS) production is triggered by stress, such as ozone or pathogen attack (1). ROS cause accumulation of salicylic acid (SA) and cell death (2), which leads to ethylene production and propagation of cell death (3). Jasmonates function in containment of cell death (4). Modified from Van Camp *et al.* (1998), Rao *et al.* (2002), Overmyer *et al.* (2003) and Tuominen *et al.* (2004).

these lesion initiation sites are predetermined, or random, is not known. What is known, however, is that the O_3 lesions are not randomly distributed throughout the leaf, but are preferentially located in the close proximity of the second- and third-degree veins (Schraudner *et al.*, 1998; Wohlgemuth *et al.*, 2002). In SAR, an analogous ROS-dependent death of a small number of cells close to the veins, termed as micro HR, is essential for the development of SAR (Alvarez *et al.*, 1998). In the systemic leaves cell death – as a response to transmittable signal from the cells in contact with the pathogen – does not spread further from these clusters of cells. This is in contrast to the primary leaves where cells around the primary initiation site die by PCD as a result of signals from the cell death initiation site. This raises the question of whether O_3 lesion formation uses the same process and whether the small number of dead cells in the leaves of O_3-tolerant plants is analogous to the micro HR in the systemic leaves and the enlargement of the lesion in the O_3-sensitive plants is analogous to the spreading of the HR lesion in the primary leaves of pathogen-infected plants. The possible similarities and processes involved are discussed here.

In plants, SA (2-hydroxybenzoic acid) is best known for its role in SAR (Raskin, 1995; Hoeberichts & Woltering, 2002) for which it seems to be essential. It is also required for the execution of HR-like cell death (Durner *et al.*, 1997). Both O_3 exposure and pathogen attack induce SA synthesis within hours of exposure (Örvar *et al.*, 1997). Studies on plant signaling during pathogen attack and oxidative stress have revealed a position for SA both up- and downstream of ROS and cell death (Chamnongpol *et al.*, 1998; Van Camp *et al.*, 1998; Durner & Klessig, 1999).

The role of SA has been mostly demonstrated through experiments utilizing transgenic SA degrading NahG plants and the SA insensitive *npr1* mutant. These studies have shown that oxidative stress-induced cell death is abolished in the absence of SA or its action (Örvar *et al.*, 1997; Rao & Davis, 1999). The integral role for SA in cell death is also supported by experiments in which exogenously applied SA significantly increased the O_3 sensitivity of otherwise tolerant genotypes (Rao *et al.*, 2000b; Mazel & Levine, 2001). Furthermore, double mutant analysis in *Arabidopsis* has shown that O_3-sensitive accessions have become tolerant when either the transgene NahG or the *npr1* mutation has been introduced to the sensitive background (Örvar *et al.*, 1997; Rao *et al.*, 2000b; unpublished results from the authors' laboratory).

All this suggests that SA has a vital role in the process of cell death and that without SA, active PCD is not initiated in O_3-exposed plants. However, Rao and Davis (1999) have shown both O_3-induced necrotic and HR-like cell death, where the mechanism of cell death was dependent on the genotype and the duration of exposure to O_3. In short-term O_3 exposures, the SA-deficient NahG plants were more tolerant to O_3 than the wild-type plants, whereas in long-term exposures, NahG plants developed necrotic lesions without the involvement of the active ROS formation typical to the lesions in short-duration O_3 exposures. In the NahG plants, the necrotic cell death was a result of the depletion of antioxidative capacity, which led to a drastic shift in the cellular redox balance and the death of the cells. Thus, it is possible that O_3, or more likely the ROS directly generated from O_3 degradation (Figure 11.1), can also directly damage components of plasma membrane and cause necrotic death. Accordingly, it has been suggested that both death by rampant oxidation and PCD may occur depending on the O_3 concentration (Pell *et al.*, 1997). This means that the possibility for mosaics of PCD and necrotic cells occurring in the same O_3-exposed tissue cannot be excluded. Accordingly, signals from the cells that have undergone necrotic cell death caused by the ROS directly formed from O_3 may trigger surrounding cells to die by PCD. This results in the beginning of lesion propagation and the formation of the visible lesion.

Lesion propagation, as discussed later, is clearly an ethylene-dependent process, but it has been suggested (Rao *et al.*, 2002) that the O_3-induced ethylene synthesis could be SA-dependent. However, the results presented in Rao *et al.* (2002) do not differentiate if the lower ethylene evolution in O_3-exposed NahG or *npr1* plants was a result of direct regulatory action of SA on ethylene synthesis, or a result of just less initial cell death due to lack of SA or SA responses in these accessions. Thus, whether, and how, SA is involved in the regulation of ethylene biosynthesis is still an open question.

11.2.2   Lesion propagation

The lesion propagation in O_3 damage is under the control of the gaseous plant hormone, ethylene (Overmyer *et al.*, 2000; Rao *et al.*, 2000b, 2002; Moeder *et al.*, 2002; Kanna *et al.*, 2003; Tuominen *et al.*, 2004). The ethylene-insensitive

Arabidopsis mutants (*etr1*, *ein2*) are O_3 tolerant, while both ethylene-overproducing mutants (*eto1*, *eto2*) and jasmonate-insensitive mutants (*jar1*, *coi1*) are O_3 sensitive due to lesion propagation – yet, due to different mechanisms. In ethylene-overproducing mutants, lesion propagation is exacerbated by increased ethylene synthesis, while jasmonate-insensitive mutants are, as described later, deficient in lesion containment due to their apparent inability to downregulate ethylene-dependent lesion propagation.

Ethylene is involved in many developmental and inducible processes (reviewed in Johnson & Ecker, 1998; Wang *et al.*, 2002). Mostly, ethylene levels in plants are extremely low, but in addition to specific developmental cues, many stresses induce rapid ethylene synthesis (Wang *et al.*, 2002). Stress ethylene production is also intimately linked to O_3 damage (Tingey *et al.*, 1976; Mehlhorn & Wellburn, 1987; Overmyer *et al.*, 2000; Rao *et al.*, 2002) and ethylene production is one of the fastest responses of plants to O_3 (Tuomainen *et al.*, 1997; Vahala *et al.*, 1998; Overmyer *et al.*, 2000). The amount of ethylene produced and the extent of cell death are well correlated (Langebartels *et al.*, 1991; Tuomainen *et al.*, 1997; Tamaoki *et al.*, 2003). Therefore, a role in promoting cell death has been assigned to ethylene in ROS-dependent PCD (de Jong *et al.*, 2002; Overmyer *et al.*, 2003).

Ethylene biosynthesis is regulated at the level of the production of its immediate precursor 1-aminocyclopropane-1-carboxylic acid (ACC), synthesized by ACC synthases (ACSs) from *S*-adenosyl-methionine. The conversion of ACC to ethylene by ACC oxidases, and the simultaneous production of CO_2 and cyanide (HCN), is considered to be less rigorously controlled (Wang *et al.*, 2002). O_3-induced ethylene biosynthesis is the result of ACS gene induction. ACSs are a gene family consisting of several members that show differential expression patterns in plant organs during development and as a response to external challenges. In *Arabidopsis*, a specific member of the gene family, *ACS6*, is rapidly induced by O_3 (Vahala *et al.*, 1998; Overmyer *et al.*, 2000; Tamaoki *et al.*, 2003). Similarly, in tomato, a bi-phasic, sequential induction of ACS genes by O_3 has been observed (Nakajima *et al.*, 2001; Moeder *et al.*, 2002): first, the *LE-ACS1B* and *LE-ACS6* genes were upregulated by O_3 very rapidly, followed by a sharp reduction in their mRNA levels and a simultaneous induction of *LE-ACS2*.

Antisense suppression of *LE-ACS2* expression resulted in an almost complete block of ethylene synthesis in developing tomato fruit (Oeller *et al.*, 1991). In O_3-exposed *LE-ACS2* antisense plants, however, ethylene biosynthesis increased rapidly during the first phase, but the increase in ethylene evolution during the second phase, when *LE-ACS2* induction normally takes place, did not occur (Moeder *et al.*, 2002). Accordingly, O_3 sensitivity of the *LE-ACS2* antisense plants did not differ from the untransformed control. On the contrary, when the tomato *LE-ACS6* was transformed to tobacco in antisense orientation, low initial rates of O_3-induced ethylene production were observed in the transgenic plants and the extent of O_3 damage was lower than in the non-transformed controls (Nakajima *et al.*, 2002). This suggests that the enzymes encoded by the sequentially induced members of the ACS gene family may have a distinct function in the process and that the first phase ethylene synthesis was sufficient to trigger the ethylene-dependent processes that resulted in O_3 lesion formation.

One further possibility presented for ethylene in the promotion of cell death is the formation of HCN – a toxic by-product of ethylene biosynthesis – formed in stoichiometric amounts with ethylene and CO_2 by ACC oxidase (Abeles *et al.*, 1992). It has been suggested (Grossmann *et al.*, 1993; Grossmann, 1996) that HCN may play a role in cell death in specific stresses, which are connected with highly elevated ethylene production. Normally, β-cyanoalanine synthase (β-CAS) detoxifies HCN to β-cyanoalanine in the cells where ethylene is formed. High ethylene production *per se* is not stimulatory for cell death, but, under specific circumstances where ethylene synthesis is highly elevated and ethylene sensitivity is compromised, detoxification of HCN by β-CAS may become compromised since the induction of β-CAS gene is strongly ethylene-dependent (Vahala *et al.*, 2003). In O_3-exposed tomato, ethylene synthesis (visualized as the activation of *LE-ACO1* gene) was evident only in small clusters of cells with similar spatial pattern as the later developing lesions (Moeder *et al.*, 2002). This suggests a high local synthesis and concentration of ethylene in these clusters, and also that the induction of ethylene receptor genes observed took place in the same area. As discussed in the next paragraph, increased synthesis of ethylene receptor results in decreased ethylene, and O_3, sensitivity. In transgenic, ethylene-insensitive birch trees, however, instead of the expected complete tolerance to O_3, insensitivity to ethylene provided only a partial protection against O_3, when the prevention of ethylene biosynthesis in the same ET-insensitive trees blocked cell death completely (Vahala *et al.*, 2003). This suggests that ethylene synthesis contributed to O_3 sensitivity also by mechanisms that were independent of ethylene perception. If the O_3-induced ethylene synthesis decreases plant ethylene sensitivity by the activation of genes encoding ethylene receptor, which at the same time also prevents the activation of β-CAS, the deficient induction of β-CAS may compromise the HCN detoxification leading to cell death.

11.2.3   Lesion containment

Without a counteracting regulatory system that limits the spread of the lesion once it is initiated, induction of spreading cell death would result in a progressive destruction of the whole organ. Two different mechanisms can be suggested to be responsible for the containment of lesion spread.

Ethylene itself is, in a way, responsible for the containment of (the ethylene-dependent) O_3 lesion propagation since ethylene synthesis also induces a desensitization of the cells to the hormone. Ethylene binding to the receptor has dissociation rates in the yeast-expressed ETR1 of more than 10 h (Bleecker, 1999). This raises the question of how ethylene responses can be downregulated after the initial association of the hormone with the receptor. Ethylene receptor acts as suppressor of ethylene signaling when it is not in contact with the hormone. It is thought that the ethylene-induced synthesis of new, unoccupied receptor molecules in response to stress is responsible for the desensitization of ethylene signaling and, thereby, rapid elimination of the ethylene responses. Induction of certain receptor isoforms has been described after ethylene treatment (Wilkinson *et al.*, 1995) and also in O_3-exposed tomato (Moeder *et al.*, 2002). Thus, the O_3-induced synthesis of new ethylene receptor proteins could, in part, lead into decreased ethylene sensitivity and downregulation of ethylene-dependent lesion spread.

However, other hormones also act in lesion containment. JA and its methyl ester, methyl jasmonate (MeJA), are the most studied of the linolenic-acid-derived signaling molecules in plants that are collectively referred to as oxylipins or jasmonates (Farmer *et al.*, 1998). Exposure to O_3, as well as pathogen attack, stimulates jasmonate biosynthesis in plants (Blechert *et al.*, 1995; Penninckx *et al.*, 1998; Rao *et al.*, 2000b; Tuominen *et al.*, 2004). This stimulation could be partly due to membrane lipid peroxidation that creates free fatty acids for JA biosynthesis, but also a receptor-mediated induction is thought to exist, since fungal elicitors and plant cell wall fragments stimulate JA production (Creelman & Mullet, 1997). In the oxidative cell death cycle, jasmonates seem to protect tissues from ROS-induced cell death and thus counteract the effects of SA and ET (Figure 11.4). Örvar *et al.* (1997) showed that pre-treating tobacco plants with jasmonates inhibited O_3-induced cell death. The same was evident in *Arabidopsis*, where jasmonate treatment also reduced the amount of SA produced in response to O_3 (Rao *et al.*, 2000b), jasmonate insensitive and deficient mutants were hypersensitive to O_3 (Overmyer *et al.*, 2000; Rao *et al.*, 2000b), and MeJA applied after O_3 halted lesion spread in the O_3-sensitive *rcd1* mutant (Overmyer *et al.*, 2000). Not all JA-insensitive mutants are, however, sensitive to O_3. The jasmonate-insensitive *jin1* is tolerant to O_3, as well as to necrotrophic pathogens (Nickstadt *et al.*, 2004), which suggests that JA has, at least, two different roles in O_3-responses. Accordingly, it was shown recently that the MYC-type transcription factor (AtMYC2) encoded by *JIN1* is required to discriminate between two different branches of jasmonate responses (Lorenzo *et al.*, 2004). Both these branches of JA signaling seem to be, however, involved in O_3-related processes, one in lesion formation and the other in lesion containment, since lesion formation seems to require AtMYC2 (JIN1) and lesion containment requires JAR1 and COI1.

11.2.4   Interactions between the hormonal signaling cascades

According to the current understanding, hormonal control seems to be a more important factor than the antioxidative capacity in determining plant O_3 sensitivity. Overall, the evidence on hormonal control of plant O_3 responses is strong and the balance between SA, ethylene and JA is very likely to be largely responsible for the processes induced (Overmyer *et al.*, 2003; Tamaoki *et al.*, 2004). The balance can be accomplished by mutual inhibition of biosynthesis, as e.g. is the case with SA and JA (Peña-Cortés *et al.*, 1993; Doares *et al.*, 1995; Örvar *et al.*, 1997). Since JA prevented the O_3-induced accumulation of SA (Örvar *et al.*, 1997), it seems that JA's antagonism to cell death is mediated at least partly through its effect on SA.

JA also antagonized ethylene signaling (Tuominen *et al.*, 2004). This interaction was mutually antagonistic, since ethylene inhibited JA-induced gene expression as well. Because ethylene levels in O_3-treated JA-insensitive *jar1* mutant were similar to wild type (Overmyer *et al.*, 2000) and MeJA application did not change ethylene levels in *eto1* (Tuominen *et al.*, 2004), JA antagonism on ethylene is not likely to take place at the level of biosynthesis. The triple response assay for ethylene suggested that JA action on ethylene was downstream of biosynthesis but upstream of CTR1, a raf-type protein kinase active in ethylene signaling just downstream of the receptor (Tuominen *et al.*, 2004). Thus, JA could affect ethylene signaling at the receptor level. Accordingly, JA-induced upregulation of an ethylene receptor isoform has been discovered in microarray studies (Schenk *et al.*, 2000).

The role of JA in these processes is, however, complicated. In birch and *Arabidopsis*, both ethylene and JA accumulation correlated with O_3-induced cell death in the O_3-sensitive accessions, whereas in the O_3-tolerant accessions, JA concentration did not increase (Vahala *et al.*, 2003; Tuominen *et al.*, 2004). The involvement of JA in lesion containment may at first seem contradictory to JA accumulation in the sensitive accessions and not in the tolerant ones. However, because JA biosynthesis appears to be limited by substrate availability (Laudert *et al.*, 2000; Ziegler *et al.*, 2001), it is also possible that the accumulation of JA is a consequence of the ET-dependent cell death: the substrate for JA synthesis (α-linolenic acid or 13-(*S*)-hydroperoxylinolenic acid) could be released from the dying cells, thus resulting in increased JA synthesis, which then halts the ET-dependent lesion propagation by affecting the ethylene sensitivity of the cells by increased ethylene receptor synthesis. On the tissue level, the balance probably shifts temporally and spatially so that in the initial cells the SA- and ET-driven processes prevail, but further away from the site of initiation the JA pathways become progressively more induced. This response overcomes the initial processes and containment of cell death follows.

11.3   Other regulators of plant O_3 responses

11.3.1   Induction of plant volatiles

Leaves normally release small quantities of volatile compounds, but when a plant is damaged by biotic or abiotic factors, the quantity and the number of volatiles increase dramatically (Croft *et al.*, 1993; Heiden *et al.*, 1999). Under various stress conditions, the interaction between ROS, ethylene, salicylate, jasmonate (and other molecules like the gaseous NO) is thought to determine the extent of lesion formation. This common mechanism forms the basis of the principal structural uniformity of the emitted compounds in various plant families, albeit with a specific 'bouquet' for individual species.

With regard to the signaling compounds of the oxidative cell death cycle, it is remarkable to note that all of them can form volatile metabolites. This is evident for the ethylene pathway where the hormone itself is the volatile compound. Methyl salicylate, which occurs as a volatile SA metabolite in TMV-infected and O_3-treated tobacco, was effective in mediating defense reactions from the infected to a control plant, *via* the air path (Shulaev *et al.*, 1997). MeJA, the volatile form of JA and a containment signal in the oxidative cell death cycle, is released from pathogen-infected plants. It is effective in triggering a subset of defense-related genes, e.g. thionins, defensin and a hevein-like protein (Reymond & Farmer, 1998). In addition to jasmonate biosynthesis, 13-lipoxygenase together with hydroperoxide lyase ultimately converts 13-hydroperoxylinolenic acid to volatile C6 aldehydes and alcohols, among which 3-hexen-1-ol and 3-hexen-1-al are major components of the 'green odor' in plants (Hatanaka, 1993). It was demonstrated by Heiden *et al.* (1999) that these defined products of the 13-lipoxygenase pathway are released from O_3-treated tobacco during post-cultivation in pollutant-free air. The spectrum of the major emitted compounds clearly pointed to a specific activation of biosynthetic pathways by the air pollutant O_3 rather than to unspecific lipid peroxidation processes.

When the spatial and temporal episodes of O_3 induction are considered *in toto*, it is evident that following O_3 exposure, and probably oxidative stress in general, intracellular signal pathways and defense reactions are activated first (episode I). In the case of pathogen attack, the responses are directed to protect the invaded cell. In episode II, local cell-to-cell signaling leads to an induction of antioxidant and antimicrobial defense responses on one hand and damage amplifying reactions on the other in the vicinity of the attacked cell. Episode III comprises of systemic movement of signal molecules in the vascular tissue (SA and additional compounds, the ethylene precursor ACC) and, in parallel, emission of volatile compounds, which may transmit the message through the gas phase to distal tissues of the leaf and the plant, and, putatively, to neighbor plants. In addition to volatile metabolites, conjugates of ACC, SA, jasmonate and cis-3-hexen-1-ol are formed in episode III. Both processes help to limit the effective time period for these highly active signal molecules by releasing them into the surrounding air (volatile compounds) or by storing them (in an inactive conjugate form) in the vacuole for rapid release of the signal in subsequent stress situations.

When the oxidative cell death cycle is operative in O_3-sensitive plants, it leads to highly increased VOC emission by two to three orders of magnitude. This is not only found for the signal molecules (e.g. 50 times for ethylene and 600 times for methyl salicylate emissions; Heiden *et al.*, 1999), but also for several sesquiterpenes that are emitted on the day following the treatment (e.g. 100 times for valencene). The O_3-triggered emission of sesquiterpenes and other compounds that possess O_3 generation potential in turn leads to back-coupling effects, and potentially elevated O_3 levels under prolonged exposure.

Various O_3-induced metabolic changes are maintained in plants over days to months. Therefore, O_3 episodes in the field may be connected to later infection periods through these 'memory' effects (Sandermann, 1998; Langebartels & Kangasjärvi, 2004). The changed metabolite status of O_3-exposed plants may lead to either enhanced or decreased likelihood of disease, depending on the individual plant, herbivore species or viral, bacterial or fungal pathogen.

11.3.2 O_3 and mitogen-activated protein kinases

Mitogen-activated protein-kinase (MAPK) cascades are a conserved signal transduction system present in all eukaryotes and their importance for plants is also well known (Jonak et al., 2002; see Chapter 7 for further discussion of MAPKs in relation to ROS). The universal structure of this kind of a signaling module consists of a MAP kinase kinase kinase (MAPKKK) that phosphorylates a MAP kinase kinase (MAPKK), which, in turn, phosphorylates a MAP kinase (MAPK). Phosphorylation generally leads to nuclear localization of, and transcription factor activation by, the MAPK.

Although *Arabidopsis* genome codes for 20 different MAPKs (MAPK Group, 2002), only a few have been studied in more detail. The two primary oxidative stress-responsive MAPKs are AtMPK6 and AtMPK3. They, and their orthologs in tobacco (SIPK and WIPK), are induced among other stresses by O_3 (Samuel et al., 2000; Samuel & Ellis, 2002; Ahlfors et al., 2004b). AtMPK6/SIPK and AtMPK3/WIPK are also activated by H_2O_2 and O_2^- (Kovtun et al., 2000; Samuel et al., 2000; Moon et al., 2003).

Although AtMPK3 and AtMPK6 are activated both in response to pathogens and oxidative stress, the MAP kinase cascades induced by these stresses seem to be different: the H_2O_2-induced MAPK cascade started with the activation of the MAPKKK ANP1, followed by the activation of AtMPK3 and AtMPK6, and finally the expression of an oxidative stress-induced gene *GST6* (Kovtun et al., 2000). The MAPKK leading to MAPK activation was not identified in this study. In contrast, the MAPKKK activated by the bacterial elicitor flg22 was MEKK1, which activated two MAPKKs, MKK4 and MKK5, which, in turn, activated AtMPK3 and AtMPK6. Additionally, MKK4 and MKK5 were not effective in inducing expression of *GST6*, a gene that responds to H_2O_2 (Asai et al., 2002).

Despite the rapid activation of both AtMPK6 and AtMPK3 by O_3, their activity under O_3 exposure is differentially regulated. AtMPK3 was upregulated by O_3 on transcriptional, translational as well as on post-translational levels whereas only post-translational stimulation of the kinase activity of AtMPK6 was detected by Ahlfors et al. (2004b). In addition, the activation of AtMPK3 lasted longer than that of AtMPK6. The plant O_3 sensitivity and the expression of antioxidant genes is affected by these kinase classes, since both suppression and overexpression of SIPK (the AtMPK6 ortholog) led to increased O_3 sensitivity and changes in the expression of *APX* and *GST* (Samuel & Ellis, 2002). Interestingly, SIPK also seems to regulate the activity of WIPK (the AtMPK3 ortholog), since in the SIPK overexpression lines, WIPK activity induced by O_3 was significantly reduced, whereas the opposite was true for the SIPK suppression line (Samuel & Ellis, 2002). However, the interaction of these MAP kinases requires further study, because, in opposition to the relationship described above, SIPK was found to be necessary for WIPK transcription and activation in TMV-inoculated tobacco (Liu *et al.*, 2003).

In addition to being activated by ROS, the activation of AtMPK6 and AtMPK3 by constitutively active forms of AtMKK4 and AtMKK5 induced endogenous H_2O_2 production and cell death in *Arabidopsis* (Ren *et al.*, 2002). Since MAP kinases are connected to both ROS and cell death, their relationship to plant hormones is interesting. MAPK involvement in ethylene, SA and ABA signaling has been established in different contexts (Zhang & Klessig, 1997; Kim *et al.*, 2003; Ouaked *et al.*, 2003; Xiong & Yang, 2003) but the question of whether plant hormones regulate MAP kinase signaling or *vice versa* in O_3-induced cell death has not been resolved. Ahlfors *et al.* (2004b) showed that the initial activation of AtMPK6 and AtMPK3 during O_3 exposure was independent or upstream of SA, ethylene or JA. However, hormone signaling is intertwined with these MAPKs since the activation of AtMPK6 was prolonged and the induction of AtMPK3 activation delayed in the ethylene insensitive *etr1* mutant. Additionally, the basal expression level of AtMPK3 was only half of the wild-type levels in SA-insensitive and deficient accessions and these lower levels of expression were also reflected in AtMPK3 activity. Kim *et al.* (2003) suggest the involvement of tobacco SIPK (AtMPK6 homolog) in the induction of ethylene biosynthesis and the downregulation of this kinase by ethylene-dependent feedback loop. This model would be in accordance with the prolonged AtMPK6 activation in *etr1* reported by Ahlfors *et al.* (2004b).

11.3.3   G-proteins

O_3 is believed to be perceived in the apoplast (Kangasjärvi *et al.*, 1994). The membrane-bound GTPases (G-proteins) that transduce extracellular signals to intracellular processes are involved in several different processes in eukaryotic organisms. In plants, G-proteins are thought to be involved in interactions with plant hormones and plant defense responses to wounding and pathogen infection (Assmann, 2002). Recently, Booker *et al.* (2004) showed that insertion mutants of *Arabidopsis* canonical G-protein α subunit gene (*GPA1*) were more tolerant to O_3 than the wild-type Col-0, suggesting that these null mutants had altered perception of O_3 or downstream signals. This further supports the notion that the action of O_3 in plants is due to processes perceived, transduced and induced by the network of regulatory processes in the cells affected.

G-proteins are also involved in ABA signaling (Wang *et al.*, 2001; Coursol *et al.*, 2003; Pandey & Assmann, 2004). T-DNA insertion mutants of the *GPA1* gene exhibited insensitivity to ABA in the inhibition of stomatal opening whereas the related process of promotion of stomatal closure was unaffected (Wang *et al.*, 2001). However, the *gpa1* mutants did have increased transpiration from detached leaves. As discussed earlier, stomatal conductivity might be one of the factors regulating plant sensitivity to O_3 but the comparison of results presented by Booker *et al.* (2004) and Wang *et al.* (2001) indicates, however, that this is not the sole determinant of O_3 sensitivity. Additionally, knocking out the putative G-protein coupled receptor GCR1 that has been shown to interact with GPA1 renders plants hypersensitive to ABA with respect to root growth inhibition, ABA-induced gene expression and stomatal responses. These plants were also more drought tolerant than wild type, probably due to their lower transpiration rates. Therefore, GCR1 is thought to be a negative regulator of ABA responses (Pandey & Assmann, 2004). However, this was not reflected in the O_3 sensitivity of the null mutants used by Booker et al. (2004) which did not differ from the wild type. Thus, the interaction between ABA signaling, G-proteins and O_3 tolerance clearly requires further studies.

The suggested involvement of G-proteins in the perception and/or early signaling of O_3 (Booker et al., 2004) creates an intriguing connection with the recent identification (Ahlfors et al., 2004a) of the gene mutated in the O_3-sensitive Arabidopsis mutant rcd1. The RCD1 protein has domains and features that suggest that it might be involved in ADP ribosylation of proteins – the trimeric G-protein is one of the known examples of proteins that can be regulated by ADP ribosylation. Whether there is a mechanistic connection between these proteins and processes in O_3-induced responses remains to be seen in the future.

11.4   Conclusions

The processes regulating the entry of O_3 in the leaf and the early apoplastic reactions, and the downstream responses after the perception of O_3/ROS have been elucidated to the degree where the interactions of the processes can be understood and the whole picture and the similarities with other similar processes have become visible. However, the perception of O_3 and ROS, in general, is still mostly unresolved and the components involved unidentified. O_3 exposure and other abiotic stresses seem to have more in common than believed so far. Also, the extent to which the signal transduction pathways are shared between pathogen response and oxidative stress is still poorly understood and the topic clearly requires further studies.

References

Abeles, F.B., Morgan, P.W. and Saltveit, M.E. (1992) Ethylene in Plant Biology, 2nd edn., Academic Press, San Diego, CA, USA.
Ahlfors, R., Lång, S., Overmyer, K. et al. (2004a) 'Arabidopsis thaliana RADICAL-INDUCED CELL DEATH 1 belongs to the WWE protein–protein interaction-domain protein family and modulates abscisic acid, ethylene, and methyl jasmonate responses', The Plant Cell 16, 1925–1937.
Ahlfors, R., Macioszek, V., Rudd, J. et al. (2004b) 'Stress hormone-independent activation and nuclear translocation of mitogen-activated protein kinases (MAPKs) in Arabidopsis thaliana plants during ozone exposure', The Plant Journal 40, 512–522.
Alvarez, M.E., Pennell, R.I., Meijer, P.-J., Ishikawa, A., Dixon, R.A. and Lamb, C. (1998) 'Reactive oxygen intermediates mediate a systemic signal network in the establishment of plant immunity', Cell 92, 773–784.
Asai, T., Tena, G., Plotnikova, J. et al. (2002) 'MAP kinase signalling cascade in Arabidopsis innate immunity', Nature 415, 977–983.

Assmann, S.M. (2002) 'Heterotrimeric and unconventional GTP binding proteins in Plant cell signaling', *The Plant Cell* **14**, S355–S373.

Beers, E.P. and McDowell, J.M. (2001) 'Regulation and execution of programmed cell death in response to pathogens, stress and developmental cues', *Current Opinion in Plant Biology* **4**, 561–567.

Berger, S. (2002) 'Jasmonate-related mutants of *Arabidopsis* as tools for studying stress signals', *Planta* **214**, 497–504.

Blechert, S., Brodschelm, W., Hölder, S. *et al.* (1995) 'The octadecanoic pathway: signal molecules for the regulation of secondary pathways', *Proceedings of the National Academy of Sciences USA* **92**, 4099–4105.

Bleecker, A.B. (1999) 'Ethylene perception and signaling: an evolutionary perspective', *Trends in Plant Science* **4**, 269–274.

Bolwell, G.P., Bindschedler, L.V., Blee, K.A. *et al.* (2002) 'The apoplastic oxidative burst in response to biotic stress in plants: a three-component system', *Journal of Experimental Botany* **53**, 1367–1376.

Booker, F.L., Burkey, K.O., Overmyer, K. and Jones, A.M. (2004) 'Differential responses of G-protein *Arabidopsis thaliana* mutants to ozone', *New Phytologist* **162**, 633–641.

Buckland, S.M., Price, A.H. and Hendry, G.A.F. (1991) 'The role of ascorbate in drought-treated *Cochlearia atlantica* (Pobed.) and *Armeria maritima* (Mill.) Willd', *The New Phytologist* **119**, 155–160.

Castillo, F.J. and Heath, R.L. (1990) 'Ca^{2+} transport in membrane vesicles from pinto bean leaves and its alteration after ozone exposure', *Plant Physiology* **94**, 788–795.

Chameides, W.L. (1989) 'The chemistry of ozone deposition to plant leaves: role of ascorbic acid', *Environmental Science and Technology* **23**, 595–600.

Chamnongpol, S., Willekens, H., Moeder, W. *et al.* (1998) 'Defense activation and enhanced pathogen tolerance induced by H_2O_2 in transgenic tobacco', *Proceedings of the National Academy of Sciences USA* **95**, 5818–5823.

Clayton, H., Knight, M.R., Knight, H., McAinsh, M.R. and Hetherington, A.M. (1999) 'Dissection of the ozone-induced calcium signature', *The Plant Journal* **17**, 575–579.

Conklin, P.L., Williams, E.H. and Last, R.L. (1996) 'Environmental stress sensitivity of an ascorbic acid-deficient *Arabidopsis* mutant', *Proceedings of the National Academy of Sciences USA* **93**, 9970–9974.

Coursol, S., Fan, L.-M., Le Stunff, H., Spiegel, S., Gilroy, S. and Assmann, S.M. (2003) 'Sphingolipid signalling in *Arabidopsis* guard cells involves heterotrimeric G proteins', *Nature* **423**, 651–654.

Creelman, R.A. and Mullet, J.E. (1997) 'Biosynthesis and action of jasmonates in plants', *Annual Review of Plant Physiology and Plant Molecular Biology* **48**, 355–381.

Croft, K.P.C., Jüttner, F. and Slusarenko, A.J. (1993) 'Volatile products of the lipoxygenase pathway evolved from *Phaseolus vulgaris* (L.) leaves inoculated with *Pseudomonas syringae* pv *phaseolicola*', *Plant Physiology* **101**, 13–24.

de Jong, A., Yakimova, E.T., Kapchina, V.M. and Woltering, E.J. (2002) 'A critical role for ethylene in hydrogen peroxide release during programmed cell death in tomato suspension cells', *Planta* **214**, 537–545.

del Rio, L.A., Corpas, F.J., Sandalio, L.M., Palma, J.M. and Barroso, J.B. (2003) 'Plant peroxisomes, reactive oxygen metabolism and nitric oxide', *IUBMB Life* **55**, 71–81.

Doares, S.H., Narváez-Vásquez, J., Conconi, A. and Ryan, C.A. (1995) 'Salicylic acid inhibits synthesis of proteinase inhibitors in tomato leaves induced by systemin and jasmonic acid', *Plant Physiology* **108**, 1741–1746.

Durner, J. and Klessig, D.F. (1999) 'Nitric oxide as a signal in plants', *Current Opinion in Plant Biology* **2**, 369–374.

Durner, J., Shah, J. and Klessig, D.F. (1997) 'Salicylic acid and disease resistance in plants', *Trends in Plant Science* **2**, 266–274.

Farmer, E.E., Weber, H. and Vollenweider, S. (1998) 'Fatty acid signaling in *Arabidopsis*', *Planta* **206**, 167–174.

Finkelstein, R.R., Gampala, S.S.R. and Rock, C.D. (2002) 'Abscisic acid signaling in seeds and seedlings', *The Plant Cell* **14**, S15–S45.

Foyer, C.H. and Halliwell, B. (1976) 'The presence of glutathione and glutathione reductase in chloroplasts: a proposed role in ascorbic acid metabolism', *Planta* **133**, 21–25.

Fryer, M.J., Oxborough, K., Mullineaux, P.M. and Baker, N.R. (2002) 'Imaging of photo-oxidative stress responses in leaves', *Journal of Experimental Botany* **53**, 1249–1254.

Fujibe, T., Saji, H., Arakawa, K., Yabe, N., Takeuchi, Y. and Yamamoto, K.T. (2004) 'A methyl viologen-resistant mutant of *Arabidopsis*, which is allelic to ozone-sensitive *rcd1*, is tolerant to supplemental UV-B irradiation', *Plant Physiology* **134**, 275–285.

Giamalva, D., Church, D.F. and Pryor, W.A. (1985) 'A comparison of the rates of ozonation of biological antioxidants and oleate and linoleate esters', *Biochemical and Biophysical Research Communications* **133**, 773–779.

Grimes, H.D., Perkins, K.K. and Boss, W.F. (1983) 'Ozone degrades into hydroxyradical under physiological conditions. A spin trapping study', *Plant Physiology* **72**, 1016–1020.

Grossmann, K. (1996) 'A role for cyanide, derived from ethylene biosynthesis, in the development of stress symptoms', *Physiologia Plantarum* **97**, 772–775.

Grossmann, K., Siefert, F., Kwiatkowski, J., Schraudner, M., Langebartels, C. and Sandermann, H. Jr. (1993) 'Inhibition of ethylene production in sunflower cell suspensions by the plant growth retardant BAS 111.W: possible relations to changes in polyamine and cytokinin contents', *Journal of Plant Growth Regulation* **12**, 5–11.

Gunderson, C.A. and Taylor, G.E. (1991) 'Ethylene directly inhibits foliar gas-exchange in *Glycine max*', *Plant Physiology* **95**, 337–339.

Halliwell, B. and Gutteridge, J.M.C., eds. (1999) *Free Radicals in Biology and Medicine*, 3rd edn., Oxford University Press, Oxford, UK.

Hatanaka A. (1993) 'The biogeneration of green odour by green leaves', *Phytochemistry* **34**, 1201–1218.

Heath, R.L. and Taylor, G.E. Jr. (1997) Physiological processes and plant responses to ozone exposure, in *Forest Decline and Ozone* (ed. H. Sandermann, A.R. Wellburn and R.L. Heath), Springer-Verlag, Berlin, Heidelberg, pp. 317–368.

Heiden, A.C., Hoffmann, T., Kahl, J. *et al.* (1999) 'Emission of volatile organic compounds from ozone-exposed plants', *Ecological Applications* **9**, 1160–1167.

Hideg, É., Csengele, B., Kálai, T., Vass, I., Hideg, K. and Asada, K. (2002) 'Detection of singlet oxygen and superoxide with fluorescent sensors in leaves under stress by photoinhibition or UV radiation', *Plant Cell Physiology* **43**, 1154–1164.

Hill, A.C. and Littlefield, N. (1969) 'Ozone. Effect on apparent photosynthesis rate of transpiration and stomatal closure in plants', *Environmental Science and Technology* **3**, 52–56.

Himmelbach, A., Yang, Y. and Grill, E. (2003) 'Relay and control of abscisic acid signaling', *Current Opinion in Plant Biology* **6**, 470–479.

Hoeberichts, F.A. and Woltering, E.J. (2002) 'Multiple mediators of plant programmed cell death: interplay of conserved cell death mechanisms and plant-specific regulators', *BioEssays* **25**, 47–57.

Horemans, N., Asard, H. and Caubergs, R.J. (1994) 'The role of ascorbate free radical as an electron acceptor to cytochrome b-mediated trans-plasma membrane electron transport in higher plants', *Plant Physiology* **104**, 1455–1458.

Igamberdiev, A.U. and Lea, P.J. (2002) 'The role of peroxisomes in the integration of metabolism and evolutionary diversity of photosynthetic organisms', *Phytochemistry* **60**, 651–674.

Jabs, T., Tschöpe, M., Colling, C., Hahlbrock, K. and Scheel, D. (1997) 'Elicitor-stimulated ion fluxes and O_2^- from the oxidative burst are essential components in triggering defense gene activation and phytoalexin synthesis in parsley', *Proceedings of the National Academy of Sciences USA* **94**, 4800–4805.

Jakob, B. and Heber, U. (1998) 'Apoplastic ascorbate does not prevent the oxidation of fluorescent amphiphilic dyes by ambient and elevated concentrations of ozone in leaves', *Plant Cell Physiology* **39**, 313–322.

Johnson, P.R. and Ecker, J.R. (1998) 'The ethylene gas signal transduction pathway: a molecular perspective' *Annual Review of Genetics* **32**, 227–254.

Jonak, C., Okresz, L., Bogre, L. and Hirt, H. (2002) 'Complexity, cross talk and integration of plant MAP kinase signalling', *Current Opinion in Plant Biology* **5**, 415–424.

Kangasjärvi, J., Talvinen, J., Utriainen, M. and Karjalainen, R. (1994) 'Plant defence systems induced by ozone', *Plant, Cell and Environment* **17**, 783–794.

Kanna, M., Tamaoki, M., Kubo, A. *et al.* (2003) 'Isolation of an ozone-sensitive and jasmonate-semi-insensitive *Arabidopsis* mutant (*oji1*)', *Plant Cell Physiology* **44**, 1301–1310.

Kerstiens, G. and Lendzian, K.J. (1989) 'Interactions between ozone and plant cuticles. 1. Ozone deposition and permeability', *The New Phytologist* **112**, 13–19.

Kim, C.Y., Liu, Y., Thorne, E.T. *et al.* (2003) 'Activation of a stress-responsive mitogen-activated protein kinase cascade induces the biosynthesis of ethylene in plants', *The Plant Cell* **15**, 2707–2718.

Koch, J.R., Creelman, R.A., Eshita, S.M., Seskar, M., Mullet, J.E. and Davis, K.R. (2000) 'Ozone sensitivity in hybrid poplar correlates with insensitivity to both salicylic acid and jasmonic acid. The role of programmed cell death in lesion formation', *Plant Physiology* **123**, 487–496.

Kollist, H. (2001) Leaf apoplastic ascorbate as ozone scavenger and its transport across the plasma membrane. Ph.D. Dissertation, University of Tartu.

Kollist, H., Moldau, H., Mortensen, L., Rasmussen, S.K. and Jorgensen, L.B. (2000) 'Ozone flux to plasmalemma in barley and wheat is controlled rather by stomata than by direct reaction of ozone with apoplastic ascorbate', *Journal of Plant Physiology* **156**, 645–651.

Kovtun, Y., Chiu, W.-L., Tena, G. and Sheen, J. (2000) 'Functional analysis of oxidative stress activated mitogen-activated protein kinase cascades in plants', *Proceedings of the National Academy of Sciences USA* **97**, 2940–2945.

Kwak, J.M., Mori, I.C., Pei, Z.-M. *et al.* (2003) 'NADPH oxidase *AtrbohD* and *AtrbohF* genes function in ROS-dependent ABA signaling in *Arabidopsis*', *EMBO Journal* **22**, 2623–2633.

Laisk, A., Kull, O. and Moldau, H. (1989) 'Ozone concentration in leaf intercellular air spaces is close to zero', *Plant Physiology* **90**, 1163–1167.

Laloi, C., Apel, K. and Danon, A. (2004) 'Reactive oxygen signalling: the latest news', *Current Opinion in Plant Biology* **7**, 323–328.

Lamb, C. and Dixon, R.A. (1997) 'The oxidative burst in plant disease resistance', *Annual Review of Plant Physiology and Plant Molecular Biology* **48**, 251–275.

Langebartels, C. and Kangasjärvi, J. (2004) Ethylene and jasmonate as regulators of cell death in disease resistance, in *Molecular Ecotoxicology of Plants* (ed. H. Sandermann), Springer-Verlag, Berlin, Heidelberg, pp. 75–110.

Langebartels, C., Kerner, K., Leonardi, S. *et al.* (1991) 'Biochemical plant responses to ozone I. Differential induction of polyamine and ethylene biosynthesis in tobacco', *Plant Physiology* **95**, 882–889.

Langebartels, C., Wohlgemuth, H., Kschieschan, S., Grün, S. and Sandermann, H. (2002) 'Oxidative burst and cell death in ozone-exposed plants', *Plant Physiology and Biochemistry* **40**, 567–575.

Laudert, D., Schaller, F. and Weiler, E.W. (2000) 'Transgenic *Nicotiana tabacum* and *Arabidopsis thaliana* plants overexpressing allene oxide synthase', *Planta* **211**, 163–165.

Levine, A., Tenhaken, R., Dixon, R. and Lamb, C. (1994) 'H_2O_2 from the oxidative burst orchestrates the plant hypersensitive disease resistance response', *Cell* **79**, 583–593.

Liu, Y., Jin, H., Yang, K.-Y., Kim, C.Y., Baker, B. and Zhang, S. (2003) 'Interaction between two mitogen-activated protein kinases during tobacco defense signaling', *The Plant Journal* **34**, 149–160.

Lorenzo, O., Chico, J.M., Sánchez-Serrano, J.J. and Solano, R. (2004) '*JASMONATE-INSENSITIVE1* encodes a MYC transcription factor essential to discriminate between different jasmonate-regulated responses in *Arabidopsis*', *The Plant Cell* **16**, 1938–1950.

MAPK Group (2002) 'Mitogen-activated protein kinase cascades in plants: a new nomenclature', *Trends in Plant Science* **7**, 301–308.

Mazel, A. and Levine, A. (2001) 'Induction of cell death in *Arabidopsis* by superoxide in combination with salicylic acid or with protein synthesis inhibitors', *Free Radical Biology and Medicine* **30**, 98–106.

McAinsh, M.R., Clayton, H., Mansfield, T.A. and Hetherington, A.M. (1996) 'Changes in stomatal behaviour and guard cell cytosolic free calcium in response to oxidative stress', *Plant Physiology* **111**, 1031–1042.

Mehlhorn, H. and Wellburn, A.R. (1987) 'Stress ethylene formation determines plant sensitivity to ozone', *Nature* **327**, 417–418.

Meinhard, M. and Grill, E. (2001) 'Hydrogen peroxide is a regulator of ABI1, a protein phosphatase 2C from *Arabidopsis*', *Federation of European Biochemical Societies Letters* **508**, 443–446.

Meinhard, M., Rodriguez, P.L. and Grill, E. (2002) 'The sensitivity of ABI2 to hydrogen peroxide links the abscisic acid-response regulator to redox signalling', *Planta* **214**, 775–782.

Menser, A. (1964) 'Response of plants to air pollutants: III. A relation between ascorbate levels and ozone susceptibility of light pre-conditioned tobacco leaves', *Plant Physiology* **39**, 564–567.

Moeder, W., Barry, C.S., Tauriainen, A.A. *et al.* (2002) 'Ethylene synthesis regulated by bi-phasic induction of ACC synthase and ACC oxidase genes is required for H_2O_2 accumulation and cell death in ozone-exposed tomato', *Plant Physiology* **130**, 1918–1926.

Moldau, H. (1998) 'Hierarchy of ozone scavenging reactions in the plant cell wall', *Physiologia Plantarum* **104**, 617–622.

Moldau, H., Sober, J. and Sober, A. (1990) 'Differential sensitivity of stomata and mesophyll to sudden exposure of bean shoots to ozone', *Photosynthetica* **24**, 446–458.

Moon, H., Lee, B., Choi, G. *et al.* (2003) 'NDP kinase 2 interacts with two oxidative stress-activated MAPKs to regulate cellular redox state and enhances multiple stress tolerance in transgenic plants', *Proceedings of the National Academy of Sciences USA* **100**, 358–363.

Mou, Z., Fan, W. and Dong, X. (2003) 'Inducers of plant systemic acquired resistance regulate NPR1 function through redox changes', *Cell* **113**, 935–944.

Murata, Y., Pei, Z.M., Mori, I.C. and Schroeder, J.I. (2001) 'Abscisic acid activation of plasma membrane Ca^{2+} channels in guard cells requires cytosolic NAD(P)H and is differentially disrupted upstream and downstream of reactive oxygen species production in *abi1-1* and *abi2-1* protein phosphatase 2C mutants', *The Plant Cell* **13**, 2513–2523.

Nakajima, N., Itoh, T., Takikawa, S. *et al.* (2002) 'Improvement in ozone tolerance of tobacco plants with an antisense DNA for 1-aminocyclopropane-1-carboxylate synthase', *Plant, Cell and Environment* **25**, 727–736.

Nakajima, N., Matsuyama, T., Tamaoki, M. *et al.* (2001) 'Effects of ozone exposure on the gene expression of ethylene biosynthetic enzymes in tomato leaves', *Plant Physiology and Biochemistry* **39**, 993–998.

Nickstadt, A., Thomma, B.P.H.J., Feussner, I. *et al.* (2004) 'The jasmonate-insensitive mutant *jin1* shows increased resistance to biotrophic as well as necrotrophic pathogens', *Molecular Plant Pathology* **5**, 425–434.

Noctor, G. and Foyer, C.H. (1998) 'Ascorbate and glutathione: keeping active oxygen under control', *Annual Review of Plant Physiology and Plant Molecular Biology* **49**, 249–279.

Oeller, P.W., Min-Wong, L., Taylor, L.P., Pike, D.A. and Theologis, A. (1991) 'Reversible inhibition of tomato fruit senescence by antisense RNA', *Science* **254**, 437–439.

Örvar, B.L. and Ellis, B.E. (1997) 'Transgenic tobacco plants expressing antisense RNA for cytosolic ascorbate peroxidase show increased susceptibility to ozone injury', *The Plant Journal* **11**, 1297–1305.

Örvar, B.L., McPherson, J. and Ellis, B.E. (1997) 'Pre-activating wounding response in tobacco prior to high-level ozone exposure prevents necrotic injury', *The Plant Journal* **11**, 203–212.

Ouaked, F., Rozhon, W., Lecourieux, D. and Hirt, H. (2003) 'A MAPK pathway mediates ethylene signaling in plants', *EMBO Journal* **22**, 1282–1288.

Overmyer, K., Brosché, M. and Kangasjärvi, J. (2003) 'Reactive oxygen species and hormonal control of cell death', *Trends in Plant Science* **8**, 335–342.

Overmyer, K., Brosché, M., Pellinen, R. *et al.* (2005) 'Ozone-induced programmed cell death in the *Arabidopsis* radical-induced cell death1 mutant', *Plant Physiology* **137**, 1092–1104.

Overmyer, K., Tuominen, H., Kettunen, R. *et al.* (2000) 'The ozone-sensitive Arabidopsis *rcd1* mutant reveals opposite roles for ethylene and jasmonate signaling pathways in regulating superoxide-dependent cell death', *The Plant Cell* **12**, 1849–1862.

Pandey, S. and Assmann, S.M. (2004) 'The *Arabidopsis* putative G protein-coupled receptor GCR1 interacts with the G protein α subunit GPA1 and regulates abscisic acid signaling', *The Plant Cell* **16**, 1616–1632.

Pasqualini, S., Piccioni, C., Reale, L., Ederli, L., Della Torre, G. and Ferranti, F. (2003) 'Ozone-induced cell death in tobacco cultivar Bel W3 plants. The role of programmed cell death in lesion formation', *Plant Physiology* **133**, 1122–1134.

Pei, Z.-M., Murata, Y., Benning, G. *et al.* (2000) 'Calcium channels activated by hydrogen peroxide mediate abscisic acid signalling in guard cells', *Nature* **406**, 731–734.

Pell, E.J., Schlagnhaufer, C.D. and Arteca, R.N. (1997) 'Ozone-induced oxidative stress: mechanisms of action and reaction', *Physiologia Plantarum* **100**, 264–273.

Pellinen, R., Palva, T. and Kangasjärvi, J. (1999) 'Subcellular localization of ozone-induced hydrogen peroxide production in birch (*Betula pendula*) leaf cells', *The Plant Journal* **20**, 349–356.

Pellinen, R.I., Korhonen, M.S., Tauriainen, A.A., Palva, E.T. and Kangasjärvi, J. (2002) 'H_2O_2 activates cell death and defense gene-expression in birch (*Betula pendula*)', *Plant Physiology* **130**, 549–560.

Peña-Cortés, H., Albrecht, T., Prat, S., Weiler, E.W. and Willmitzer, L. (1993) 'Aspirin prevents wound-induced gene expression in tomato leaves by blocking jasmonic acid biosynthesis', *Planta* **191**, 123–128.

Penninckx, I.A.M.A., Thomma, B.P.H.J., Buchala, A., Métraux, J.-P. and Broekaert, W.F. (1998) 'Concomitant activation of jasmonate and ethylene response pathways is required for induction of a plant defensin gene in *Arabidopsis*', *The Plant Cell* **10**, 2103–2113.

Pignocchi, C. and Foyer, C.H. (2003) 'Apoplastic ascorbate metabolism and its role in the regulation of cell signaling', *Current Opinion in Plant Biology* **6**, 379–389.

Ranieri, A., Castagna, A., Padu, E., Moldau, H., Rahi, M. and Soldatini, G.F. (1999) 'The decay of O_3 through direct reaction with cell wall ascorbate is not sufficient to explain the different degrees of O_3 sensitivity in two poplar clones', *Journal of Plant Physiology* **154**, 250–255.

Rao, M.V. and Davis, K.R. (1999) 'Ozone-induced cell death occurs via two distinct mechanisms in *Arabidopsis*: the role of salicylic acid', *The Plant Journal* **17**, 603–614.

Rao, M.V. and Davis, K.R. (2001) 'The physiology of ozone-induced cell death', *Planta* **213**, 682–690.

Rao, M.V., Koch, J.R. and Davis, K.R. (2000a) 'Ozone: a tool for probing programmed cell death in plants', *Plant Molecular Biology* **44**, 345–358.

Rao, M.V., Lee, H.I., Creelman, R.A., Mullet, J.A. and Davis, K.R. (2000b) 'Jasmonic acid signaling modulates ozone-induced hypersensitive cell death', *The Plant Cell* **12**, 1633–1646.

Rao, M.V., Lee, H.-I. and Davis, K.R. (2002) 'Ozone-induced ethylene production is dependent on salicylic acid, and both salicylic acid and ethylene act in concert to regulate ozone-induced cell death', *The Plant Journal* **32**, 447–456.

Raskin, I. (1995) Salicylic acid, in *Plant Hormones Physiology, Biochemistry and Molecular Biology*, 2nd edn. (ed. P.J. Davies), Kluwer, Dordrecht, The Netherlands, pp. 188–205.

Ren, D., Yang, H. and Zhang, S. (2002) 'Cell death mediated by MAPK is associated with hydrogen peroxide production in *Arabidopsis*', *The Journal of Biological Chemistry* **277**, 559–565.

Reymond, P. and Farmer, E.E. (1998) 'Jasmonate and salicylate as global signals for defense gene expression', *Current Opinion in Plant Biology* **1**, 404–411.

Rougee, M. and Bensasson, R.V. (1986) 'Determination of the decay-rate constant of singlet oxygen (1-Δ-G) in presence of biomolecules', *Comptes Rendus de l'Academie des Sciences Serie II* **302**, 1223–1226.

Salinas, J. (2002) Molecular mechanisms of signal transduction in cold acclimation, in *Plant Signal Transduction* (ed. D. Scheel and C. Wasternack), Oxford University Press, New York, pp. 116–139.

Samuel, M.A. and Ellis, B.E. (2002) 'Double jeopardy: both overexpression and suppression of a redox-activated plant mitogen-activated protein kinase render tobacco plants ozone sensitive', *The Plant Cell* **14**, 2059–2069.

Samuel, M.A., Miles, G.P. and Ellis, B.E. (2000) 'Ozone treatment rapidly activates MAP kinase signalling in plants', *The Plant Journal* **22**, 367–376.

Sandermann, H. Jr. (1998) 'Ozone: an air pollutant acting as a plant-signaling molecule', *Naturwissenschaften* **85**, 369–375.

Sandermann, H. and Matyssek, R. (2004) Scaling up from molecular to ecological processes, in *Molecular Ecotoxicology of Plants* (ed. H. Sandermann), Springer-Verlag, Berlin, Heidelberg, pp. 207–226.

Sandermann, H. Jr., Ernst, D., Heller, W. and Langebartels, C. (1998) 'Ozone: an abiotic elicitor of plant defence reactions', *Trends in Plant Science* **3**, 47–50.

Sanmartin, M., Drogoudi, P.D., Lyons, T., Pateraki, I., Barnes, J. and Kanellis, A.K. (2003) 'Overexpression of ascorbate oxidase in the apoplast of transgenic tobacco results in altered ascorbate and glutathione redox states and increased sensitivity to ozone', *Planta* **216**, 918–928.

Schenk, P.M., Kazan, K., Wilson, I. *et al.* (2000) 'Coordinated plant defense responses in *Arabidopsis* revealed by microarray analysis', *Proceedings of the National Academy of Sciences USA* **97**, 11655–11660.

Schraudner, M., Moeder, W., Wiese, C. *et al.* (1998) 'Ozone-induced oxidative burst in the ozone biomonitor plant, tobacco Bel W3', *The Plant Journal* **16**, 235–245.

Sharma, Y.K., León, J., Raskin, I. and Davis, K.R. (1996) 'Ozone-induced responses in *Arabidopsis thaliana*: the role of salicylic acid in the accumulation of defense-related transcripts and induced resistance', *Proceedings of the National Academy of Sciences USA* **93**, 5099–5104.

Shulaev, V., Silverman, P. and Raskin, I. (1997) 'Airborne signalling by methyl salicylate in plant pathogen resistance', *Nature* **385**, 718–721.

Simon-Plas, F., Elmayan, T. and Blein, J.P. (2002) 'The plasma membrane oxidase NtrbohD is responsible for AOS production in elicited tobacco cells', *The Plant Journal* **31**, 137–147.

Stockwell, W.R., Kramm, G., Scheel, H.-E., Mohnen, V.A. and Seiler, W. (1997) Ozone formation, destruction and exposure in Europe and the United States, in *Forest Decline and Ozone* (ed. H. Sandermann, Jr., A.R. Wellburn and R.L. Heath), Springer-Verlag, Berlin, Heidelberg, New York, pp. 1–38.

Tamaoki, M., Matsuyama, T., Kanna, M. *et al.* (2003) 'Differential ozone sensitivity among *Arabidopsis* accessions and its relevance to ethylene synthesis', *Planta* **216**, 552–560.

Tamaoki, M., Nakajima, N., Kubo, A., Aono, M., Matsuyama, T. and Saji, H. (2004) 'Transcriptome analysis of O_3-exposed *Arabidopsis* reveals that multiple signal pathways act mutually antagonistically to induce gene expression', *Plant Molecular Biology* **53**, 443–456.

Timonen, U., Huttunen, S. and Manninen, S. (2004) 'Ozone sensitivity of wild field layer plant species of northern Europe. A review', *Plant Ecology* **172**, 27–39.

Tingey, D.T., Standley, C. and Field, R.W. (1976) 'Stress ethylene evolution: a measure of ozone effects on plants', *Atmospheric Environment* **10**, 969–974.

Torres, M.A., Dangl, J.L. and Jones, J.D.G. (2002) '*Arabidopsis* gp91[phox] homologues *AtrbohD* and *AtrbohF* are required for accumulation of reactive oxygen intermediates in the plant defense response', *Proceedings of the National Academy of Sciences USA* **99**, 517–522.

Torsethaugen, G., Pell, E.J. and Assmann, S.M. (1999) 'Ozone inhibits guard cell K^+ channels implicated in stomatal opening', *Proceedings of the National Academy of Sciences USA* **96**, 13577–13582.

Tuomainen, J., Betz, C., Kangasjärvi, J. *et al.* (1997) 'Ozone induction of ethylene emission in tomato plants: regulation by differential transcript accumulation for the biosynthetic enzymes', *The Plant Journal* **12**, 1151–1162.

Tuominen, H., Overmyer, K., Keinänen, M., Kollist, H. and Kangasjärvi, J. (2004) 'Mutual antagonism of ethylene and jasmonic acid regulates ozone-induced spreading cell death in *Arabidopsis*', *The Plant Journal* **39**, 59–69.

Vahala, J., Ruonala, R., Keinänen, M., Tuominen, H. and Kangasjärvi, J. (2003) 'Ethylene insensitivity modulates ozone-induced cell death in birch (*Betula pendula*)', *Plant Physiology* **132**, 185–195.

Vahala, J., Schlagnhaufer, C.D. and Pell, E.J. (1998) 'Induction of an ACC synthase cDNA by ozone in light-grown *Arabidopsis thaliana* leaves', *Physiologia Plantarum* **103**, 45–50.

Van Camp, W., Van Montagu, M. and Inzé, D. (1998) 'H_2O_2 and NO: redox signals in disease resistance', *Trends in Plant Science* **3**, 330–334.

Vranová, E., Inzé, D. and Van Breusegem, F. (2002) 'Signal transduction during oxidative stress', *Journal of Experimental Botany* **53**, 1227–1236.

Wang, K.L.C., Li, H. and Ecker, J.R. (2002) 'Ethylene biosynthesis and signaling networks mediating responses to stress', *The Plant Cell* **14**, S131–S151.

Wang, X.Q., Ullah, H., Jones, A.M. and Assmann, S.M. (2001) 'G protein regulation of ion channels and abscisic acid signaling in *Arabidopsis* guard cells', *Science* **292**, 2070–2072.

Wilkinson, J.Q., Lanahan, M.B., Yen, H.-C., Giovannoni, J.J. and Klee, H.J. (1995) 'An ethylene-inducible component of signal transduction encoded by *Never-ripe*', *Science* **270**, 1807–1809.

Wohlgemuth, H., Mittelstrass, K., Kschieschan, S. *et al.* (2002) 'Activation of an oxidative burst is a general feature of sensitive plants exposed to the air pollutant ozone', *Plant, Cell and Environment* **25**, 717–726.

Xiong, L. and Yang, Y. (2003) 'Disease resistance and abiotic stress tolerance in rice are inversely modulated by an abscisic acid-inducible mitogen-activated protein kinase', *The Plant Cell* **15**, 745–759.

Zhang, S. and Klessig, D.F. (1997) 'Salicylic acid activates a 48-kD MAP kinase in tobacco', *The Plant Cell* **9**, 809–824.

Zhang, X., Zhang, L., Dong, F., Gao, J., Galbraith, D.W. and Song, C.-P. (2001) 'Hydrogen peroxide is involved in abscisic acid-induced stomatal closure in *Vicia faba*', *Plant Physiology* **126**, 1438–1448.

Ziegler, J., Keinänen, M. and Baldwin, I.T. (2001) 'Herbivore-induced allene oxide synthase transcripts and jasmonic acid in *Nicotiana attenuata*', *Phytochemistry* **58**, 729–738.

Index